Lecture Notes in Mathematics 1716

Springer
Berlin
Heidelberg
New York
Barcelona
Hong Kong
London
Milan
Paris
Singapore
Tokyo

J. Coates R. Greenberg
K. A. Ribet K. Rubin

Arithmetic Theory of Elliptic Curves

Lectures given at the 3rd Session of the
Centro Internazionale Matematico Estivo
(C.I.M.E.) held in Cetraro, Italy,
July 12–19, 1997

Editor: C. Viola

Fondazione
C.I.M.E.

Springer

Authors

John H. Coates
Department of Pure Mathematics
and Mathematical Statistics
University of Cambridge
16 Mill Lane
Cambridge CB2 1SB, UK

Ralph Greenberg
Department of Mathematics
University of Washington
Seattle, WA 98195, USA

Kenneth A. Ribet
Department of Mathematics
University of California
Berkeley CA 94720, USA

Karl Rubin
Department of Mathematics
Stanford University
Stanford CA 94305, USA

Editor

Carlo Viola
Dipartimento di Matematica
Università di Pisa
Via Buonarroti 2
56127 Pisa, Italy

Cataloging-in-Publication Data applied for

Die Deutsche Bibliothek - CIP-Einheitsaufnahme

Arithmetic theory of elliptic curves : held in Cetraro, Italy, July 12 - 19, 1997 / Fondazione CIME. J. Coates ... Ed.: C. Viola. - Berlin ; Heidelberg ; New York ; Barcelona ; Hong Kong ; London ; Milan ; Paris ; Singapore ; Tokyo : Springer, 1999
(Lectures given at the ... session of the Centro Internazionale Matematico Estivo (CIME) ... ; 1997,3) (Lecture notes in mathematics ; Vol. 1716 : Subseries: Fondazione CIME)
ISBN 3-540-66546-3

Mathematics Subject Classification (1991):
11G05, 11G07, 11G15, 11G18, 11G40, 11R18, 11R23, 11R34, 14G10, 14G35

ISSN 0075-8434
ISBN 3-540-66546-3 Springer-Verlag Berlin Heidelberg New York

Springer-Verlag Berlin Heidelberg New York
a member of BertelsmannSpringer Science+Business Media GmbH

Printed in Germany

Typesetting: Camera-ready TEX output by the authors
SPIN: 10774740 41/3111-54321 – Printed on acid-free paper

Preface

The C.I.M.E. Session "Arithmetic Theory of Elliptic Curves" was held at Cetraro (Cosenza, Italy) from July 12 to July 19, 1997.

The arithmetic of elliptic curves is a rapidly developing branch of mathematics, at the boundary of number theory, algebra, arithmetic algebraic geometry and complex analysis. After the pioneering research in this field in the early twentieth century, mainly due to H. Poincaré and B. Levi, the origin of the modern arithmetic theory of elliptic curves goes back to L. J. Mordell's theorem (1922) stating that the group of rational points on an elliptic curve is finitely generated. Many authors obtained in more recent years crucial results on the arithmetic of elliptic curves, with important connections to the theories of modular forms and L-functions. Among the main problems in the field one should mention the Taniyama–Shimura conjecture, which states that every elliptic curve over $\mathbb{Q}$ is modular, and the Birch and Swinnerton–Dyer conjecture, which, in its simplest form, asserts that the rank of the Mordell–Weil group of an elliptic curve equals the order of vanishing of the L-function of the curve at 1. New impetus to the arithmetic of elliptic curves was recently given by the celebrated theorem of A. Wiles (1995), which proves the Taniyama–Shimura conjecture for semistable elliptic curves. Wiles' theorem, combined with previous results by K. A. Ribet, J.-P. Serre and G. Frey, yields a proof of Fermat's Last Theorem. The most recent results by Wiles, R. Taylor and others represent a crucial progress towards a complete proof of the Taniyama–Shimura conjecture. In contrast to this, only partial results have been obtained so far about the Birch and Swinnerton–Dyer conjecture.

The fine papers by J. Coates, R. Greenberg, K. A. Ribet and K. Rubin collected in this volume are expanded versions of the courses given by the authors during the C.I.M.E. session at Cetraro, and are broad and up-to-date contributions to the research in all the main branches of the arithmetic theory of elliptic curves. A common feature of these papers is their great clarity and elegance of exposition.

Much of the recent research in the arithmetic of elliptic curves consists in the study of modularity properties of elliptic curves over $\mathbb{Q}$, or of the structure of the Mordell–Weil group $E(K)$ of K-rational points on an elliptic curve E defined over a number field K. Also, in the general framework of Iwasawa theory, the study of $E(K)$ and of its rank employs algebraic as well as analytic approaches.

Various algebraic aspects of Iwasawa theory are deeply treated in Greenberg's paper. In particular, Greenberg examines the structure of the p-primary Selmer group of an elliptic curve E over a $\mathbb{Z}_p$-extension of the field K, and gives a new proof of Mazur's control theorem. Rubin gives a detailed and thorough description of recent results related to the Birch and Swinnerton–Dyer conjecture for an elliptic curve defined over an imaginary quadratic field K, with complex multiplication by K. Coates' contribution is mainly concerned with the construction of an analogue of Iwasawa theory for elliptic curves without complex multiplication, and several new results are included in his paper. Ribet's article focuses on modularity properties, and contains new results concerning the points on a modular curve whose images in the Jacobian of the curve have finite order.

The great success of the C.I.M.E. session on the arithmetic of elliptic curves was very rewarding to me. I am pleased to express my warmest thanks to Coates, Greenberg, Ribet and Rubin for their enthusiasm in giving their fine lectures and for agreeing to write the beautiful papers presented here. Special thanks are also due to all the participants, who contributed, with their knowledge and variety of mathematical interests, to the success of the session in a very co-operative and friendly atmosphere.

Carlo Viola

Table of Contents

Fragments of the GL_2 Iwasawa Theory of Elliptic Curves without Complex Multiplication

John Coates

"Fearing the blast
Of the wind of impermanence,
I have gathered together
The leaflike words of former mathematicians
And set them down for you."

Thanks to the work of many past and present mathematicians, we now know a very complete and beautiful Iwasawa theory for the field obtained by adjoining all p-power roots of unity to $\mathbb{Q}$, where p is any prime number. Granted the ubiquitous nature of elliptic curves, it seems natural to expect a precise analogue of this theory to exist for the field obtained by adjoining to $\mathbb{Q}$ all the p-power division points on an elliptic curve E defined over $\mathbb{Q}$. When E admits complex multiplication, this is known to be true, and Rubin's lectures in this volume provide an introduction to a fairly complete theory. However, when E does not admit complex multiplication, all is shrouded in mystery and very little is known. These lecture notes are aimed at providing some fragmentary evidence that a beautiful and precise Iwasawa theory also exists in the non complex multiplication case. The bulk of the lectures only touch on one initial question, namely the study of the cohomology of the Selmer group of E over the field of all p-power division points, and the calculation of its Euler characteristic when these cohomology groups are finite. But a host of other questions arise immediately, about which we know essentially nothing at present.

Rather than tempt uncertain fate by making premature conjectures, let me illustrate two key questions by one concrete example. Let E be the elliptic curve $X_1(11)$, given by the equation

$$y^2 + y = x^3 - x^2.$$

Take p to be the prime 5, let K be the field obtained by adjoining the 5-division points on E to $\mathbb{Q}$, and let F_∞ be the field obtained by adjoining all 5-power division points to $\mathbb{Q}$. We write Ω for the Galois group of F_∞ over K. The action of Ω on the group of all 5-power division points allows us to identify Ω with a subgroup of $GL_2(\mathbb{Z}_5)$, and a celebrated theorem of Serre tells us that Ω is an open subgroup. Now it is known that the Iwasawa algebra $\Lambda(\Omega)$ (see (14)) is left and right Noetherian and has no divisors of zero. Let $C(E/F_\infty)$ denote the compact dual of the Selmer group of E over F_∞ (see (12)), endowed with its natural structure as a left $\Lambda(\Omega)$-module. We prove in these lectures that $C(E/F_\infty)$ is large in the sense that

$$\dim_{\mathbb{Q}_5}\left(C(E/F_\infty) \otimes_{\mathbb{Z}_5} \mathbb{Q}_5\right) = \infty.$$

But we also prove that every element of $C(E/F_\infty)$ has a non-zero annihilator in $\Lambda(\Omega)$. We strongly suspect that $C(E/F_\infty)$ has a deep and interesting arithmetic structure as a representation of $\Lambda(\Omega)$. For example, can one say anything about the irreducible representations of $\Lambda(\Omega)$ which occur in $C(E/F_\infty)$? Is there some analogue of Iwasawa's celebrated main conjecture on cyclotomic fields, which, in this case, should relate the $\Lambda(\Omega)$-structure of $C(E/F_\infty)$ to a 5-adic L-function formed by interpolating the values at $s = 1$ of the twists of the complex L-function of E by all Artin characters of Ω? I would be delighted if these lectures could stimulate others to work on these fascinating non-abelian problems.

In conclusion, I want to warmly thank R. Greenberg, S. Howson and Sujatha for their constant help and advice throughout the time that these lectures were being prepared and written. Most of the material in Chapters 3 and 4 is joint work with S. Howson. I also want to thank Y. Hachimori, K. Matsuno, Y. Ochi, J.-P. Serre, R. Taylor, and B. Totaro for making important observations to us while this work was evolving. Finally, it is a great pleasure to thank Carlo Viola and C.I.M.E. for arranging for these lectures to take place at an incomparably beautiful site in Cetraro, Italy.

1 Statement of Results

1.1 Serre's theorem

Throughout these notes, F will denote a finite extension of the rational field $\mathbb{Q}$, and E will denote an elliptic curve defined over F, which will always be assumed to satisfy the hypothesis:

Hypothesis. *The endomorphism ring of E over $\overline{\mathbb{Q}}$ is equal to $\mathbb{Z}$, i.e. E does not admit complex multiplication.*

Let p be a prime number. For all integers $n \geqslant 0$, we define

$$E_{p^{n+1}} = \mathrm{Ker}\left(E(\overline{\mathbb{Q}}) \xrightarrow{p^{n+1}} E(\overline{\mathbb{Q}})\right), \quad E_{p^\infty} = \bigcup_{n \geqslant 0} E_{p^{n+1}}.$$

We define the corresponding Galois extensions of F

$$F_n = F(E_{p^{n+1}}), \quad F_\infty = F(E_{p^\infty}). \tag{1}$$

Write

$$\Sigma_n = G(F_\infty/F_n), \quad \Sigma = G(F_\infty/F) \tag{2}$$

for the Galois groups of F_∞ over F_n, and F_∞ over F, respectively. Now the action of Σ on E_{p^∞} defines a canonical injection

$$i : \Sigma \hookrightarrow \mathrm{Aut}(E_{p^\infty}) \simeq \mathrm{GL}_2(\mathbb{Z}_p). \tag{3}$$

When there is no danger of confusion, we shall drop the homomorphism i from the notation, and identify Σ with a subgroup of $\mathrm{GL}_2(\mathbb{Z}_p)$. Note that i maps Σ_n into the subgroup of $\mathrm{GL}_2(\mathbb{Z}_p)$ consisting of all matrices which are congruent to the identity modulo p^{n+1}. In particular, it follows that Σ_0 is always a pro-p-group. However, it is not in general true that Σ is a pro-p-group. The following fundamental result about the size of Σ is due to Serre [26].

Theorem 1.1.

(i) *Σ is open in $\mathrm{GL}_2(\mathbb{Z}_p)$ for all primes p, and*

(ii) *$\Sigma = \mathrm{GL}_2(\mathbb{Z}_p)$ for all but a finite number of primes p.*

Serre's method of proof in [26] of Theorem 1.1 is effective, and he gives many beautiful examples of the calculations of Σ for specific elliptic curves and specific primes p. We shall use some of these examples to illustrate the theory developed in these lectures. For convenience, we shall always give the name of the relevant curves in Cremona's tables [9].

Example. Consider the curves of conductor 11

$$11(A1): \quad y^2 + y = x^3 - x^2 - 10x - 20 \tag{4}$$

$$11(A3): \quad y^2 + y = x^3 - x^2. \tag{5}$$

The first curve corresponds to the modular group $\Gamma_0(11)$ and is often denoted by $X_0(11)$, and the second curve corresponds to the group $\Gamma_1(11)$, and is often denoted by $X_1(11)$. Neither curve admits complex multiplication (for example, their j-invariants are non-integral). Both curves have a $\mathbb{Q}$-rational point of order 5, and they are linked by a $\mathbb{Q}$-isogeny of degree 5. For both curves, Serre [26] has shown that $\Sigma = \mathrm{GL}_2(\mathbb{Z}_p)$ for all primes $p \geqslant 7$. Subsequently, Lang and Trotter [21] determined Σ for the curve $11(A1)$ and the primes $p = 2, 3, 5$.

We now briefly discuss Σ-Euler characteristics, since this will play an important role in our subsequent work. By virtue of Theorem 1.1, Σ is a p-adic Lie group of dimension 4. By results of Serre [28] and Lazard [22], Σ will have p-cohomological dimension equal to 4 provided Σ has no p-torsion.

Since Σ is a subgroup of $\mathrm{GL}_2(\mathbb{Z}_p)$, it will certainly have no p-torsion provided $p \geqslant 5$. Whenever we talk about Σ-Euler characteristics in these notes, we shall always assume that $p \geqslant 5$. Let W be a discrete p-primary Σ-module. We shall say that W has finite Σ-Euler characteristic if all of the cohomology groups $H^i(\Sigma, W)$ $(i = 0, \ldots, 4)$ are finite. When W has finite Σ-Euler characteristic, we define its Euler characteristic $\chi(\Sigma, W)$ by the usual formula

$$\chi(\Sigma, W) = \prod_{i=0}^{4} (\#(H^i(\Sigma, W)))^{(-1)^i}. \tag{6}$$

Example. Take $W = E_{p^\infty}$. Serre [29] proved that E_{p^∞} has finite Σ-Euler characteristic, and recently he determined its value in [30].

Theorem 1.2. *If $p \geqslant 5$, then $\chi(\Sigma, E_{p^\infty}) = 1$ and $H^4(\Sigma, E_{p^\infty}) = 0$.*

This result will play an important role in our later calculations of the Euler characteristics of Selmer groups. Put

$$h_i(E) = \#(H^i(\Sigma, E_{p^\infty})). \tag{7}$$

We now give a lemma which is often useful for calculating the $h_i(E)$. Let μ_{p^n} denote the group of p^n-th roots of unity, and put

$$\mu_{p^\infty} = \bigcup_{n \geqslant 1} \mu_{p^n}, \quad T_p(\mu) = \varprojlim \mu_{p^n}. \tag{8}$$

By the Weil pairing, $F(\mu_{p^\infty}) \subset F(E_{p^\infty})$ and so we can view Σ as acting in the natural fashion on the two modules (8). As usual, define

$$E_{p^\infty}(-1) = E_{p^\infty} \otimes_{\mathbb{Z}_p} T_p(\mu)^{\otimes(-1)}, \tag{9}$$

where

$$T_p(\mu)^{\otimes(-1)} = \mathrm{Hom}(T_p(\mu), \mathbb{Z}_p);$$

here Σ acts on both groups again in the natural fashion.

Lemma 1.3. *Let p be any prime number. Then*

(i) *$h_0(E)$ divides $h_1(E)$.*
(ii) *If Σ has no p-torsion, we have $h_3(E) = \#H^0(\Sigma, E_{p^\infty}(-1))$.*

Corollary 1.4. *If $p \geqslant 5$, and $h_3(E) > 1$, then $h_2(E) > 1$.*

Indeed, Theorem 1.2 shows that

$$h_2(E) = \frac{h_1(E)}{h_0(E)} \times h_3(E),$$

whence the assertion of the Corollary is clear from (i) of Lemma 1.3. The corollary is useful because it does not seem easy to compute $h_2(E)$ in a direct manner.

We now turn to the proof of (i) of Lemma 1.3. Let K_∞ denote the cyclotomic $\mathbb{Z}_p$-extension of F, and let $E_{p^\infty}(K_\infty)$ be the subgroup of E_{p^∞} which is rational over K_∞. We claim that $E_{p^\infty}(K_\infty)$ is finite. Granted this claim, it follows that

$$h_0(E) = \#(H^0(\Gamma, E_{p^\infty}(K_\infty))) = \#(H^1(\Gamma, E_{p^\infty}(K_\infty))),$$

where Γ denotes the Galois group of K_∞ over F. But $H^1(\Gamma, E_{p^\infty}(K_\infty))$ is a subgroup of $H^1(\Sigma, E_{p^\infty})$ under the inflation map, and so (i) is clear. To show that $E_{p^\infty}(K_\infty)$ is finite, let us note that it suffices to show that $E_{p^\infty}(H_\infty)$ is finite, where $H_\infty = F(\mu_{p^\infty})$. Let $\Omega = G(F_\infty/H_\infty)$. By virtue of the Weil pairing, we have $\Omega = \Sigma \cap \mathrm{SL}_2(\mathbb{Z}_p)$, for any embedding $i : \Sigma \hookrightarrow \mathrm{GL}_2(\mathbb{Z}_p)$ given by choosing any $\mathbb{Z}_p$-basis e_1, e_2 of $T_p(E)$. If $E_{p^\infty}(H_\infty)$ was infinite, we could choose e_1 so that it is fixed by Ω. But then the embedding i would inject Ω into the subgroup of $\mathrm{SL}_2(\mathbb{Z}_p)$ consisting of all matrices of the form $\begin{pmatrix} 1 & x \\ 0 & 1 \end{pmatrix}$, where x runs over $\mathbb{Z}_p$. But this is impossible since Ω must be open in $\mathrm{SL}_2(\mathbb{Z}_p)$ as Σ is open in $\mathrm{GL}_2(\mathbb{Z}_p)$. To prove assertion (ii) of Lemma 1.3, we need the fact that Σ is a Poincaré group of dimension 4 (see Corollary 4.8, [25], p. 75). Moreover, as was pointed out to us by B. Totaro, the dualizing module for Σ is isomorphic to $\mathbb{Q}_p/\mathbb{Z}_p$ with the trivial action for Σ (see Lazard [22], Theorem 2.5.8, p. 184 when Σ is pro-p, and the same proof works in general for any open subgroup of $\mathrm{GL}_2(\mathbb{Z}_p)$ which has no p-torsion). Moreover, the Weil pairing gives a Σ-isomorphism

$$E_{p^n}(-1) \simeq \mathrm{Hom}(E_{p^n}, \mathbb{Z}/p^n\mathbb{Z}).$$

Using that Σ is a Poincaré group of dimension 4, it follows that $H^3(\Sigma, E_{p^n})$ is dual to $H^1(\Sigma, E_{p^n}(-1))$ for all integers $n \geqslant 1$. As usual, let

$$T_p(E) = \varprojlim E_{p^n}.$$

Passing to the limit as $n \to \infty$, we conclude that

$$H^3(\Sigma, E_{p^\infty}) = \varinjlim H^3(\Sigma, E_{p^n})$$

is dual to

$$H^1(\Sigma, T_p(E)(-1)) = \varprojlim H^1(\Sigma, E_{p^n}(-1)). \tag{10}$$

Write $V_p(E) = T_p(E) \otimes \mathbb{Q}_p$. Then we have the exact sequence of Σ-modules

$$0 \longrightarrow T_p(E)(-1) \longrightarrow V_p(E)(-1) \longrightarrow E_{p^\infty}(-1) \longrightarrow 0.$$

Now $V_p(E)(-1)^{\Sigma} = 0$ since $E_{p^\infty}(H_\infty)$ is finite. Moreover, (10) is finite by the above duality argument, and so it must certainly map to 0 in the $\mathbb{Q}_p$-vector space $H^1(\Sigma, V_p(E)(-1))$. Thus, taking Σ-cohomology of the above exact sequence, we conclude that

$$H^1(\Sigma, T_p(E)(-1)) = H^0(\Sigma, E_{p^\infty}(-1)). \tag{11}$$

As (11) is dual to $H^3(\Sigma, E_{p^\infty})$, this completes the proof of (ii) of Lemma 1.3.

Example. Take $F = \mathbb{Q}$, E to be the curve $X_0(11)$ given by (4), and $p = 5$. The point $(5, 5)$ is a rational point of order 5 on E. As remarked earlier, Lang-Trotter [21] (see Theorem 8.1 on p. 55) have explicitly determined Σ in this case. In particular, they show that

$$E_5 = \mathbb{Z}/5\mathbb{Z} \oplus \mu_5$$

as Σ-modules. Moreover, although we do not give the details here, it is not difficult to deduce from their calculations that

$$h_0(E) = h_3(E) = 5, \text{ and } h_1(E) \geqslant 5^2.$$

It also then follows from Theorem 1.2 that $h_2(E) = h_1(E)$.

1.2 The basic Iwasawa module

Iwasawa theory can be fruitfully applied in the following rather general setting. Let H_∞ denote a Galois extension of F whose Galois group $\Omega = G(H_\infty/F)$ is a p-adic Lie group of positive dimension. By analogy with the classical situation over F, we define the Selmer group $\mathcal{S}(E/H_\infty)$ of E over H_∞ by

$$\mathcal{S}(E/H_\infty) = \mathrm{Ker}\big(H^1(H_\infty, E_{p^\infty}) \longrightarrow \prod_w H^1(H_{\infty,w}, E)\big), \tag{12}$$

where w runs over all finite primes of H_∞, and, as usual for infinite extensions, $H_{\infty,w}$ denotes the union of the completions at w of all finite extensions of F contained in H_∞. Of course, the Galois group Ω has a natural left action on $\mathcal{S}(E/H_\infty)$, and the central idea of the Iwasawa theory of elliptic curves is to exploit this Ω-action to obtain deep arithmetic information about E. This Ω-action makes $\mathcal{S}(E/H_\infty)$ into a discrete p-primary left Ω-module. It will often be convenient to study its compact dual

$$\mathcal{C}(E/H_\infty) = \mathrm{Hom}\big(\mathcal{S}(E/F_\infty), \mathbb{Q}_p/\mathbb{Z}_p\big), \tag{13}$$

which is endowed with the left action of Ω given by $(\sigma f)(x) = f(\sigma^{-1}x)$ for f in $\mathcal{C}(E/H_\infty)$ and σ in Ω. Clearly $\mathcal{S}(E/H_\infty)$ and $\mathcal{C}(E/H_\infty)$ are continuous modules over the ordinary group ring $\mathbb{Z}_p[\Omega]$ of Ω with coefficients in $\mathbb{Z}_p$. But, as Iwasawa was the first to observe in the case of the cyclotomic theory, it is more useful to view them as modules over a larger algebra, which we denote by $\Lambda(\Omega)$ and call the Iwasawa algebra of Ω, and which is defined by

$$\Lambda(\Omega) = \varprojlim \mathbb{Z}_p[\Omega/W], \tag{14}$$

where W runs over all open normal subgroups of Ω. Now if A is any discrete p-primary left Ω-module and $X = \mathrm{Hom}(A, \mathbb{Q}_p/\mathbb{Z}_p)$ is its Pontrjagin dual, then we have

$$A = \bigcup_W A^W, \quad X = \varprojlim X_W,$$

where W again runs over all open normal subgroups of Ω, and X_W denotes the largest quotient of X on which W acts trivially. It is then clear how to extend the natural action of $\mathbb{Z}_p[\Omega]$ on A and X by continuity to an action of the whole Iwasawa algebra $\Lambda(\Omega)$.

In Greenberg's lectures in this volume, the extension H_∞ is taken to be the cyclotomic $\mathbb{Z}_p$-extension of F. In Rubin's lectures, H_∞ is taken to be the field generated over F by all $\mathfrak{p}$-power division points on E, where $\mathfrak{p}$ is now a prime ideal in the ring of endomorphisms of E (Rubin assumes that E admits complex multiplication). In these lectures, we shall be taking $H_\infty = F_\infty = F(E_{p^\infty})$, and recall our hypothesis that E does not admit complex multiplication. Thus, in our case, $\Omega = \Sigma$ is an open subgroup of $\mathrm{GL}_2(\mathbb{Z}_p)$ by Theorem 1.1.

The first question which arises is how big is $\mathcal{S}(E/F_\infty)$? The following result, whose proof will be omitted from these notes, was pointed out to me by Greenberg.

Theorem 1.5. *For all primes p, we have*

$$\dim_{\mathbb{Q}_p}\big(\mathcal{C}(E/F_\infty) \otimes_{\mathbb{Z}_p} \mathbb{Q}_p\big) = \infty.$$

Example. Take $F = \mathbb{Q}$, $E = X_1(11)$, and $p = 5$. It was pointed out to me some years back by Greenberg that

$$\mathcal{C}(E/\mathbb{Q}(\mu_{5^\infty})) = 0 \tag{15}$$

(see his article in this volume, or [7], Chapter 4 for a detailed proof). On the other hand, we conclude from Theorem 1.5 that

$$\dim_{\mathbb{Q}_5}\big(\mathcal{C}(E/\mathbb{Q}(E_{5^\infty})) \otimes_{\mathbb{Z}_5} \mathbb{Q}_5\big) = \infty. \tag{16}$$

This example is a particularly interesting one, and we make the following observations now. Since E has a non-trivial rational point of order 5, we have the exact sequence of $G(\overline{\mathbb{Q}}/\mathbb{Q})$-modules

$$0 \longrightarrow \mathbb{Z}/5\mathbb{Z} \longrightarrow E_5 \longrightarrow \mu_5 \longrightarrow 0. \tag{17}$$

This exact sequence is not split. Indeed, since the j-invariant of E has order -1 at 11, and the curve has split multiplicative reduction at 11, the 11-adic Tate period q_E of E has order 1 at 11. Hence

$$\mathbb{Q}_{11}(E_5) = \mathbb{Q}_{11}(\mu_5, \sqrt[5]{q_E}),$$

and so we see that 5 must divide the absolute ramification index of every prime dividing 11 in any global splitting field for the Galois module E_5. It follows, in particular, that $[F_0 : \mathbb{Q}(\mu_5)] = 5$, where $F_0 = \mathbb{Q}(E_5)$. Moreover, 11 splits completely in $\mathbb{Q}(\mu_5)$, and then each of the primes of $\mathbb{Q}(\mu_5)$ dividing 11 are totally ramified in the extension $F_0/\mathbb{Q}(\mu_5)$. In view of (15) and the fact that $F_0/\mathbb{Q}(\mu_5)$ is cyclic of degree 5, we can apply the work of Hachimori and Matsuno [15] (see Theorem 3.1) to it to conclude that the following assertions are true for the $\Lambda(\Gamma)$-module $C(E/F_0(\mu_{5^\infty}))$, where Γ denotes the Galois group of $F_0(\mu_{5^\infty})$ over F_0: (i) $C(E/F_0(\mu_{5^\infty}))$ is $\Lambda(\Gamma)$-torsion, (ii) the μ-invariant of $C(E/F_0(\mu_{5^\infty}))$ is 0, and (iii) we have

$$\dim_{\mathbb{Q}_5}\big(C(E/F_0(\mu_{5^\infty})) \otimes_{\mathbb{Z}_5} \mathbb{Q}_5\big) = 16. \qquad (18)$$

However, I do not know at present whether E has a point of infinite order which is rational over F_0. Finally, we remark that one can easily deduce (16) from Theorem 3.1 of [15], on noting that $F_n/\mathbb{Q}(\mu_5)$ is a Galois 5-extension for all integers $n \geqslant 0$.

We now return to the discussion of the size of $C(E/F_\infty)$ as a left $\Lambda(\Sigma)$-module. It is easy to see (Theorem 2.7) that $C(E/F_\infty)$ is a finitely generated left $\Lambda(\Sigma)$-module. Recall that $F_n = F(E_{p^{n+1}})$, and that $\Sigma_n = G(F_\infty/F_n)$. We define Φ to be Σ_1 if $p = 2$, and to be Σ_0 if $p > 2$. The following result is a well known special case of a theorem of Lazard (see [10]).

Theorem 1.6. *The Iwasawa algebra $\Lambda(\Phi)$ is left and right Noetherian and has no divisors of* 0.

Now it is known (see Goodearl and Warfield [11], Chapter 9) that Theorem 1.6 implies that $\Lambda(\Phi)$ admits a skew field of fractions, which we denote by $K(\Phi)$. If X is any left $\Lambda(\Sigma)$-module, we define the $\Lambda(\Sigma)$-rank of X by the formula

$$\Lambda(\Sigma)\text{-rank of } X = \frac{1}{[\Sigma : \Phi]} \dim_{K(\Phi)}\big(K(\Phi) \otimes_{\Lambda(\Phi)} X\big). \qquad (19)$$

This $\Lambda(\Sigma)$-rank will not in general be an integer.

It is not difficult to see that the $\Lambda(\Sigma)$-rank is additive with respect to short exact sequences of finitely generated left $\Lambda(\Sigma)$-modules. Also, we say that X is $\Lambda(\Sigma)$-torsion if every element of X has a non-zero annihilator in $\Lambda(\Phi)$. Then X is $\Lambda(\Sigma)$-torsion if and only if X has $\Lambda(\Sigma)$-rank equal to 0.

It is natural to ask what is the $\Lambda(\Sigma)$-rank of the dual $C(E/F_\infty)$ of the Selmer group of E over F_∞. The conjectural answer to this problem depends on the nature of the reduction of E at the places v of F dividing p. We recall that E is said to have potential supersingular reduction at a prime v of F if there exists a finite extension L of the completion F_v of F at v such that E has good supersingular reduction over L. We then define the integer $\tau_v(E/F)$ to be 0 or $[F_v : \mathbb{Q}_v]$, according as E does not or does have potential supersingular reduction at v. Put

$$\tau_p(E/F) = \sum_{v|p} \tau_v(E/F), \tag{20}$$

where the sum on the right is taken over all primes v of F dividing p. Note that $\tau_p(E/F) \leqslant [F : \mathbb{Q}]$.

Conjecture 1.7. *For every prime p, the $\Lambda(\Sigma)$-rank of $C(E/F_\infty)$ is equal to $\tau_p(E/F)$.*

It is interesting to note that Conjecture 1.7 is entirely analogous to the conjecture made in the cyclotomic case in Greenberg's lectures. Specifically, if K_∞ denotes the cyclotomic $\mathbb{Z}_p$-extension of F, and if $\Gamma = G(K_\infty/F)$, then it is conjectured that the $\Lambda(\Gamma)$-rank of $C(E/K_\infty)$ is equal to $\tau_p(E/F)$ for all primes p.

Example. Consider the curve of conductor 50

$$E = 50(A1): \quad y^2 + xy + y = x^3 - x - 2. \tag{21}$$

Take $F = \mathbb{Q}$. This curve has multiplicative reduction at 2, so that $\tau_2(E/\mathbb{Q}) = 0$. It has potential supersingular reduction at 5, since it can be shown to achieve good supersingular reduction over the field $\mathbb{Q}_5(\mu_3, \sqrt[3]{-2\cdot 5^4})$. Hence $\tau_5(E/\mathbb{Q}) = 1$. It has good ordinary reduction at $3, 7, 11, 13, 17, 19, 23, 31, \ldots$, and so $\tau_p(E/\mathbb{Q}) = 0$ for all such primes p. It has good supersingular reduction at $29, 59, \ldots$, and $\tau_p(E/\mathbb{Q}) = 1$ for these primes.

Theorem 1.8. *Let $t_p(E/F)$ denote the $\Lambda(\Sigma)$-rank for $C(E/F_\infty)$. Then, for all primes $p \geqslant 5$, we have*

$$\tau_p(E/F) \leqslant t_p(E/F) \leqslant [F : \mathbb{Q}]. \tag{22}$$

We remark that the lower bound for $t_p(E/F)$ given in (22) is entirely analogous to what is known in the cyclotomic case (see Greenberg's lectures [13]). However, the upper bound for $t_p(E/F)$ in (22) still has not been proven unconditionally in the cyclotomic theory. We also point out that we do not at present know that $t_p(E/F)$ is an integer.

Corollary 1.9. *Conjecture 1.7 is true for all odd primes p such that E has potential supersingular reduction at all places v of F dividing p.*

This is clear since $\tau_p(E/F) = [F : \mathbb{Q}]$ when E has potential supersingular reduction at all places v of F dividing p. For example, if we take E to be the curve $50(A1)$ above and $F = \mathbb{Q}$, we conclude that $C(E/F_\infty)$ has $\Lambda(\Sigma)$-rank equal to 1 for $p = 5$, and for all primes $p = 29, 59, \ldots$ where E has good supersingular reduction.

We long tried unsuccessfully to prove examples of Conjecture 1.7 when $\tau_p(E/F) = 0$, and we are very grateful to Greenberg for making a suggestion which at last enables us to do this using recent work of Hachimori and Matsuno [15]. As before, let K_∞ denote the cyclotomic $\mathbb{Z}_p$-extension of F, and let $\Gamma = G(K_\infty/F)$. Let Y denote a finitely generated torsion $\Lambda(\Gamma)$-module. We recall that Y is said to have μ-invariant 0 if $(Y)_\Gamma$ is a finitely generated $\mathbb{Z}_p$-module, where $(Y)_\Gamma$ denotes the largest quotient of Y on which Γ acts trivially.

Theorem 1.10. *Let p be a prime such that* (i) $p \geqslant 5$, (ii) $\Sigma = G(F_\infty/F)$ *is a pro-p-group, and* (iii) *E has good ordinary reduction at all places v of F dividing p. Assume that $C(E/K_\infty)$ is $\Lambda(\Gamma)$-torsion and has μ-invariant* 0. *Then $C(E/F_\infty)$ is $\Lambda(\Sigma)$-torsion.*

Example. Take $E = X_1(11)$, $F = \mathbb{Q}(\mu_5)$, and $p = 5$. Then E has good ordinary reduction at the unique prime of F above 5. The cyclotomic $\mathbb{Z}_5$-extension of $\mathbb{Q}(\mu_5)$ is the field $\mathbb{Q}(\mu_{5^\infty})$. As was remarked earlier, F_n/F is a 5-extension for all $n \geqslant 0$, because F_0/F is a cyclic extension of degree 5, and F_n/F_0 is clearly a 5-extension. Hence Σ is pro-5 in this case. Hence (15) shows that the hypotheses of Theorem 1.10 hold in this case, and so it follows that $C(E/F_\infty)$ is $\Lambda(\Sigma)$-torsion.

The next result proves a rather surprising vanishing theorem for the cohomology of $\mathcal{S}(E/F_\infty)$. If $p \geqslant 5$, we recall that both Σ and every open subgroup Σ' of Σ have p-cohomological dimension equal to 4.

Theorem 1.11. *Assume that* (i) $p \geqslant 5$, *and* (ii) *$C(E/F_\infty)$ has $\Lambda(\Sigma)$-rank equal to $\tau_p(E/F)$. Then, for every open subgroup Σ' of Σ, we have*

$$H^i(\Sigma', \mathcal{S}(E/F_\infty)) = 0 \tag{23}$$

for all $i \geqslant 2$.

For example, the vanishing assertion (23) holds for $E = 50(A1)$ and $p = 5, 29, 59, \ldots$, and for $E = X_1(11)$ and $p = 5$, with $F = \mathbb{Q}$ in both cases.

1.3 The Euler characteristic formula

Exact formulae play an important part in the Iwasawa theory of elliptic curves. For example, if the Selmer group $\mathcal{S}(E/F_\infty)$ is to eventually be useful for studying the arithmetic of E over the base field F, we must be able to recover the basic arithmetic invariants of E over F from some exact formula related to the Σ-structure of $\mathcal{S}(E/F_\infty)$. The natural means of obtaining such an exact formula is via the calculation of the Σ-Euler characteristic of $\mathcal{S}(E/F_\infty)$. When do we expect this Σ-Euler characteristic to be finite?

Conjecture 1.12. *For each prime $p \geqslant 5$, $\chi(\Sigma, \mathcal{S}(E/F_\infty))$ is finite if and only if both $\mathcal{S}(E/F)$ is finite and $\tau_p(E/F) = 0$.*

We shall show later that even the finiteness of $H^0(\Sigma, \mathcal{S}(E/F_\infty))$ implies that $\mathcal{S}(E/F)$ is finite and $\tau_p(E/F) = 0$. However, the implication of the conjecture in the other direction is difficult and unknown. The second natural question to ask is what is the value of $\chi(\Sigma, \mathcal{S}(E/F_\infty))$ when it is finite? We will now describe a conjectural answer to this question given by Susan Howson and myself (see [5], [6]). Let $\text{Ш}(E/F)$ denote the Tate-Shafarevich group of E over F. For each finite prime v of F, let $E_0(F_v)$ be the subgroup of $E(F_v)$ consisting of the points with non-singular reduction, and put

$$c_v = [E(F_v) : E_0(F_v)]. \tag{24}$$

If A is any abelian group, $A(p)$ will denote its p-primary subgroup. Let $|\ |_p$ be the p-adic valuation of $\mathbb{Q}$, normalized so that $|p|_p = p^{-1}$. We then define

$$\rho_p(E/F) = \frac{\#(\text{Ш}(E/F)(p))}{\left(\#(E(F)(p))\right)^2} \times \left|\prod_v c_v\right|_p^{-1}, \tag{25}$$

where it is assumed that $\text{Ш}(E/F)(p)$ is finite. If v is a finite place of F, write k_v for the residue field of v and $\widetilde{E}_v$ for the reduction of E modulo v. Let j_E denote the classical j-invariant of our curve E. We define

$$\mathfrak{M} = \{\text{finite places } v \text{ of } F \text{ such that } \operatorname{ord}_v(j_E) < 0\}. \tag{26}$$

In other words, $\mathfrak{M}$ is the set of places of F where E has potential multiplicative reduction. For each $v \in \mathfrak{M}$, let $L_v(E, s)$ be the Euler factor of E at v. Thus $L_v(E, s)$ is equal to 1, $(1 - (Nv)^{-s})^{-1}$ or $(1 + (Nv)^{-s})^{-1}$, according as E has additive, split multiplicative, or non-split multiplicative reduction at v. The following conjecture is made in [6]:

Conjecture 1.13. *Assume that p is a prime such that* (i) $p \geqslant 5$, (ii) *E has good ordinary reduction at all places v of F dividing p, and* (iii) $\mathcal{S}(E/F)$

is finite. Then $H^i(\Sigma, \mathcal{S}(E/F_\infty))$ *is finite for* $i = 0, 1$, *and equal to* 0 *for* $i = 2, 3, 4$, *and*

$$\chi(\Sigma, \mathcal{S}(E/F_\infty)) = \rho_p(E/F) \times \left| \prod_{v|p} (\#(\widetilde{E}_v(k_v)))^2 \times \prod_{v \in \mathfrak{M}} L_v(E,1)^{-1} \right|_p^{-1}. \quad (27)$$

We remark in passing that Conjecture 1 made in our earlier note [5] is not correct because it does not contain the term coming from the Euler factors in $\mathfrak{M}$. We are very grateful to Richard Taylor for pointing this out to us.

Example. Take $F = \mathbb{Q}$ and E to be one of the two curves $X_0(11)$ and $X_1(11)$ given by (4) and (5). The conjecture applies to the primes $p = 5, 7, 13, 17, 23, 31, \ldots$ where these two isogenous curves admit good ordinary reduction.

We shall simply denote either curve by E when there is no need to distinguish between them. We have

$$\mathfrak{M} = \{11\}, \quad L_{11}(E,s) = (1 - 11^{-s})^{-1}, \quad (28)$$

and

$$\#(\widetilde{E}_5(\mathbb{F}_5)) = 5, \quad \widetilde{E}_p(\mathbb{F}_p)(p) = 0 \text{ for all } p \geqslant 7 \text{ with } p \neq 11. \quad (29)$$

This last statement is true because of Hasse's bound for the order of $\widetilde{E}_p(\mathbb{F}_p)$ and the fact that 5 must divide the order of $\widetilde{E}_p(\mathbb{F}_p)$ for all primes $p \neq 5, 11$. We also have $c_q = 1$ for all $q \neq 11$, and

$$c_{11}(X_0(11)) = 5, \quad c_{11}(X_1(11)) = 1.$$

As is explained in Greenberg's article in this volume, a 5-descent on either curve shows that

$$Ш(E/\mathbb{Q})(5) = 0, \quad E(\mathbb{Q}) = \mathbb{Z}/5\mathbb{Z}.$$

Hence we see that Conjecture 1.13 for $p = 5$ predicts that

$$\chi(\Sigma, \mathcal{S}(X_0(11)/F_\infty)) = 5^2, \quad \chi(\Sigma, \mathcal{S}(X_1(11)/F_\infty)) = 5. \quad (30)$$

In Chapter 4 of these notes (see Proposition 4.10), we prove Conjecture 1.13 for both of the elliptic curves $X_0(11)$ and $X_1(11)$ with $F = \mathbb{Q}$ and $p = 5$. Hence the values (30) are true. Now assume p is a prime $\geqslant 7$. We claim that

$$Ш(E/\mathbb{Q})(p) = 0. \quad (31)$$

Indeed, the conjecture of Birch and Swinnerton-Dyer predicts that $Ш(E/\mathbb{Q}) = 0$, and Kolyvagin's theorem tells us that $Ш(E/\mathbb{Q})$ is finite since $L(E,1) \neq 0$. In fact, Kolyvagin's method (see Gross [14], in particular Proposition 2.1) shows that (31) holds if we can find an imaginary quadratic field K, in which 11 splits, such that the Heegner point attached to K in $E(K)$ is not divisible by p; here we are using Serre's result [26] that $G(F_0/\mathbb{Q}) = \mathrm{GL}_2(\mathbb{F}_p)$ for all primes $p \neq 5$. The determination of such a field K is well known by computation, but unfortunately the details of such a computation do not seem to have been published anywhere. Granted (31), we deduce from (28) and (29) that Conjecture 1.13 predicts that

$$\chi(\Sigma, \mathcal{S}(E/F_\infty)) = 1 \tag{32}$$

for all primes $p \geqslant 7$ where E has good ordinary reduction. At present, we cannot prove (32) for a single prime $p \geqslant 7$.

In these notes, we shall prove two results in the direction of Conjecture 1.13, both of which are joint work with Susan Howson.

Theorem 1.14. *In addition to the hypotheses of Conjecture* 1.13, *let p be such that $C(E/F_\infty)$ is $\Lambda(\Sigma)$-torsion. Then Conjecture* 1.13 *is valid for p.*

Of course, Theorem 1.14 is difficult to apply in practice, since we only have rather weak results (see Theorem 1.10) for showing that $C(E/F_\infty)$ is $\Lambda(\Sigma)$-torsion. The next result avoids making this hypothesis, but only establishes a partial result. Put

$$\xi_p(E/F) = \rho_p(E/F) \times \left| \prod_{v|p} (\#(\widetilde{E}_v(k_v)))^2 \times \prod_{v \in \mathfrak{M}} L_v(E,1)^{-1} \right|_p^{-1}. \tag{33}$$

Theorem 1.15. *Let E be a modular elliptic curve over $\mathbb{Q}$ such that $L(E,1) \neq 0$. Let p be a prime $\geqslant 5$ where E has good ordinary reduction. As before, let $F_\infty = \mathbb{Q}(E_{p^\infty})$. Then* (i) *$H^1(\Sigma, \mathcal{S}(E/F_\infty))$ is finite and its order divides $\#(H^3(\Sigma, E_{p^\infty}))$, and* (ii) *$H^0(\Sigma, \mathcal{S}(E/F_\infty))$ is finite of exact order $\xi_p(E/\mathbb{Q}) \times \#(H^3(\Sigma, E_{p^\infty}))$.*

We recall that we conjecture that $H^j(\Sigma, \mathcal{S}(E/F_\infty)) = 0$ for $j = 2,3,4$ for all $p \geqslant 5$, but we cannot prove at present that these cohomology groups are even finite under the hypotheses of Theorem 1.15. Note also that the order of $H^3(\Sigma, E_{p^\infty})$ can easily be calculated using Lemma 1.3. As an example of Theorem 1.15, we see that for E given either by $X_0(11)$ or $X_1(11)$, we have

$$H^0(\Sigma, \mathcal{S}(E/F_\infty)) = H^1(\Sigma, \mathcal{S}(E/F_\infty)) = 0 \tag{34}$$

for all primes $p \geqslant 7$ where E has good ordinary reduction. Indeed, we have $H^3(\Sigma, E_{p^\infty}) = 0$ for all primes $p \neq 5$ because of Lemma 1.3 and Serre's result that $G(F_0/\mathbb{Q}) = \mathrm{GL}_2(\mathbb{F}_p)$ for all $p \neq 5$.

2 Basic Properties of the Selmer Group

2.1 Nakayama's lemma

If G is an arbitrary profinite group, we recall that its Iwasawa algebra $\Lambda(G)$ is defined by

$$\Lambda(G) = \varprojlim_{H} \mathbb{Z}_p[G/H], \tag{35}$$

where H runs over all open normal subgroups of G. We endow $\mathbb{Z}_p[G/H]$ with the p-adic topology for every open normal subgroup H. This induces a topology on $\Lambda(G)$ which makes it a compact $\mathbb{Z}_p$-algebra, in which the ordinary group ring $\mathbb{Z}_p[G]$ is a dense sub-algebra. If X is a compact left $\Lambda(G)$-module, our aim is to establish a version of Nakayama's lemma giving a sufficient condition for X to be finitely generated over $\Lambda(G)$. Balister and Howson (see [1], §3) have pointed out that there are unexpected subtleties in this question for arbitrary compact X. Fortunately, we will need only the case when X is pro-finite, and these difficulties do not occur. We define the augmentation ideal $I(G)$ of $\Lambda(G)$ by

$$I(G) = \mathrm{Ker}\big(\Lambda(G) \longrightarrow \mathbb{Z}_p = \mathbb{Z}_p[G/G]\big). \tag{36}$$

Proposition 2.1. *Assume that G is a pro-p-group, and that X is a pro-p-abelian group, which is a left $\Lambda(G)$-module. Then $X = 0$ if and only if $X/I(G)X = 0$.*

Proof. We have $X = \mathrm{Hom}(A, \mathbb{Q}_p/\mathbb{Z}_p)$, where A is a discrete p-primary abelian group. Moreover, $X/I(G)X$ is dual to A^G. Hence we must show that $A^G = 0$ if and only if $A = 0$. One implication being trivial, we assume that $A^G = 0$. Suppose, on the contrary, that $A \neq 0$. Since A is a discrete G-module, it follows that $A^U \neq 0$ for some open normal subgroup U of G. Hence there exists a non-zero finite G-submodule B of A^U. But then $B^{G/U} = 0$ since $A^G = 0$, and so, as G/U is a finite p-group, we have $B = 0$ by the standard result for finite p-groups. This is the desired contradiction, and the proof is complete.

Corollary 2.2. *Assume that G is a pro-p-group, and that X is a pro-p-abelian group, which is a left $\Lambda(G)$-module. If $X/I(G)X$ is a finitely generated $\mathbb{Z}_p$-module, then X is a finitely generated $\Lambda(G)$-module.*

Proof. Let $x_1, \ldots, x_s$ be lifts to X of any finite set of $\mathbb{Z}_p$-generators of $X/I(G)X$. Define Y to be the left $\Lambda(G)$-submodule of X generated by $x_1, \ldots, x_s$. Then Y is a closed subgroup of X and X/Y is also a pro-p-abelian group. But

$$I(G)(X/Y) = (I(G)X + Y)/Y = X/Y.$$

Hence $X = Y$ by Proposition 2.1, completing the proof.

We make the following remark (see [1]). Let X be a pro-p-abelian group which is a finitely generated left $\Lambda(G)$-module. If G is isomorphic to $\mathbb{Z}_p$, the structure theory for finitely generated $\Lambda(G)$-modules implies that, if $X/I(G)X$ is finite, then X is certainly torsion over the Iwasawa algebra $\Lambda(G)$. However, contrary to what is asserted in Harris [16], Balister and Howson [1] show that the analogue of this assertion breaks down completely if we take G to be any pro-p open subgroup of $\mathrm{GL}_2(\mathbb{Z}_p)$.

2.2 The fundamental diagram

We now return to our elliptic curve E which is defined over a finite extension F of $\mathbb{Q}$, and does not admit complex multiplication. We use without comment all the notation of Chapter 1. Thus p denotes an arbitrary prime number, $F_\infty = F(E_{p^\infty})$, and Σ the Galois group of F_∞ over F. The crucial ingredient in studying the Selmer group $\mathcal{S}(E/F_\infty)$ as a module over the Iwasawa algebra $\Lambda(\Sigma)$ is the single natural commutative diagram (43) given below. Because of its importance, we shall henceforth call it the *fundamental diagram*.

Let T denote any finite set of primes of F which contains all the primes dividing p, and all primes where E has bad reduction. Let F_T denote the maximal extension of F which is unramified outside of T and all the archimedean primes of F, and let

$$G_T = G(F_T/F)$$

be the corresponding Galois group. We mention in passing that very little is known about the Galois group G_T beyond the fact that its profinite order is divisible by infinitely many distinct prime numbers. For example, it is unknown whether G_T is topologically finitely generated even in the simplest case when $F = \mathbb{Q}$ and $T = \{2\}$, as was remarked to us by Serre.

By our choice of T, we clearly have $F_\infty \subset F_T$. If H denotes any intermediate field with $F \subset H \subset F_T$, we put

$$G_T(H) = G(F_T/H). \tag{37}$$

Suppose now that L is a finite extension of F. For each finite place v of F, we define

$$J_v(L) = \bigoplus_{w|v} H^1(L_w, E)(p), \tag{38}$$

where w runs over all primes of L dividing v. We then have the localization map

$$\lambda_T(L) : H^1(G_T(L), E_{p^\infty}) \longrightarrow \bigoplus_{v \in T} J_v(L). \tag{39}$$

If H is an infinite extension of F, we define

$$J_v(H) = \varinjlim_{L} J_v(L),$$

where the inductive limit is taken with respect to the restriction maps, and L runs over all finite extensions of F contained in H. We also define $\lambda_T(H)$ to be the inductive limit of the localization maps $\lambda_T(L)$. The following lemma is classical:

Lemma 2.3. *For every algebraic extension H of F, we have $\mathcal{S}(E/H) =$ $\mathrm{Ker}(\lambda_T(H))$.*

Proof. It clearly suffices to prove Lemma 2.3 for every finite extension L of F, and we quickly sketch the proof in this case. By definition, $\mathcal{S}(E/L)$ is given by the exactness of the sequence

$$0 \longrightarrow \mathcal{S}(E/L) \longrightarrow H^1(L, E_{p^\infty}) \longrightarrow \prod_{w} H^1(L_w, E),$$

where w ranges over all finite places of L. Let

$$\varphi_T(L) : H^1(L, E_{p^\infty}) \longrightarrow \prod_{w \nmid T} H^1(L_w, E)$$

denote the map given by localization at all finite primes w of L which do not lie above T. Clearly Lemma 2.3 is equivalent to the assertion

$$H^1(G_T(L), E_{p^\infty}) = \mathrm{Ker}(\varphi_T(L)). \tag{40}$$

The proof of (40) follows easily from two standard classical facts about the arithmetic of E over local fields. Let w denote any finite prime of L which does not lie above T. Then E has good reduction at w, and so

$$H^1(G(L_w^{nr}/L_w), E(L_w^{nr})) = 0, \tag{41}$$

where L_w^{nr} denotes the maximal unramified extension of L_w. It follows immediately from (41) that the left hand side of (40) is contained in the right hand side. Next, let P be any point in $E(L_w)$, and let P_n be any point in $E(\overline{L}_w)$ such that $p^n P_n = P$ for some integer $n \geqslant 0$. Then the second fact is that, because w does not divide p and E has good reduction at w, we have that the extension $L_w(P_n)/L_w$ is unramified. The inclusion of the right hand side of (40) in the left hand side follows immediately from this second fact and local Kummer theory on E. This completes the proof of Lemma 2.3.

In view of Lemma 2.3, we have the exact sequence

$$0 \longrightarrow \mathcal{S}(E/F_\infty) \longrightarrow H^1(G_T(F_\infty), E_{p^\infty}) \xrightarrow{\lambda_T(F_\infty)} \bigoplus_{v \in T} J_v(F_\infty). \tag{42}$$

Taking Σ-invariants of (42), we obtain the fundamental diagram

$$\begin{array}{ccccccc} 0 \longrightarrow & \mathcal{S}(E/F_\infty)^\Sigma & \longrightarrow & H^1(G_T(F_\infty), E_{p^\infty})^\Sigma & \xrightarrow{\psi_T(F_\infty)} & \bigoplus_{v\in T} J_v(F_\infty)^\Sigma & \\ & \uparrow \alpha & & \uparrow \beta & & \uparrow \gamma & \\ 0 \longrightarrow & \mathcal{S}(E/F) & \longrightarrow & H^1(G_T, E_{p^\infty}) & \xrightarrow{\lambda_T(F)} & \bigoplus_{v\in T} J_v(F), & \end{array} \tag{43}$$

where the rows are exact, and the vertical maps are the obvious restriction maps. We emphasize that all of our subsequent arguments revolve around analysing this diagram. Since γ is a direct sum

$$\gamma = \bigoplus_{v\in T} \gamma_v,$$

where γ_v denotes the restriction map from $J_v(F)$ to $J_v(F_\infty)$, we see that the analysis of $\text{Ker}(\gamma)$ and $\text{Coker}(\gamma)$ is a purely local question, whose answer at the primes v dividing p uses the theory of deeply ramified p-adic fields developed in [4].

2.3 Finite generation over $\Lambda(\Sigma)$

As a first application of the fundamental diagram (43), we shall prove that, for all primes p, the Pontrjagin duals of both $H^1(G_T(F_\infty), E_{p^\infty})$ and $\mathcal{S}(E/F_\infty)$ are finitely generated left $\Lambda(\Sigma)$-modules. We begin with a very well known lemma. If A is a discrete p-primary abelian group, we recall that the Pontrjagin dual $\widehat{A}$ of A is defined by

$$\widehat{A} = \text{Hom}(A, \mathbb{Q}_p/\mathbb{Z}_p).$$

Lemma 2.4. *The Pontrjagin dual of $H^1(G_T, E_{p^\infty})$ is a finitely generated $\mathbb{Z}_p$-module.*

Proof. Taking G_T-cohomology of the exact sequence

$$0 \longrightarrow E_p \longrightarrow E_{p^\infty} \xrightarrow{p} E_{p^\infty} \longrightarrow 0,$$

and putting $A = H^1(G_T, E_{p^\infty})$, we obtain a surjection

$$H^1(G_T, E_p) \twoheadrightarrow (A)_p, \tag{44}$$

where $(A)_p$ denotes the elements of A of order dividing p. But $H^1(G_T, M)$ is well known to be finite for any finite p-primary G_T-module M. Hence the fact that (44) is a surjection implies that $(A)_p$ is finite. Let X be the Pontrjagin dual of A. Now $(A)_p$ is dual to X/pX, and so this latter group is finite. But then, by Nakayama's lemma, X must be a finitely generated $\mathbb{Z}_p$-module. This completes the proof of Lemma 2.4.

Lemma 2.5. *The Pontrjagin dual of* $\mathrm{Ker}(\gamma)$ *is a finitely generated* $\mathbb{Z}_p$-*module of rank at most* $[F:\mathbb{Q}]$.

Proof. We recall that the $\mathbb{Z}_p$-rank of a $\mathbb{Z}_p$-module X is defined by

$$\mathrm{rank}_{\mathbb{Z}_p}(X) = \dim_{\mathbb{Q}_p}(X \otimes_{\mathbb{Z}_p} \mathbb{Q}_p).$$

Also, if A is an abelian group, we write

$$A^* = \varprojlim A/p^n A \tag{45}$$

for the p-adic completion of A. Let Y be the Pontrjagin dual of $\mathrm{Ker}(\gamma)$, let Z_v be the dual of $J_v(F)$, and let $Z = \bigoplus_{v \in T} Z_v$. Since $\mathrm{Ker}(\gamma)$ is a subgroup of $\bigoplus_{v\in T} J_v(F)$, it follows that Y is a quotient of Z. By Tate duality, $H^1(F_v, E)$ is dual to $E(F_v)$, and so $J_v(F) = H^1(F_v, E)(p)$ is dual to $E(F_v)^*$. But, by the theory of the formal group of E at v, we have that $E(F_v)^*$ is finite if v does not divide p, and that $E(F_v)^*$ is a finitely generated $\mathbb{Z}_p$-module of rank equal to $[F_v : \mathbb{Q}_p]$ if v does divide p. Hence Z is a finitely generated $\mathbb{Z}_p$-module of rank equal to $[F:\mathbb{Q}]$, and so Y must be a finitely generated $\mathbb{Z}_p$-module of rank at most $[F:\mathbb{Q}]$ because it is a quotient of Z. This completes the proof of Lemma 2.5.

Lemma 2.6. *For all primes p, we have* (i) $\mathrm{Ker}(\beta)$ *and* $\mathrm{Coker}(\beta)$ *are finite,* (ii) $\mathrm{Ker}(\alpha)$ *is finite, and* (iii) $\mathrm{Coker}(\alpha)$ *is dual to a finitely generated* $\mathbb{Z}_p$-*module of* $\mathbb{Z}_p$-*rank at most* $[F:\mathbb{Q}]$.

Proof. By the inflation-restriction sequence, we have that $\mathrm{Ker}(\beta) = H^1(\Sigma, E_{p^\infty})$ and that $\mathrm{Coker}(\beta)$ injects into $H^2(\Sigma, E_{p^\infty})$. Assertion (i) follows immediately because $H^i(\Sigma, E_{p^\infty})$ is finite for all $i \geqslant 0$ (see [29]). Assertion (ii) is then plain because $\mathrm{Ker}(\alpha)$ injects into $\mathrm{Ker}(\beta)$. Finally, it is clear from (43) that we have the exact sequence

$$\mathrm{Im}(\lambda_T(F)) \cap \mathrm{Ker}(\gamma) \longrightarrow \mathrm{Coker}(\alpha) \longrightarrow \mathrm{Coker}(\beta),$$

and so (iii) follows from (i) and Lemma 2.4. This completes the proof of the lemma.

Theorem 2.7. *The Pontrjagin duals of both* $H^1(G_T(F_\infty), E_{p^\infty})$ *and* $\mathcal{S}(E/F_\infty)$ *are finitely generated* $\Lambda(\Sigma)$-*modules.*

Proof. As in Chapter 1, let

$$F_0 = F(E_p), \quad \Sigma_0 = G(F_\infty/F_0).$$

While Σ is not in general a pro-p-group, Σ_0 always is pro-p since it is isomorphic under the injection (3) to an open subgroup of the kernel of the reduction map from $\mathrm{GL}_2(\mathbb{Z}_p)$ to $\mathrm{GL}_2(\mathbb{F}_p)$. Let X and Y denote the Pontrjagin duals of

$H^1(G_T(F_\infty), E_{p^\infty})$ and $\mathcal{S}(E/F_\infty)$, respectively. Since $\Lambda(\Sigma_0)$ is a sub-algebra of $\Lambda(\Sigma)$, it clearly suffices to show that X and Y are finitely generated left $\Lambda(\Sigma_0)$-modules. We shall establish this using Corollary 2.2. Note first that X and Y are pro-finite. To prove this, we must show that $H^1(G_T(F_\infty), E_{p^\infty})$ and $\mathcal{S}(E/F_\infty)$ are inductive limits of finite groups. But clearly $H^1(G_T(F_\infty), E_{p^\infty})$ is the inductive limit of the finite groups $H^1(G_T(L), E_{p^n})$ where L runs over all finite extensions of F contained in F_∞ and n runs over all integers $\geqslant 1$. Similarly, $\mathcal{S}(E/F_\infty)$ is the inductive limit of the finite groups

$$\mathcal{S}(E/L, p^n) = \mathrm{Ker}(H^1(G_T(L), E_{p^n}) \longrightarrow \bigoplus_{v \in T} J_v(L)),$$

where L and n run over the same sets. We now appeal to the fundamental diagram (43), but with the base field F replaced by F_0, and consequently Σ replaced by Σ_0. Since $\mathrm{Ker}(\beta)$ and $\mathrm{Coker}(\beta)$ are finite, it follows immediately from Lemma 2.4 for F_0 and (43) that the dual of

$$H^1(G_T(F_\infty), E_{p^\infty})^{\Sigma_0} \tag{46}$$

is a finitely generated $\mathbb{Z}_p$-module. Hence X is a finitely generated $\Lambda(\Sigma_0)$-module by Corollary 2.2, since $X/I(\Sigma_0)X$ is dual to (46). We next claim that the dual of

$$\mathcal{S}(E/F_\infty)^{\Sigma_0} \tag{47}$$

is a finitely generated $\mathbb{Z}_p$-module. By virtue of (43), it suffices to show that both the image of α and the cokernel of α are dual to finitely generated $\mathbb{Z}_p$-modules. Now (iii) of Lemma 2.6 for F_0 gives that $\mathrm{Coker}(\alpha)$ is dual to a finitely generated $\mathbb{Z}_p$-module. Also $\mathcal{S}(E/F_0)$ is contained in $H^1(G_T(F_0), E_{p^\infty})$, and so we deduce from Lemma 2.4 that $\mathrm{Im}(\alpha)$ is also dual to a finitely generated $\mathbb{Z}_p$-module. Having proved the above claim, it again follows from Corollary 2.2 that Y is a finitely generated $\Lambda(\Sigma_0)$-module. This completes the proof of Theorem 2.7.

2.4 Decomposition of primes in F_∞

We need hardly remind the reader that no precise reciprocity law is known for giving the decomposition of finite primes of F in F_∞. Nevertheless, we collect together here some coarser elementary results in this direction, which will be used later in these notes. We have omitted discussing the primes v dividing p where E has potential supersingular reduction at v, since this involves the notion of formal complex multiplication (see Serre [27]), and this case will not be needed in our subsequent arguments.

Let v denote any finite prime of F. For simplicity, we write $D(v)$ for the decomposition group in Σ of any fixed prime of F_∞ above v. Thus $D(v)$ is only determined up to conjugation in Σ by v. Now $D(v)$ is itself a p-adic Lie group, since it is a closed subgroup of the 4-dimensional p-adic Lie group Σ. We can easily determine the dimension of $D(v)$ in many cases.

Lemma 2.8. (i) *If v does not divide p, then $D(v)$ has dimension* 1 *or* 2, *according as E has potential good or potential multiplicative reduction at v.* (ii) *Assume that v does divide p. Then $D(v)$ has dimension* 2 *if E has potential multiplicative reduction at v, and dimension* 3 *if E has potential ordinary reduction at v.*

Since the global Galois group $\Sigma = G(F_\infty/F)$ is a p-adic Lie group of dimension 4, we immediately obtain the following corollary.

Corollary 2.9. *There are infinitely many primes of F_∞ lying above each finite prime of F which does not divide p. There are also infinitely many primes of F_∞ lying above each prime of F, which divides p, and where E has potential ordinary or potential multiplicative reduction.*

We now prove Lemma 2.8. Let L denote the completion of F at v, and let $L_\infty = L(E_{p^\infty})$. We can then identify $D(v)$ with local Galois group $\Omega = G(L_\infty/L)$. We first remark that the dimension of Ω does not change if we replace L by a finite extension L', i.e. if we put $L'_\infty = L'(E_{p^\infty})$ and $\Omega' = G(L'_\infty/L)$, then Ω and Ω' have the same dimension as p-adic Lie groups. This is clear since restriction to L_∞ defines an isomorphism from Ω' onto an open subgroup of Ω. Thus we may assume that E has either good reduction or split multiplicative reduction over L. We first dispose of the easy case when v does not divide p. If E/L has good reduction, then $L(E_{p^\infty})/L$ is unramified, and Ω plainly has dimension 1. If E/L has split multiplicative reduction, we write q_E for the v-adic Tate period of E. Then L_∞ is clearly obtained by adjoining all p^m-th ($m = 1, 2, \ldots$) roots of q_E to $L(\mu_{p^\infty})$, and it is then clear that Ω has dimension 2 as a p-adic Lie group. We now turn to the two cases when v divides p.

Case 1. Assume that v divides p, and that E/L has good ordinary reduction. Let $\widetilde{E}_{p^\infty}$ denote the reduction of E_{p^∞} modulo v, and let $\widehat{E}_{p^\infty}$ be the kernel of reduction modulo v. As usual, if A is an abelian group, we write $T_p(A) = \varprojlim (A)_{p^n}$, where $(A)_{p^n}$ denotes the kernel of multiplication by p^n on A. Now we have the exact sequence of Ω-modules

$$0 \longrightarrow T_p(\widehat{E}_{p^\infty}) \longrightarrow T_p(E) \longrightarrow T_p(\widetilde{E}_{p^\infty}) \longrightarrow 0, \tag{48}$$

where the two end groups are free of rank 1 over $\mathbb{Z}_p$ by our ordinary hypothesis. Let η and ε denote the characters of Ω with values in $\mathbb{Z}_p^\times$ giving its action on $T_p(\widehat{E}_{p^\infty})$ and $T_p(\widetilde{E}_{p^\infty})$, respectively. By the Weil pairing $\eta\varepsilon$ is the character giving the action of Ω on $T_p(\mu)$. Choosing a basis of $T_p(E_{p^\infty})$

whose first element is a basis of $T_p(\widehat{E}_{p^\infty})$, we deduce from (48) an injection $\rho : \Omega \hookrightarrow \mathrm{GL}_2(\mathbb{Z}_p)$ such that, for all $\sigma \in \Omega$, we have

$$\rho(\sigma) = \begin{pmatrix} \eta(\sigma) & a(\sigma) \\ 0 & \varepsilon(\sigma) \end{pmatrix}, \tag{49}$$

where $a(\sigma) \in \mathbb{Z}_p$. Since E does not have complex multiplication, it is known (see Serre [27]) that (48) does not split as an exact sequence of Ω-modules, and so we have that $a(\sigma)$ is non-zero for some $\sigma \in \Omega$. Now let H_∞ denote the maximal unramified extension of L contained in L_∞, and put $M_\infty = H_\infty(\mu_{p^\infty})$. Thus we have the tower of fields

$$L \subset H_\infty \subset M_\infty = H_\infty(\mu_{p^\infty}) \subset L_\infty = L(E_{p^\infty}). \tag{50}$$

We claim that the three Galois groups $G(H_\infty/L)$, $G(M_\infty/H_\infty)$ and $G(L_\infty/M_\infty)$ are each p-adic Lie groups of dimension 1, whence it follows immediately that $\Omega = G(L_\infty/L)$ is a p-adic Lie group of dimension 3, as required. Since $\widetilde{E}_{p^\infty}$ is rational over H_∞, it is clear that H_∞ must be a finite extension of the unramified $\mathbb{Z}_p$-extension of L, whence $G(H_\infty/L)$ has dimension 1. The action of $G(M_\infty/H_\infty)$ on μ_{p^∞} defines an injection of this Galois group onto an open subgroup of $\mathbb{Z}_p^\times$, whence it also has dimension 1. Finally, since ε and η are both trivial on $G(L_\infty/M_\infty)$, the map $\sigma \mapsto a(\sigma)$ defines an injection of $G(L_\infty/M_\infty)$ into $\mathbb{Z}_p$. But the image of this map cannot be 0 as we remarked above, and so we conclude that $G(L_\infty/M_\infty)$ is isomorphic to $\mathbb{Z}_p$. This completes the proof that Ω has dimension 3 in this case.

Case 2. Assume that v divides p, and that E/L has split multiplicative reduction at v. The argument that Ω has dimension 2 is entirely parallel to that given when E does not divide p. Indeed, let q_E denote the Tate period of E. Then L_∞ is again obtained by adjoining to $L(\mu_{p^\infty})$ the p^m-th roots $(m = 1, 2, \dots)$ of q_E. This completes the proof of Lemma 2.8.

2.5 The vanishing of $H^2(G_T(F_\infty), E_{p^\infty})$

We are grateful to Y. Ochi ([24]) for pointing out to us the following basic fact about the cohomology of E over F_∞.

Theorem 2.10. *For all odd primes p, we have*

$$H^2(G_T(F_\infty), E_{p^\infty}) = 0. \tag{51}$$

This is a rare example of a statement which is easier to prove for the extension F_∞ rather than the cyclotomic $\mathbb{Z}_p$-extension of F. Indeed, if K_∞ denotes the cyclotomic $\mathbb{Z}_p$-extension of F, then it has long been conjectured that

$$H^2(G_T(K_\infty), E_{p^\infty}) = 0$$

for all odd primes p. However, at present this latter assertion has only been proven in some rather special cases.

We now give the proof of Theorem 2.10. Since E_{p^∞} is rational over F_∞, the Galois group $G_T(F_\infty) = G(F_T/F_\infty)$ operates trivially on E_{p^∞}, which is therefore isomorphic to $(\mathbb{Q}_p/\mathbb{Z}_p)^2$ as a $G_T(F_\infty)$-module. Hence it suffices to show that

$$H^2(G_T(F_\infty), \mathbb{Q}_p/\mathbb{Z}_p) = 0 \tag{52}$$

for all primes p. But F_∞ is the union of the fields $F_n = F(E_{p^{n+1}})$ ($n = 0, 1, \ldots$), and thus it is also the union of the fields $K_{n,\infty} = F_n(\mu_{p^\infty})$ ($n = 0, 1, \ldots$), since $\mu_{p^\infty} \subset F_\infty$ by the Weil pairing. Hence we have

$$H^2(G_T(F_\infty), \mathbb{Q}_p/\mathbb{Z}_p) = \varinjlim H^2(G_T(K_{n,\infty}), \mathbb{Q}_p/\mathbb{Z}_p),$$

where the inductive limit is taken with respect to the restriction maps. But each of the cohomology groups in the inductive limit on the right vanishes, thanks to the following general result due essentially to Iwasawa [19]. Let K be any finite extension of $\mathbb{Q}$, K_∞ the cyclotomic $\mathbb{Z}_p$-extension of K, and K_T the maximal extension of K unramified outside T and the archimedean primes of K, where T is an arbitrary finite set of primes of K containing all primes dividing p. Then we claim that

$$H^2(G(K_T/K_\infty), \mathbb{Q}_p/\mathbb{Z}_p) = 0 \tag{53}$$

for all odd primes p. Here is an outline of the proof of (53). Let $\Gamma = G(K_\infty/K)$, and write $\Lambda(\Gamma)$ for the Iwasawa algebra of Γ. Let $\delta_i(K)$ denote the $\Lambda(\Gamma)$-rank of the Pontrjagin dual of $H^i(G(K_T/K_\infty), \mathbb{Q}_p/\mathbb{Z}_p)$ ($i = 1, 2$). By a basic Euler characteristic calculation (see [12], Proposition 3), we have

$$\delta_1(K) - \delta_2(K) = r_2(K), \tag{54}$$

where $r_2(K)$ denotes the number of complex places of K. On the other hand, we have

$$H^1(G(K_T/K_\infty), \mathbb{Q}_p/\mathbb{Z}_p) = \mathrm{Hom}(G(M_\infty/K_\infty), \mathbb{Q}_p/\mathbb{Z}_p),$$

where M_∞ denotes the maximal abelian p-extension of K_∞ which is unramified outside T and the archimedean primes. But it follows easily from one of the principal results of Iwasawa [19] that $G(M_\infty/K_\infty)$ has $\Lambda(\Gamma)$-rank exactly equal to $r_2(K)$. It follows from (54) that we must have $\delta_2(K) = 0$. But the dual of $H^2(G(K_T/K_\infty), \mathbb{Q}_p/\mathbb{Z}_p)$ is a free $\Lambda(\Gamma)$-module for all odd primes p (see [12], Proposition 4), and so (53) follows. This completes the proof of Theorem 2.10.

Although we will not have time to give the proof in the present notes, we mention in passing that Y. Ochi [24] and S. Howson [18] have proven that the analogue of the Euler characteristic formula also holds for the $\Lambda(\Sigma)$-ranks of the Pontrjagin duals of the $H^i(G_T(F_\infty), E_{p^\infty})$ $(i = 1, 2)$. Let $\varepsilon_i(F)$ denote the $\Lambda(\Sigma)$-rank of the Pontrjagin dual of $H^i(G_T(F_\infty), E_{p^\infty})$, where we recall that the $\Lambda(\Sigma)$-rank is defined by (19). Although it is far from obvious, we again have

$$\varepsilon_1(F) - \varepsilon_2(F) = [F : \mathbb{Q}], \tag{55}$$

in exact parallel with the result given in Proposition 3 of [12] when F_∞ is replaced by the cyclotomic $\mathbb{Z}_p$-extension K_∞ of F. Granted (55), we obtain the following consequence of Theorem 2.10.

Corollary 2.11. *For all odd primes p, the Pontrjagin dual of*

$$H^1(G_T(F_\infty), E_{p^\infty})$$

has $\Lambda(\Sigma)$-rank equal to $[F : \mathbb{Q}]$.

Since the Selmer group $\mathcal{S}(E/F_\infty)$ is a submodule of $H^1(G_T(F_\infty), E_{p^\infty})$, we see that the upper bound for the $\Lambda(\Sigma)$-rank of the dual of $\mathcal{S}(E/F_\infty)$ asserted in Theorem 1.8 is an immediate consequence of Corollary 2.11.

3 Local cohomology calculations

3.1 Strategy

As always, E denotes an elliptic curve defined over a finite extension F of $\mathbb{Q}$, which does not admit complex multiplication. Throughout this chapter, p will denote an arbitrary prime number, $F_\infty = F(E_{p^\infty})$, and Σ will denote the Galois group of F_∞ over F. The aim of this chapter is to study the Σ-cohomology of the local terms $J_v(F_\infty)$, for any $v \in T$, which occur in the fundamental diagram (43). We recall that

$$J_v(F_\infty) = \varinjlim_n \bigoplus_{w|v} H^1(F_{n,w}, E)(p),$$

where $F_n = F(E_{p^{n+1}})$, w runs over all places of F_n dividing our given v in T, and the inductive limit is taken with respect to the restriction maps. Knowledge of the Σ-cohomology of the $J_v(F_\infty)$ will, in particular, play a crucial role in the calculation of the Σ-Euler characteristic of $\mathcal{S}(E/F_\infty)$ in the next chapter. These questions are purely local, thanks to the following well known principle. If w is a place of F_∞, we recall that $F_{\infty,w} = \bigcup_{n \geqslant 0} F_{n,w}$.

Lemma 3.1. *For each prime v in T, let w denote a fixed prime of F_∞ above v. Let $\Sigma_w \subset \Sigma$ denote the decomposition group of w over v. Then, for all $i \geqslant 0$, we have a canonical isomorphism*

$$H^i(\Sigma, J_v(F_\infty)) \simeq H^i(\Sigma_w, H^1(F_{\infty,w}, E)(p)).$$

Indeed, this is a simple and well known consequence of Shapiro's lemma. Let $\Delta_n = G(F_n/F)$, and let $\Delta_{n,w} \subset \Delta_n$ denote the decomposition group of the restriction of w to F_n. Then, for all $n \geqslant 0$, Shapiro's lemma gives a canonical isomorphism

$$H^i(\Delta_n, J_v(F_n)) \simeq H^i(\Delta_{n,w}, H^1(F_{n,w}, E)(p)) \quad (i \geqslant 0).$$

Passing to the inductive limit via the restriction maps as $n \to \infty$ immediately gives the assertion of Lemma 3.1.

When v does not divide p, we shall see that well known classical methods suffice to compute the cohomology. However, when v divides p, we will make essential use of the results about the cohomology of elliptic curves over deeply ramified extensions which are established in [4], noting that in this case $F_{\infty,w}$ is indeed deeply ramified because it contains the deeply ramified field $F_v(\mu_{p^\infty})$. All of the material discussed in this chapter is joint work with Susan Howson.

3.2 A vanishing theorem

Theorem 3.2. *Let p be any prime $\geqslant 5$. Then*

$$H^i(\Sigma, J_v(F_\infty)) = 0 \tag{56}$$

for all $i \geqslant 1$ and all primes v of F.

We break the proof of Theorem 3.2 up into a series of lemmas.

As before, let j_E denote the classical j-invariant of E. Thus, for any prime v of F, we have $\mathrm{ord}_v(j_E) < 0$ if and only if E has potential multiplicative reduction at v.

Lemma 3.3. *Let p be any prime, and let v be a place of F such that* $\mathrm{ord}_v(j_E) < 0$ *and v does not divide p. Then $J_v(F_\infty) = 0$.*

Proof. In view of the definition of $J_v(F_\infty)$, we must show that, under the hypothesis of Lemma 3.3, we have

$$H^1(F_{\infty,w}, E)(p) = 0 \tag{57}$$

for all places w of F_∞ lying above p. Since v does not divide p, we have

$$E(F_{\infty,w}) \otimes \mathbb{Q}_p/\mathbb{Z}_p = 0.$$

Hence local Kummer theory on E over $F_{\infty,w}$ shows that

$$H^1(F_{\infty,w}, E_{p^\infty}) \simeq H^1(F_{\infty,w}, E)(p). \tag{58}$$

Thus (57) will certainly follow if we can show that the Galois group of $\overline{F}_v$ over $F_{\infty,w}$ has p-cohomological dimension zero. Let M_∞ denote the maximal pro-p-extension of F_v. We will show that $F_{\infty,w}$ contains M_∞, which will certainly show that $G(\overline{F}_v/F_{\infty,w})$ has p-cohomological dimension zero. Now recall that, by the Weil pairing, $F_{\infty,w}$ contains the field $H_\infty = F_v(\mu_{p^\infty})$, and this latter field is an unramified extension of F_v which contains the unique unramified $\mathbb{Z}_p$-extension of F_v. Let F_v^{nr} denote the maximal unramified extension of F_v. It is well known (see Serre [26]) that the maximal tamely ramified extension of F_v has a topologically cyclic Galois group over F_v^{nr}. Now M_∞ is a tamely ramified extension of F_v because v does not divide p. It follows easily from these remarks that any Galois extension of H_∞, whose profinite degree over H_∞ is divisible by p^∞, must automatically contain M_∞. But this latter condition holds for $F_{\infty,w}$ thanks to our hypothesis that $\mathrm{ord}_v(j_E) < 0$. Indeed, to see this, assume first that E has split multiplicative reduction at v, so that E is isomorphic over F_v to a Tate curve. Let q_E denote the Tate period of E over F_v. Then $F_{\infty,w}$ is obtained by adjoining to H_∞ the p^n-th roots of q_E for $n = 1, 2, \ldots$, and thus it is clear that the Galois group of $F_{\infty,w}$ over H_∞ is isomorphic to $\mathbb{Z}_p$. If E does not have split multiplicative reduction at v, there exists a finite extension L of F_v such that E has split multiplicative reduction over L. But then our previous argument shows that the profinite degree of $LF_{\infty,w}$ over LH_∞ is divisible by p^∞, whence the same must be true for the profinite degree of $F_{\infty,w}$ over H_∞ since L is of finite degree over F_v. This completes the proof of Lemma 3.3.

Lemma 3.4. *Let p be a prime $\geqslant 5$, and let v be a place of F such that* $\mathrm{ord}_v(j_E) \geqslant 0$ *and v does not divide p. Then $H^i(\Sigma, J_v(F_\infty)) = 0$ for all $i \geqslant 1$.*

Proof. Let w be a fixed prime of F_∞ above v. Since v does not divide p, the isomorphism (58) is again valid. Combining (58) with Lemma 3.1, we see that the assertion of Lemma 3.4 is equivalent to

$$H^i(\Sigma_w, H^1(F_{\infty,w}, E_{p^\infty})) = 0 \tag{59}$$

for all $i \geqslant 1$. We will first show that Σ_w has p-cohomological dimension equal to 1, which will establish (59) for all $i \geqslant 2$. Now E has potential good reduction at v since $\mathrm{ord}_v(j_E) \geqslant 0$, and we appeal to the results of Serre-Tate [31]. It follows from [31] that (i) E has good reduction over the field $F_{0,w}$ (here we need $p \neq 2$), and (ii) the inertial subgroup of the Galois group of $F_{0,w}$ over F_v has order dividing 24. We deduce from (i) that $F_{\infty,w}$ is an unramified extension of $F_{0,w}$. Let $L_{\infty,w}$ denote the maximal unramified extension of F_v contained in $F_{\infty,w}$. By virtue of (ii) and our hypothesis that $p \geqslant 5$, we see that $[F_{\infty,w} : L_{\infty,w}]$ is finite and of order prime to p. Moreover, $L_{\infty,w}$ contains the unramified $\mathbb{Z}_p$-extension of F_v because $L_{\infty,w} \supset F_v(\mu_{p^\infty})$. It is now plain that Σ_w has p-cohomological dimension equal to 1.

We are left proving (59) for $i = 1$. We begin by observing that $H^2(L, E_{p^\infty}) = 0$ for all finite extensions L of F_v, because $H^2(L, E_{p^\infty})$ is dual by Tate local duality to $H^0(L, T_p(E))$, and this latter group is clearly zero because the torsion subgroup of $E(L)$ is finite. On allowing L to range over all finite extensions of F_v contained in $F_{\infty,w}$, we deduce that

$$H^2(F_{\infty,w}, E_{p^\infty}) = 0. \tag{60}$$

In view of (60), we conclude from the Hochschild-Serre spectral sequence ([17], Theorem 3) applied to the extension $F_{\infty,w}$ over F_v that we have the exact sequence

$$H^2(F_v, E_{p^\infty}) \longrightarrow H^1(\Sigma_w, H^1(F_{\infty,w}, E_{p^\infty})) \longrightarrow H^3(\Sigma_w, E_{p^\infty}). \tag{61}$$

But the group on the left of (61) is zero by the above remark, and the group on the right is zero because Σ_w has p-cohomological dimension equal to 1. Hence the group in the middle of (61) is zero, and the proof of Lemma 3.4 is complete.

Lemma 3.5. *Let p be a prime $\geqslant 5$, and let v be any place of F dividing p. Then $H^i(\Sigma, J_v(F_\infty)) = 0$ for all $i \geqslant 1$.*

Proof. We must show that

$$H^i(\Sigma_w, H^1(F_{\infty,w}, E)(p)) = 0$$

for all $i \geqslant 1$, where w is some fixed prime of F_∞ above v. However, the question is now much subtler than in the proof of Lemma 3.4 for two reasons, both arising from the fact that v now divides p. Firstly, Σ_w will now have p-cohomological dimension greater than 1 because it will now be a p-adic Lie group of dimension greater than 1 (see §2.4). Secondly and more seriously there is no longer any simple way like that given by the isomorphism (58) for identifying $H^1(F_{\infty,w}, E)(p)$ with $H^1(F_{\infty,w}, A)$ for an appropriate discrete p-primary Galois module A. Happily, the ramification-theoretic methods developed in [4] give a complete answer to this latter problem, which we now explain. Write G_v for the Galois group of $\overline{F}_v$ over F_v, and I_v for the inertial subgroup of G_v. As is explained in [4] (see p. 150), it is easy to see that there is a canonical exact sequence of G_v-modules

$$0 \longrightarrow C \longrightarrow E_{p^\infty} \longrightarrow D \longrightarrow 0, \tag{62}$$

which is characterized by the fact that C is divisible and that D is the maximal quotient of E_{p^∞} by a divisible subgroup such that I_v acts on D via a finite quotient. We recall that $F_{\infty,w}$ is deeply ramified in the sense of [4] because it contains the deeply ramified field $F_v(\mu_{p^\infty})$. Hence, combining Propositions 4.3 and 4.8 of [4], we obtain a canonical Σ_w-isomorphism

$$H^1(F_{\infty,w}, E)(p) \simeq H^1(F_{\infty,w}, D). \tag{63}$$

Now $D = 0$ if and only if E has potential supersingular reduction at v, and so we see that Lemma 3.5 is certainly true in view of (63) when E has potential supersingular reduction at v.

We note that $G(\overline{F}_v/F_{\infty,w})$ has p-cohomological dimension equal to 1 because the profinite degree of $F_{\infty,w}$ over F_v is divisible by p^∞ (see [25]). It follows that

$$H^i(F_{\infty,w}, D) = 0 \tag{64}$$

for all $i \geqslant 2$. In view of (64), we conclude from the Hochschild-Serre spectral sequence ([17], Theorem 3) applied to the extension $F_{\infty,w}$ over F_v that we have the exact sequence

$$H^{j+1}(F_v, D) \longrightarrow H^j(\Sigma_w, H^1(F_{\infty,w}, D)) \longrightarrow H^{j+2}(\Sigma_w, D) \tag{65}$$

for all $j \geqslant 1$. We proceed to show that the cohomology groups at both ends of (65) are zero, which will establish Lemma 3.5 in view of (63). To prove our claim that

$$H^i(F_v, D) = 0 \tag{66}$$

for all $i \geqslant 2$, we recall that $G(\overline{F}_v/F_v)$ has cohomological dimension 2. This implies firstly that (66) is valid for all $i \geqslant 3$, and secondly that, on taking cohomology of the exact sequence (62), we obtain a surjection from $H^2(F_v, E_{p^\infty})$ onto $H^2(F_v, D)$. But, as was explained in the proof of Lemma 2.4, we have $H^2(F_v, E_{p^\infty}) = 0$, and so (66) also follows for $i = 2$. Next we claim that

$$H^i(\Sigma_w, D) = 0 \tag{67}$$

for all $i \geqslant 3$. If E has potential multiplicative reduction at v, then Σ_w has p-cohomological dimension equal to 2, because Σ_w is a p-adic Lie group of dimension 2 (see Lemma 2.8) which has no p-torsion because $p \geqslant 5$. Hence (67) follows in this case. Suppose finally that E has potential ordinary reduction at v. Then Σ_w has p-cohomological dimension equal to 3, because Σ_w is a p-adic Lie group of dimension 3 (see Lemma 2.8) and has no p-torsion. This implies that (67) is valid for all $i \geqslant 4$, and also, on taking Σ_w-cohomology of the exact sequence (62), that there is a surjection of $H^3(\Sigma_w, E_{p^\infty})$ onto $H^3(\Sigma_w, D)$. Hence it suffices to show that $H^3(\Sigma_w, E_{p^\infty}) = 0$. But Σ_w is a Poincaré group of dimension 3 since it is a p-adic Lie group of dimension 3. Moreover, since Σ_w is a closed subgroup of Σ, it is known (see Wingberg [33]) that the dualizing module for Σ_w is just the dualizing module for Σ viewed as a Σ_w-module, i.e. $\mathbb{Q}_p/\mathbb{Z}_p$ with the trivial action of Σ_w. But, arguing as in the proof of Lemma 1.3, we conclude that $H^3(\Sigma_w, E_{p^\infty})$ is dual to $H^0(\Sigma_w, T_p(E)(-1))$. On the other hand, since E has potential good reduction at v, a result of Imai [20] shows that the p-power torsion subgroup of E in the field $F_v(\mu_{p^\infty})$ is finite. Hence it is clear that $H^0(\Sigma_w, T_p(E)(-1))$ is zero, and the proof of Lemma 3.5 is complete. This also completes the proof of Theorem 3.2.

3.3 Analysis of the local restriction maps

A crucial element of the analysis of the fundamental diagram (43) is to study the kernel and cokernel of the restriction maps

$$\gamma_v : J_v(F) = H^1(F_v, E)(p) \longrightarrow J_v(F_\infty)^\Sigma = H^1(F_{\infty,w}, E)(p)^{\Sigma_w}. \tag{68}$$

Here v denotes any finite place of F, w is some fixed place of F_∞ above v, and $\Sigma_w \subset \Sigma$ is the decomposition group of w.

We first discuss γ_v when v does not divide p. We recall from §1.3 that $c_v = [E(F_v) : E_0(F_v)]$, and that $L_v(E, s)$ denotes the Euler factor of the complex L-function of E at v. The following lemma is very well known (see [3], Lemma 7).

Lemma 3.6. *Let v be any finite prime of F which does not divide p. Then $J_v(F) = H^1(F_v, E)(p)$ is finite, and its order is the exact power of p dividing $c_v/L_v(E, 1)$.*

Proof. If A is an abelian group, we write

$$A^* = \varprojlim_n A/p^n A$$

for its p-adic completion. By Tate duality, $H^1(F_v, E)$ is canonically dual to $E(F_v)$, from which it follows immediately that $H^1(F_v, E)(p)$ is dual to $E(F_v)^*$. Thus we must show that $E(F_v)^*$ is finite of order the exact power of p dividing $c_v/L_v(E, 1)$. Let k_v be the residue field of v, and let $\widetilde{E}_v$ denote the reduction of E modulo v. We write $\widetilde{E}_{ns}(k_v)$ for the group of non-singular points in the set $\widetilde{E}_v(k_v)$. We have the exact sequence

$$0 \longrightarrow E_1(F_v) \longrightarrow E_0(F_v) \longrightarrow \widetilde{E}_{ns}(k_v) \longrightarrow 0,$$

where $E_1(F_v)$ is the kernel of reduction modulo v. Now $E_1(F_v)$ can be identified with the points of the formal group of E at v with coordinates in the maximal ideal of the ring of integers of F_v. As v does not divide p, multiplication by p is an automorphism of $E_1(F_v)$, and thus $E_1(F_v)^* = 0$. Hence the above exact sequence yields an isomorphism from $E_0(F_v)^*$ to $\widetilde{E}_{ns}(k_v)^*$. Define B_v by $B_v = E(F_v)/E_0(F_v)$. Since B_v is finite, we see easily that the induced map from $E_0(F_v)^*$ to $E(F_v)^*$ is injective, and that we have the exact sequence

$$0 \longrightarrow E_0(F_v)^* \longrightarrow E(F_v)^* \longrightarrow B_v^* \longrightarrow 0.$$

Hence $E(F_v)^*$ is finite, and its order is the exact power of p dividing $c_v \cdot \#(\widetilde{E}_{ns}(k_v))$. Lemma 3.6 now follows immediately from the well known fact (see [32]) that

$$(Nv)L_v(E, 1)^{-1} = \#(\widetilde{E}_{ns}(k_v)).$$

Lemma 3.7. *Let v be any finite prime of F which does not divide p. Let M_∞ denote an arbitrary Galois extension of F_v, and write $\Omega_\infty = G(M_\infty/F_v)$. Let*

$$\tau_v : H^1(F_v, E)(p) \longrightarrow H^1(M_\infty, E)(p)^{\Omega_\infty}$$

denote the restriction map. Then

$$\begin{aligned} \mathrm{Ker}(\tau_v) &= H^1(\Omega_\infty, E_{p^\infty}(M_\infty)), \\ \mathrm{Coker}(\tau_v) &= H^2(\Omega_\infty, E_{p^\infty}(M_\infty)). \end{aligned} \tag{69}$$

Proof. We have the commutative diagram

$$\begin{array}{ccc} H^1(F_v, E_{p^\infty}) & \xrightarrow{s_v} & H^1(M_\infty, E_{p^\infty})^{\Omega_\infty} \\ \downarrow & & \downarrow \\ H^1(F_v, E)(p) & \xrightarrow{\tau_v} & H^1(M_\infty, E)(p)^{\Omega_\infty}, \end{array}$$

where s_v is also the restriction map, and the vertical maps are the surjections derived from Kummer theory on E. But both vertical maps are isomorphisms, since

$$E(F_v) \otimes \mathbb{Q}_p/\mathbb{Z}_p = E(M_\infty) \otimes \mathbb{Q}_p/\mathbb{Z}_p = 0,$$

because v does not divide p. Thus we can identify $\mathrm{Ker}(\tau_v)$ with $\mathrm{Ker}(s_v)$, and $\mathrm{Coker}(\tau_v)$ with $\mathrm{Coker}(s_v)$. But, as was already remarked in the proof of Lemma 3.4, we have $H^2(F_v, E_{p^\infty}) = 0$ by the Tate duality. Hence the Hochschild-Serre spectral sequence shows that the assertion (69) holds for s_v instead of τ_v, and the proof of the lemma is complete.

Lemma 3.8. *Let v be any finite prime of F which does not divide p. Let K_∞ denote the cyclotomic $\mathbb{Z}_p$-extension of F_v, and put $\Gamma_\infty = G(K_\infty/F_v)$. Then $H^1(\Gamma_\infty, E_{p^\infty}(K_\infty))$ is a finite group whose order is the exact power of p dividing c_v.*

Proof. Let F_v^{nr} be the maximal unramified extension of F_v, and put

$$W_v = H^1(G(F_v^{nr}/F_v), E(F_v^{nr})).$$

It is well known (see [23]) that W_v is the exact orthogonal complement of $E_0(F_v)$ under the dual Tate pairing of $H^1(F, E)$ and $E(F_v)$. Hence $W_v(p)$ is finite and its order is the exact power of p dividing c_v. Since v does not divide p, we can apply Lemma 3.7 with $\Omega_\infty = F_v^{nr}$, and we conclude that

$$W_v(p) = H^1(G(F_v^{nr}/F_v), E_{p^\infty}(F_v)). \tag{70}$$

But, again because v does not divide p, K_∞ is contained in F_v^{nr}, and the profinite degree of F_v^{nr} over K_∞ is prime to p. Hence the group on the right of (70) can be identified under inflation with $H^1(\Gamma_\infty, E_{p^\infty}(K_\infty))$. This completes the proof of the lemma.

Proposition 3.9. *Let v be any finite prime of F which does not divide p. If $\mathrm{ord}_v(j_E) < 0$, then γ_v is the zero map, and the order of its kernel $J_v(F)$ is the exact power of p dividing $c_v/L_v(E,1)$. If $\mathrm{ord}_v(j_E) \geqslant 0$, and $p \geqslant 5$, then γ_v is an isomorphism.*

Proof. The first assertion of the proposition is clear from Lemmas 3.3 and 3.6. Also, applying Lemma 3.7 with $M_\infty = F_{\infty,w}$, we see that

$$\begin{aligned}\mathrm{Ker}(\gamma_v) &= H^1(\Sigma_w, E_{p^\infty}),\\ \mathrm{Coker}(\gamma_v) &= H^2(\Sigma_w, E_{p^\infty}).\end{aligned} \tag{71}$$

Suppose next that $\mathrm{ord}_v(j_E) \geqslant 0$ and $p \geqslant 5$. By Lemma 2.8, Σ_w is a p-adic Lie group of dimension 1, and has no p-torsion because $p \geqslant 5$, whence it has p-cohomological dimension equal to 1. It follows from (71) that $\mathrm{Coker}(\gamma_v) = 0$. Let K_∞ denote the cyclotomic $\mathbb{Z}_p$-extension of F_v, and put $\Gamma_\infty = G(K_\infty/F_v)$, $\Phi_\infty = G(F_{\infty,w}/K_\infty)$. Then we have the exact sequence

$$H^1(\Gamma_\infty, E_{p^\infty}(K_\infty)) \longrightarrow \mathrm{Ker}(\gamma_v) \longrightarrow H^1(\Phi_\infty, E_{p^\infty}). \tag{72}$$

We claim that the terms at both ends of this exact sequence are zero. Indeed, the order of the group on the left of (72) is the exact power of p dividing c_v by Lemma 3.8. But it is well known [32] that our hypothesis that $\mathrm{ord}_v(j_E) \geqslant 0$ implies that the only primes which divide c_v lie in the set $\{2,3\}$. The term on the right of (72) will vanish if we can show that the profinite degree of Φ_∞ is prime to p. But, as was already explained in the proof of Lemma 3.4, the results of [31] show that E has good reduction over $F_{0,w} = F_v(E_p)$, and that the order of the inertial subgroup of the Galois group of $F_{0,w}$ over F_v divides 24. Thus, if L_∞ denotes the maximal unramified extension of F_v contained in $F_{\infty,w}$, the degree of $F_{\infty,w}$ over L_∞ must divide 24. But K_∞ is the unique unramified $\mathbb{Z}_p$-extension of F_v, and thus the profinite degree of L_∞ over K_∞ is prime to p. Since $p \geqslant 5$, it follows that the profinite degree of $F_{\infty,w}$ over K_∞ is prime to p. This completes the proof of Proposition 3.9.

We next consider the situation when our finite prime v of F divides p. In this case, Tate duality shows that $J_v(F)$ is dual to

$$E(F_v)^* \simeq \mathbb{Z}_p^{d_v} \times A_v,$$

where $d_v = [F_v : \mathbb{Q}_p]$, and A_v is the group of p-power torsion in $E(F_v)$. In particular, $J_v(F)$ is always infinite. When E has potential multiplicative reduction at v dividing p, it is conjectured that $\mathrm{Ker}(\gamma_v)$ is always finite. However, this is only known at present when $F = \mathbb{Q}$, and its proof in this case depends on the beautiful transcendence result of [2].

Lemma 3.10. *Let v be any prime of F dividing p. If E has potential supersingular reduction at v, then γ_v is the zero map, whence, in particular,*

$\mathrm{Ker}(\gamma_v)$ *is infinite. If* E *has potential ordinary reduction at* v*, then* $\mathrm{Ker}(\gamma_v)$ *is finite.*

Proof. The proof makes use of the theory of deeply ramified extensions discussed in the proof of Lemma 3.5. Indeed, if E has potential supersingular reduction at v, the Galois module D appearing in the exact sequence (62) is zero, and hence (63) shows that $J_v(F_\infty)$ is zero. This proves the first assertion of the lemma. To prove the second assertion, it is simplest to use the well known (see, for example, [4], §5) interpretation of the dual of $\mathrm{Ker}(\gamma_v)$ in terms of universal norms, namely that the exact orthogonal complement of $H^1(\Sigma_w, E(F_{\infty,w}))$ in the dual Tate pairing between $H^1(F_v, E)$ and $E(F_v)$ is the group $E_U(F_{\infty,w})$ defined by

$$E_U(F_{\infty,w}) = \bigcap_L N_{L/F_v}(E(L)),$$

where L runs over all finite extensions of F_v contained in $F_{\infty,w}$, and N_{L/F_v} denotes the norm map from L to F_v. Hence $\mathrm{Ker}(\gamma_v)$ is dual to $E(F_v)^*/E_U(F_{\infty,w})^*$. Suppose now that E has potential ordinary reduction at v. Thus there exists a finite extension M of F_v such that E has good ordinary reduction over M. The field $M_\infty F_v(E_{p^\infty})$ is again deeply ramified because it contains $F_{\infty,w}$. Applying Proposition 3.11, which will be proven next by an independent argument, to E over M, we conclude that the kernel of the restriction map $\gamma_v(M)$ from $H^1(M, E)(p)$ to $H^1(M_\infty, E)(p)$ is finite. Thus, by Tate duality, $E_U(M_\infty)^*$ is of finite index in $E(M)^*$. But clearly

$$N_{M/F_U}(E_U(M_\infty)^*) \subset E_U(F_{\infty,w})^* \subset E(F_v)^*.$$

However, it is well known that the norm map sends an open subgroup of $E(M)^*$ to an open subgroup of $E(F_v)^*$. Thus $E_U(F_{\infty,w})^*$ is also of finite index in $E(F_v)^*$. This completes the proof of Lemma 3.10.

In preparation for our study in Chapter 4 of the Σ-Euler characteristic of the Selmer group $S(E/F_\infty)$ in the case when E has good ordinary reduction at all primes of F dividing p, we now make a more detailed study of $\mathrm{Ker}(\gamma_v)$ and $\mathrm{Coker}(\gamma_v)$ when E has good ordinary reduction at a prime v of F dividing p. We again write k_v for the residue field of v, and $\widetilde{E}_v$ for the reduction of E modulo v. Here is the principal result which we will establish.

Proposition 3.11. *Assume that* $p \geqslant 3$*. Let* v *be a prime of* F *dividing* p*, where* E *has good ordinary reduction. Then both* $\mathrm{Ker}(\gamma_v)$ *and* $\mathrm{Coker}(\gamma_v)$ *are finite, and*

$$\frac{\#(\mathrm{Ker}(\gamma_v))}{\#(\mathrm{Coker}(\gamma_v))} = \left(\#(\widetilde{E}_v(k_v)(p))\right)^2. \tag{73}$$

The proof is rather long, and will be broken up into a series of lemmas.

Let O_v be the ring of integers of F_v, and let $\widehat{E}_v$ be the formal group defined over O_v giving the kernel of reduction modulo v on E. For each finite extension L of F_v, let k_L denote the residue field of L, and $\mathfrak{m}_L$ the maximal ideal of the ring of integers of L. Then reduction modulo v gives the exact sequence

$$0 \longrightarrow \widehat{E}_v(\mathfrak{m}_L) \longrightarrow E(L) \longrightarrow \widetilde{E}_v(k_L) \longrightarrow 0.$$

Passing to the inductive limit over all finite extensions L of F_v which are contained in $F_{\infty,w}$, we obtain the exact sequence

$$0 \longrightarrow \widehat{E}_v(\mathfrak{m}_\infty) \longrightarrow E(F_{\infty,w}) \longrightarrow \widetilde{E}_v(k_\infty) \longrightarrow 0,$$

where $\mathfrak{m}$ denotes the maximal ideal of the ring of integers of $F_{\infty,w}$, and k_∞ is the residue field of $F_{\infty,w}$.

Lemma 3.12. *For all $i \geqslant 1$, we have*

$$H^i(\Sigma_w, \widehat{E}_v(\mathfrak{m}_\infty)) = H^i(F_v, \widehat{E}_v(\overline{\mathfrak{m}})),$$

where $\overline{\mathfrak{m}}$ denotes the maximal ideal of the ring of integers of $\overline{F}_v$.

Proof. By one of the principal results of [4] (Corollary 3.2), we have

$$H^i(F_{\infty,w}, \widehat{E}_v(\overline{\mathfrak{m}})) = 0 \tag{74}$$

for all $i \geqslant 1$, because $F_{\infty,w}$ is deeply ramified. By the Hochschild-Serre spectral sequence, this vanishing implies that we have the exact sequence

$$0 \longrightarrow H^i(\Sigma_w, \widehat{E}_v(\mathfrak{m}_\infty)) \longrightarrow H^i(F_v, \widehat{E}_v(\mathfrak{m})) \longrightarrow H^i(F_{\infty,w}, \widehat{E}_v(\mathfrak{m}))^{\Sigma_w}$$

for all $i \geqslant 1$. But the group on the right is zero by (74) again, and the proof of the lemma is complete.

To lighten notation, let us define

$$e_v = \#(\widetilde{E}_v(k_v)(p)). \tag{75}$$

Lemma 3.13. *Let v be a prime of F dividing p, where E has good ordinary reduction. Then $H^1(F_v, \widehat{E}_v(\overline{\mathfrak{m}}))$ is finite of order e_v. Moreover, for all $i \geqslant 2$, we have $H^i(F_v, \widehat{E}_v(\overline{\mathfrak{m}})) = 0$.*

Proof. We only sketch the proof (see [7] for more details). For all $n \geqslant 1$, we have the exact sequence

$$0 \longrightarrow \widehat{E}_{v,p^n} \longrightarrow \widehat{E}_v(\overline{\mathfrak{m}}) \xrightarrow{p^n} \widehat{E}_v(\overline{\mathfrak{m}}) \longrightarrow 0.$$

This gives rise to a surjection

$$H^i(F_v, \widehat{E}_{v,p^n}) \longrightarrow (H^i(F_v, \widehat{E}_v(\overline{\mathfrak{m}})))_{p^n}$$

for all $i \geqslant 1$; here, if A is an abelian group, $(A)_{p^n}$ denotes the kernel of multiplication by p^n on A. Passing to the inductive limit as $n \to \infty$, we obtain the surjection

$$H^i(F_v, \widehat{E}_{v,p^\infty}) \longrightarrow H^i(F_v, \widehat{E}_v(\overline{\mathfrak{m}}))$$

for all $i \geqslant 1$. Hence the final assertion of the lemma will follow if we can show that

$$H^i(F_v, \widehat{E}_{v,p^\infty}) = 0 \tag{76}$$

for all $i \geqslant 2$. This is automatic from cohomological dimension when $i \geqslant 3$. For $i = 2$, we use the well known fact that $\widehat{E}_{v,p^n}$ is its own orthogonal complement in the Weil pairing of $E_{p^n} \times E_{p^n}$ into μ_{p^n}. Hence, by Tate local duality, $H^2(F_v, \widehat{E}_{v,p^\infty})$ is dual to $H^0(F_v, T_p(\widetilde{E}_v))$, and this latter group is zero since only finitely many elements of $\widetilde{E}_{v,p^\infty}$ are rational over F_v. This completes the proof of (76).

We now turn to the first assertion of Lemma 3.13. Taking $G(\overline{F}_v/F_v)$-cohomology of the exact sequence at the beginning of the proof, and then taking the inductive limit as $n \to \infty$, we obtain the exact sequence

$$0 \longrightarrow \widehat{E}_v(\mathfrak{m}_F) \otimes \mathbb{Q}_p/\mathbb{Z}_p \longrightarrow H^1(F_v, \widehat{E}_{v,p^\infty}) \longrightarrow H^1(F_v, \widehat{E}_v(\overline{\mathfrak{m}})) \longrightarrow 0. \tag{77}$$

On the other hand, since $H^2(F_v, \widehat{E}_{v,p^\infty}) = 0$, it follows easily from Tate's Euler characteristic theorem that the dual of $H^1(F_v, \widehat{E}_{v,p^\infty})$ is a finitely generated $\mathbb{Z}_p$-module of $\mathbb{Z}_p$-rank equal to $d_v = [F_v : \mathbb{Q}_p]$. Put $W = H^1(F_v, \widehat{E}_{v,p^\infty})$, and let W_{div} be the maximal divisible subgroup of W. Since W_{div} has $\mathbb{Z}_p$-corank equal to d_v, and since the elementary theory of the formal group tells us that the group on the left of (77) is divisible of $\mathbb{Z}_p$-corank equal to d_v, we must have

$$W_{\text{div}} = \widehat{E}_v(\mathfrak{m}_F) \otimes \mathbb{Q}_p/\mathbb{Z}_p.$$

Thus, as W/W_{div} is finite, we conclude that $H^1(F_v, \widehat{E}_v(\overline{\mathfrak{m}}))$ is finite. We now introduce the $\mathbb{Q}_p$-vector space

$$V_p(\widehat{E}_v) = T_p(\widehat{E}_v) \otimes_{\mathbb{Z}_p} \mathbb{Q}_p.$$

Clearly the continuous cohomology groups $H^i(F_v, V_p(\widehat{E}_v))$ are also $\mathbb{Q}_p$-vector spaces and so in particular divisible for all $i \geqslant 0$. Also, since $\widehat{E}_{v,p^n}$ is its own orthogonal complement under the Weil pairing for all $n \geqslant 1$, Tate local duality implies that $H^2(F_v, T_p(\widehat{E}_v))$ is dual to $H^0(F_v, \widetilde{E}_{v,p^\infty})$, and this latter group is finite of order e_v. Hence taking cohomology of the exact sequence

$$0 \longrightarrow T_p(\widehat{E}_v) \longrightarrow V_p(\widehat{E}_v) \longrightarrow \widehat{E}_{v,p^\infty} \longrightarrow 0,$$

we deduce from the above remarks that there is an isomorphism

$$W/W_{\text{div}} \simeq H^2(F_v, T_p(\widehat{E}_v)).$$

The proof of the Lemma is now complete since the group on the left is isomorphic to $H^1(F_v, \widehat{E}_v(\overline{\mathfrak{m}}))$, and the group on the right has order e_v.

Lemma 3.14. *Let v be a prime of F dividing p, where E has good ordinary reduction. Then, for all $i \geqslant 2$, we have an isomorphism*

$$H^i(\Sigma_w, E(F_{\infty,w}))(p) \simeq H^i(\Sigma_w, \widetilde{E}_{v,p^\infty}). \tag{78}$$

We also have the exact sequence

$$0 \longrightarrow H^1(F_v, \widehat{E}_v(\overline{\mathfrak{m}})) \longrightarrow H^1(\Sigma_w, E(F_{\infty,w}))(p) \longrightarrow H^1(\Sigma_w, \widetilde{E}_{v,p^\infty}) \longrightarrow 0. \tag{79}$$

Proof. We take the Σ_w-cohomology of the exact sequence

$$0 \longrightarrow \widehat{E}_v(\mathfrak{m}_\infty) \longrightarrow E(F_{\infty,w}) \longrightarrow \widetilde{E}_v(k_\infty) \longrightarrow 0,$$

then take the p-primary part of the corresponding long exact sequence, and finally apply Lemmas 3.12 and 3.13. This completes the proof of the lemma.

We continue to assume that E has good ordinary reduction at a prime v dividing p. Then, as we saw in Lemma 2.8, Σ_w is a p-adic Lie group of dimension 3. Assuming that $p \geqslant 3$, we shall show that Σ_w has no p-torsion, and hence it will follow that Σ_w has a p-cohomological dimension equal to 3. If A is a discrete p-primary Σ_w-module, we say that A has finite Σ_w-Euler characteristic if $H^i(\Sigma_w, A)$ is finite for all $i \geqslant 0$, and, when this is the case, we define

$$\chi(\Sigma_w, A) = \prod_{i=0}^{3} (\#(H^i(\Sigma_w, A)))^{(-1)^i}.$$

I am very grateful to Sujatha for suggesting to me that the following result should be true (see also Corollary 5.13 of [18]).

Lemma 3.15. *Let v be a prime of F dividing p, where E has good ordinary reduction. Assume that $p \geqslant 3$. Then $\widetilde{E}_{v,p^\infty}$ has finite Σ_w-Euler characteristic, and*

$$\chi(\Sigma_w, \widetilde{E}_{v,p^\infty}) = 1. \tag{80}$$

Moreover, $H^3(\Sigma_w, \widetilde{E}_{v,p^\infty}) = 0$.

Before proving Lemma 3.15, let us note that Proposition 3.11 follows from it and Lemmas 3.13 and 3.14. Indeed, on applying the Hochschild-Serre spectral sequence to the extension $F_{\infty,w}$ over F_v, and recalling that $H^2(F_v, E_{p^\infty}) = 0$, we deduce immediately that

$$\mathrm{Ker}(\gamma_v) = H^1(\Sigma_w, E(F_{\infty,w}))(p),$$
$$\mathrm{Coker}(\gamma_v) = H^2(\Sigma_w, E(F_{\infty,w}))(p).$$

But, by (78), we have

$$\#(\mathrm{Coker}(\gamma_v)) = h_2,$$

where we write $h_i = \#(H^i(\Sigma_w, \widetilde{E}_{v,p^\infty}))$ for $i \geqslant 0$. By (79) and Lemma 3.13, we have

$$\#(\mathrm{Ker}(\gamma_v)) = h_0 h_1.$$

Since $h_0 = e_v$, we see that (73) follows immediately from (80) and the fact that $h_3 = 1$. This completes the proof of Proposition 3.11.

We now turn to the proof of Lemma 3.15. We recall the following elementary facts which will be used repeatedly in the proof. Let G be a profinite abelian group which is the direct product of $\mathbb{Z}_p$ with a finite abelian group of order prime to p, and assume that G is topologically generated by a single element. Let γ denote a topological generator of G. Then G has p-cohomological dimension equal to 1. If A is a discrete p-primary G-module, we have

$$H^1(G, A) = A/(\gamma - 1)A.$$

Consider now the special case when $A = \mathbb{Q}_p/\mathbb{Z}_p(\psi)$, where $\psi : G \to \mathbb{Z}_p^\times$ is a continuous homomorphism (by $\mathbb{Q}_p/\mathbb{Z}_p(\psi)$ we mean $\mathbb{Q}_p/\mathbb{Z}_p$ endowed with the action of G given by $\sigma(x) = \psi(\sigma)x$ for $\sigma \in G$). Then

$$H^1(G, \mathbb{Q}_p/\mathbb{Z}_p(\psi)) = 0 \quad \text{if } \psi \neq 1. \tag{81}$$

We recall the proof of (81). We have the exact sequence of G-modules

$$0 \longrightarrow \mathbb{Z}_p(\psi) \longrightarrow \mathbb{Q}_p(\psi) \longrightarrow \mathbb{Q}_p/\mathbb{Z}_p(\psi) \longrightarrow 0.$$

If $\psi \neq 1$, then $\psi(\gamma) \neq 1$, and so $\gamma - 1$ is an automorphism of $\mathbb{Q}_p(\psi)$. But this implies that $\gamma - 1$ must be surjective on $\mathbb{Q}_p/\mathbb{Z}_p(\psi)$, proving (81).

Let H_∞ denote the maximal unramified extension of F_v which is contained in $F_{\infty,w}$, and put $M_\infty = H_\infty(\mu_{p^\infty})$. Put

$$G_1 = G(F_{\infty,w}/M_\infty),$$
$$G_2 = G(M_\infty/H_\infty),$$
$$G_3 = G(H_\infty/F_v).$$

Thus we have the tower of fields

$$\begin{array}{ll} F_{\infty,w} & \\ \mid &) \ G_1 \\ M_\infty & \\ \mid &) \ G_2 \\ H_\infty & \\ \mid &) \ G_3 \\ F_v & \end{array}$$

As in the proof of Lemma 2.8, we choose a basis of $T_p(E)$ whose first element is a basis of $T_p(\widehat{E}_{v,p^\infty})$. Then the representation ρ of Σ_w on $T_p(E)$ has the form

$$\rho(\sigma) = \begin{pmatrix} \eta(\sigma) & a(\sigma) \\ 0 & \varepsilon(\sigma) \end{pmatrix},$$

where $\eta : \Sigma_w \to \mathbb{Z}_p^\times$ is the character giving the action of Σ_w on $T_p(\widehat{E}_{v,p^\infty})$, and $\varepsilon : \Sigma_w \to \mathbb{Z}_p^\times$ is the character giving the action of Σ_w on $T_p(\widetilde{E}_{v,p^\infty})$. Now we first remark that each of G_1, G_2 and G_3 is the direct product of $\mathbb{Z}_p$ with a finite abelian group of order prime to p, and is topologically generated by a single element. This is true for G_3 because H_∞ contains the unique unramified $\mathbb{Z}_p$-extension of F_v. It is true for G_2 because of our hypothesis that $p \geqslant 3$. Finally, it holds for G_1 because the fact that E does not have complex multiplication implies that the map $\sigma \mapsto a(\sigma)$ defines an isomorphism from G_1 onto $\mathbb{Z}_p$. It now follows by an easy argument with successive quotients that Σ_w has no p-torsion. Hence H_w has p-cohomological dimension equal to 3, as required.

To simplify notation, let us put $W = \widetilde{E}_{v,p^\infty}$. Now $H^2(G_3, W) = 0$ because G_3 has p-cohomological dimension equal to 1. Moreover, G_3 acts on W via the character ε, and this action is non-trivial. Hence (81) implies that $H^1(G_3, W) = 0$. Hence the inflation-restriction sequence gives

$$H^1(\Sigma_w, W) = \mathrm{Hom}_{G_3}(X, W), \tag{82}$$

where $X = G(F_{\infty,w}/H_\infty)$. Again, we have $H^2(G_2, W) = 0$ because G_2 has p-cohomological dimension equal to 1. On the other hand, since G_2 acts trivially on W, and is topologically generated by one element, we have $H^1(G_2, W) = W$. Thus, applying the inflation-restriction sequence to $\mathrm{Hom}(X, W)$, we obtain the exact sequence

$$0 \longrightarrow W \longrightarrow \mathrm{Hom}(X, W) \longrightarrow \mathrm{Hom}_{G_2}(G_1, W) \longrightarrow 0.$$

Taking G_3 invariants of this sequence, and recalling that $H^1(G_3, W) = 0$, we obtain the exact sequence

$$0 \longrightarrow W^{G_3} \longrightarrow \mathrm{Hom}_{G_3}(X, W) \longrightarrow \mathrm{Hom}_Y(G_1, W) \longrightarrow 0, \tag{83}$$

where $Y = G(M_\infty/F_v)$. To calculate the group on the right of this exact sequence, we need the following explicit description of the action of Y on G_1, namely that, for $\tau \in G_1$ and $\sigma \in Y$, we have

$$\sigma \cdot \tau = \eta(\sigma)/\varepsilon(\sigma)\tau. \tag{84}$$

Indeed, recalling that $\sigma \cdot \tau = \widetilde{\sigma}\tau\widetilde{\sigma}^{-1}$, where $\widetilde{\sigma}$ denotes any lifting of σ to Σ_w, (84) is clear from the matrix calculation

$$\rho(\widetilde{\sigma})\rho(\tau)\rho(\widetilde{\sigma})^{-1} = \begin{pmatrix} 1 & \eta(\sigma)/\varepsilon(\sigma)a(\tau) \\ 0 & 1 \end{pmatrix}.$$

Since G_1 is isomorphic to $\mathbb{Z}_p$ with the action of Σ given by (84), we conclude that

$$\mathrm{Hom}(G_1, W) = \mathbb{Q}_p/\mathbb{Z}_p(\varepsilon^2/\eta). \tag{85}$$

Put $\chi = \varepsilon^2/\eta$. We claim that χ is not the trivial character of $Y = G(M_\infty/F_v)$. Let ψ denote the character giving the action of Y on μ_{p^∞}. By the Weil pairing, we have $\psi = \varepsilon\eta$. Hence, if $\chi = 1$, then we would have $\psi = \varepsilon^3$, which is clearly impossible since it would imply that ψ is an unramified character factoring through G_3. But then $\mathrm{Hom}_Y(G_1, W)$ must be finite, since it is annihilated by $\chi(\sigma_0) - 1$, where σ_0 is any element of Y such that $\chi(\sigma_0) \neq 1$. In view of (82) and (83), this proves the finiteness of $H^1(\Sigma_w, W)$.

We next turn to study $H^2(\Sigma_w, W)$. We have $H^2(G_1, W) = 0$ because G_1 has p-cohomological dimension equal to 1. Hence the Hochschild-Serre spectral sequence gives the exact sequence

$$H^2(Y, W) \longrightarrow H^2(\Sigma_w, W) \longrightarrow H^1(Y, H^1(G_1, W)) \longrightarrow H^3(Y, W). \tag{86}$$

Now Y is a p-adic Lie group of dimension 2 without p-torsion, and thus Y has p-cohomological dimension equal to 2. It follows that $H^3(Y, W) = 0$. We also claim that $H^2(Y, W) = 0$. Indeed, $H^2(G_2, W) = 0$ because G_2 has p-cohomological dimension equal to 1. Applying the Hochschild-Serre spectral sequence, we obtain the exact sequence

$$H^2(G_3, W) \longrightarrow H^2(Y, W) \longrightarrow H^1(G_3, H^1(G_2, W)).$$

But $H^2(G_3, W) = 0$ because G_3 has p-cohomological dimension equal to 1. On the other hand, G_3 acts trivially on G_2 since M_∞ is abelian over F_v, whence we have an isomorphism of G_3-modules

$$H^1(G_2, W) = \mathbb{Q}_p/\mathbb{Z}_p(\varepsilon).$$

Since ε is certainly not the trivial character of G_3, it follows from (81) that

$$H^1(G_3, \mathbb{Q}_p/\mathbb{Z}_p(\varepsilon)) = 0,$$

completing the proof that $H^2(Y,W) = 0$. Recalling (85), we deduce from (85) and (86) that

$$H^2(\Sigma_w, W) = H^1(Y, \mathbb{Q}_p/\mathbb{Z}_p(\chi)), \tag{87}$$

where $\chi = \varepsilon^2/\eta$. We now apply inflation-restriction to the group on the right of (87). Since G_3 has p-cohomological dimension equal to 1, we obtain the exact sequence

$$0 \longrightarrow H^1(G_3, U) \longrightarrow H^1(Y, \mathbb{Q}_p/\mathbb{Z}_p(\chi)) \longrightarrow H^1(G_2, \mathbb{Q}_p/\mathbb{Z}_p(\chi))^{G_3} \longrightarrow 0, \tag{88}$$

where $U = (\mathbb{Q}_p/\mathbb{Z}_p(\chi))^{G_2}$. But the restriction of χ to G_2 is equal to η^{-1} restricted to G_2, and so is certainly not the trivial character of G_2. It follows that U is finite, and that $H^1(G_2, \mathbb{Q}_p/\mathbb{Z}_p(\chi)) = 0$. But, since U is finite, it follows that $H^1(G_3, U)$ has the same order as $H^0(G_3, U) = \mathrm{Hom}_Y(G_1, W)$. Thus (87) and (88) imply that $H^2(\Sigma_w, W)$ is finite, and

$$\#(H^2(\Sigma_w, W)) = \#(\mathrm{Hom}_Y(G_1, W)). \tag{89}$$

Hence our Euler-characteristic formula (80) will follow from (83) and (89) provided that we can show

$$H^3(\Sigma_w, W) = 0. \tag{90}$$

To prove (90), we apply entirely similar arguments to those used above. We have $H^i(G_1, W) = 0$ for $i \geqslant 2$ since G_1 has p-cohomological dimension 1, and $H^i(Y, W) = 0$ for $i \geqslant 3$, since Y has p-cohomological dimension 2. Hence the Hochschild-Serre spectral sequence yields an isomorphism

$$H^3(\Sigma_w, W) = H^2(Y, H^1(G_1, W)). \tag{91}$$

We again apply the Hochschild-Serre spectral sequence to the right hand side of (91). Since G_2 and G_3 have p-cohomological dimension 1, we deduce using (85) that

$$H^2(Y, H^1(G_1, W)) = H^1(G_3, H^1(G_2, \mathbb{Q}_p/\mathbb{Z}_p(\chi))),$$

where $\chi = \varepsilon^2/\eta$. But, as remarked above, χ is not the trivial character of G_2, and so $H^1(G_2, \mathbb{Q}_p/\mathbb{Z}_p(\chi)) = 0$. In view of (91), we have now proven (90), and the proof of Lemma 3.15 is at last complete.

Lemma 3.16. *Assume that $p \geqslant 3$. Let v be a prime of F dividing p such that v is unramified in $F/\mathbb{Q}$, and E has good ordinary reduction at v. Then γ_v is surjective.*

Proof. By virtue of (78), we must show that, under the hypotheses of the lemma, we have

$$H^2(\Sigma_w, \widetilde{E}_{v,p^\infty}) = 0. \tag{92}$$

Now by (87), the group on the left of (92) is equal to

$$H^1(G(M_\infty/F_v), \mathbb{Q}_p/\mathbb{Z}_p(\chi)),$$

where χ is the character ε^2/η of $G(M_\infty/F_v)$. As explained immediately after (88), we have

$$\#(H^1(G(M_\infty/F_v), \mathbb{Q}_p/\mathbb{Z}_p(\chi))) = \#((\mathbb{Q}_p/\mathbb{Z}_p(\chi))^{G(M_\infty/F_v)}). \tag{93}$$

But M_∞ is the composite of the two fields H_∞ and $F_v(\mu_{p^\infty})$, and the intersection of these two fields is clearly F_v in view of our hypothesis that v is unramified in $F/\mathbb{Q}$. Hence we can choose σ in $G(M_\infty/F)$ such that $\varepsilon(\sigma) = 1$ and $\chi(\sigma)$ is a non-trivial $(p-1)$-th root of unity. But $\chi(\sigma) - 1$ annihilates the group on the right of (93), and so this group must be trivial since $\chi(\sigma)$ is not congruent to 1 mod p. This completes the proof of Lemma 3.16.

4 Global Calculations

4.1 Strategy

Again, E will denote an elliptic curve defined over a finite extension F of $\mathbb{Q}$, which does not admit complex multiplication; and $F_\infty = F(E_{p^\infty})$. We shall assume throughout that $p \geqslant 5$, thereby ensuring that $\Sigma = G(F_\infty/F)$ has p-cohomological dimension equal to 4, and that all the local cohomology results of Chapter 3 are valid. Recall that T denotes any finite set of primes of F, which contains both the primes where E has bad reduction and all the primes dividing p. We then have the localization sequence defining $\mathcal{S}(E/F_\infty)$ (see (42)), namely

$$0 \longrightarrow \mathcal{S}(E/F_\infty) \longrightarrow H^1(G_T(F_\infty), E_{p^\infty}) \xrightarrow{\lambda_T(F_\infty)} \bigoplus_{v\in T} J_v(F_\infty),$$

where $J_v(F_\infty) = \varinjlim J_v(L)$, as L runs over all finite extensions of F contained in F_∞, and

$$J_v(L) = \bigoplus_{w|v} H^1(L_w, E)(p).$$

We believe that the map $\lambda_T(F_\infty)$ should be surjective for all odd primes p, but we are only able to prove this surjectivity in some special, but non-trivial, cases using the results of Hachimori and Matsuno [15]. We then investigate consequences of the surjectivity of $\lambda_T(F_\infty)$ for the calculation of the Σ-Euler characteristic of the Selmer group $\mathcal{S}(E/F_\infty)$. In the last part of the chapter, we relate the surjectivity of $\lambda_T(F_\infty)$ to the calculation of the $\Lambda(\Sigma)$-rank of the dual $C(E/F_\infty)$ of $\mathcal{S}(E/F_\infty)$. Again, all the material discussed in this chapter is joint work with Susan Howson.

4.2 The surjectivity of $\lambda_T(F_\infty)$

In this section, we first calculate the Σ-cohomology of $H^1(G_T(F_\infty), E_{p^\infty})$. We recall that the Galois group $G_T(F) = G(F_T/F)$ has p-cohomological dimension equal to 2 for all odd primes p, and so, by a well known result, every closed subgroup of $G_T(F)$ has p-cohomological dimension at most 2.

Lemma 4.1. *Assume that $p \geqslant 5$. Then*

$$H^i(\Sigma,\, H^1(G_T(F_\infty), E_{p^\infty})) = 0 \tag{94}$$

for all $i \geqslant 2$. Moreover, if $S(E/F)$ is finite, we have

$$H^1(\Sigma,\, H^1(G_T(F_\infty), E_{p^\infty})) = H^3(\Sigma, E_{p^\infty}). \tag{95}$$

Proof. We begin by noting that

$$H^k(G_T(F_\infty), E_{p^\infty}) = 0$$

for all $k \geqslant 2$. Indeed, this is the assertion of Theorem 2.10 for $k = 2$, and it follows for $k > 2$ because $G_T(F_\infty)$ has p-cohomological dimension at most 2, since it is a closed subgroup of $G_T(F)$. Also, we clearly have

$$H^k(G_T(F), E_{p^\infty}) = 0$$

for all $k \geqslant 3$. Hence, for all $i \geqslant 1$, the Hochschild-Serre spectral sequence ([17], Theorem 3) gives the exact sequence

$$H^{i+1}(G_T(F), E_{p^\infty}) \to H^i(\Sigma,\, H^1(G_T(F_\infty), E_{p^\infty})) \to H^{i+2}(\Sigma, E_{p^\infty}) \to 0. \tag{96}$$

Assertion (94) follows, on recalling that $H^4(\Sigma, E_{p^\infty}) = 0$ by Theorem 1.2. Moreover, the next lemma shows that the hypothesis that $S(E/F)$ is finite implies that $H^2(G_T(F), E_{p^\infty}) = 0$. Hence (95) also follows on taking $i = 1$ in (96). This completes the proof of Lemma 4.1.

The following lemma about the arithmetic of E over the base field F is very well known (see Greenberg's article in this volume, or [7], Chapter 1). Recall that

$$\widehat{E(F)(p)} = \mathrm{Hom}(E(F)(p), \mathbb{Q}_p/\mathbb{Z}_p).$$

Lemma 4.2. *Let p be an odd prime, and assume that $S(E/F)$ is finite. Then $H^2(G_T(F), E_{p^\infty}) = 0$, and* $\mathrm{Coker}\,(\lambda_T(F)) = \widehat{E(F)(p)}$.

Let Σ' denote any open subgroup of Σ. Applying Theorem 3.2 when the base field F is replaced by the fixed field of Σ', we conclude that

$$H^i(\Sigma', J_v(F_\infty)) = 0 \tag{97}$$

for all $i \geqslant 1$. Similarly, we conclude from Lemma 4.1 that

$$H^i(\Sigma', H^1(G_T(F_\infty), E_{p^\infty})) = 0 \tag{98}$$

for all $i \geqslant 2$. The following result gives a surprising cohomological property of the Selmer group $\mathcal{S}(E/F_\infty)$.

Proposition 4.3. *Assume that $p \geqslant 5$, and that the map $\lambda_T(F_\infty)$ in (42) is surjective. Then, for every open subgroup Σ' of Σ, we have*

$$H^i(\Sigma', \mathcal{S}(E/F_\infty)) = 0 \tag{99}$$

for all $i \geqslant 2$.

Proof. This is immediate on taking Σ'-cohomology of the exact sequence

$$0 \longrightarrow \mathcal{S}(E/F_\infty) \longrightarrow H^1(G_T(F_\infty), E_{p^\infty}) \xrightarrow{\lambda_T(F_\infty)} \bigoplus_{v \in T} J_v(F_\infty) \longrightarrow 0, \tag{100}$$

and using (97) and (98). This completes the proof.

We now turn to the question of proving the surjectivity of the localization map $\lambda_T(F_\infty)$. There is one case which is easy to handle, and is already discussed in [8].

Proposition 4.4. *Assume that p is an odd prime, and that E has potential supersingular reduction at all primes v of F dividing p. Then $\lambda_T(F_\infty)$ is surjective.*

Proof. Let v be any prime of F dividing p, and let w be some fixed prime of F_∞ above v. As has been explained in the proof of Lemma 3.5, the fact that $F_{\infty,w}$ is deeply ramified enables us to apply one of the principal results of [4] to conclude that (63) is valid. But now $D = 0$ because, by hypothesis, E has potential supersingular reduction at v. It follows that

$$J_v(F_\infty) = 0 \tag{101}$$

for all primes v of F dividing p. Let T' denote the set of v in T which do not divide p. Now it is shown in [8] (see Theorem 2) that the localization map

$$\lambda'_T(F_\infty) : H^1(G_T(F_\infty), E_{p^\infty}) \longrightarrow \bigoplus_{v \in T'} J_v(F_\infty)$$

is surjective for all odd primes p. In view of (101), we conclude that $\lambda'_T(F_\infty) = \lambda_T(F_\infty)$, and the proof of Proposition 4.4 is complete.

Example. Proposition 4.4 applies to the curve $E = 50(A1)$ given by (21) and $F = \mathbb{Q}$, with p either 5 (where E has potential supersingular reduction) or one of the infinite set $\{29, 59, \ldots\}$ of primes where E has good supersingular reduction. It follows that, if T is any finite set of primes containing $\{2, 5, p\}$, then $\lambda_T(F_\infty)$ is surjective and (99) holds.

It seems to be a difficult and highly interesting problem to prove the surjectivity of $\lambda_T(F_\infty)$ when there is at least one prime v of F above p, where E does not have potential supersingular reduction. We are very grateful to Greenberg for pointing out to us that one can establish a first result in this direction using recent work of Hachimori and Matsuno [15]. Let K_∞ denote the cyclotomic $\mathbb{Z}_p$-extension of K. Put $\Gamma = G(K_\infty/K)$, and let $\Lambda(\Gamma)$ denote the Iwasawa algebra of Γ. We recall that $\mathcal{S}(E/K_\infty)$ denotes the Selmer group of E over K_∞, and $\mathcal{C}(E/K_\infty)$ denotes the Pontrjagin dual of $\mathcal{S}(E/K_\infty)$.

Theorem 4.5. *Let p be a prime number such that* (i) $p \geqslant 5$, (ii) $\Sigma = G(F_\infty/F)$ *is a pro-p-group,* (iii) *E has good ordinary reduction at all primes v of F dividing p, and* (iv) *$\mathcal{C}(E/K_\infty)$ is a torsion $\Lambda(\Gamma)$-module and has μ-invariant equal to* 0. *Then $\lambda_T(F_\infty)$ is surjective.*

Proof. The argument is strikingly simple. Let n be an integer $\geqslant 0$. Recall that $F_n = K(E_{p^{n+1}})$. Put

$$H_{n,\infty} = F_n(\mu_{p^\infty}), \quad \Omega_n = G(H_{n,\infty}/F_n).$$

Since $\mu_p \subset F_n$ by the Weil pairing, we see that $H_{n,\infty}$ is the cyclotomic $\mathbb{Z}_p$-extension of F_n. Now F_n is a finite Galois p-extension of F by our hypothesis that $\Sigma = G(F_\infty/F)$ is a pro-p-group. Hence, by the fundamental result of Hachimori and Matsuno [15], the fact that $\mathcal{C}(E/K_\infty)$ is $\Lambda(\Gamma)$-torsion and has μ-invariant equal to 0 implies that $\mathcal{C}(E/H_{n,\infty})$ is $\Lambda(\Omega_n)$-torsion, and has μ-invariant equal to 0. Let

$$\lambda_T(H_{n,\infty}) : H^1(G_T(H_{n,\infty}), E_{p^\infty}) \longrightarrow \bigoplus_{v \in T} J_v(H_{n,\infty})$$

be the localization map for the field $H_{n,\infty}$. Since F_∞ is plainly the union of the fields $H_{n,\infty}$ $(n = 0, 1, \ldots)$, it is clear that

$$\lambda_T(F_\infty) = \varinjlim_n \lambda_T(H_{n,\infty}), \tag{102}$$

where the inductive limit is taken with respect to the restriction maps. But it is very well known (see for example Lemma 4.6 in Greenberg's article in this volume) that the fact that $\mathcal{C}(E/H_{n,\infty})$ is $\Lambda(\Omega_n)$-torsion implies that the map $\lambda_T(H_{n,\infty})$ is surjective. Hence $\lambda_T(F_\infty)$ is also surjective because it is an inductive limit of surjective maps. This completes the proof of Theorem 4.5.

Remark. One can replace hypothesis (iv) of Theorem 4.5 by the following weaker assumption: (iv)′ E is isogenous over F to an elliptic curve E' such that $\mathcal{C}(E'/K_\infty)$ is $\Lambda(\Gamma)$-torsion, and has μ-invariant 0. Indeed, assuming (iv)$'$, the above argument shows that $\mathcal{C}(E'/H_{n,\infty})$ is $\Lambda(\Omega_n)$-torsion for all $n \geqslant 0$. But it is well known that the fact that $\mathcal{C}(E'/H_{n,\infty})$ is $\Lambda(\Omega_n)$-torsion implies that $\mathcal{C}(E/H_{n,\infty})$ is $\Lambda(\Omega_n)$-torsion (however, it will not necessarily be true that $\mathcal{C}(E/H_{n,\infty})$ has μ-invariant 0). Hence we again conclude that $\lambda_T(H_{n,\infty})$ is surjective for all $n \geqslant 0$, and thus again $\lambda_T(F_\infty)$ is surjective.

Examples. As was explained in Chapter 1, the hypotheses of Theorem 4.5 are satisfied for $E = X_1(11)$ given by equation (5), $F = \mathbb{Q}(\mu_5)$, and $p = 5$. Hence we conclude that the map $\lambda_T(F_\infty)$ is surjective in this case, where T is any finite set of primes containing 5 and 11. Moreover, the above remark enables us to conclude that, for T any finite set of primes containing 5 and 11, $\lambda_T(F_\infty)$ is surjective for $F_\infty = \mathbb{Q}(E_{5^\infty})$, and E the curve $X_0(11)$ given by (4) or the third curve of conductor 11 given by

$$y^2 + y = x^3 - x^2 - 7820x - 263580, \tag{103}$$

which is $11(A2)$ in Cremona's table [9]. This is because both of these curves are isogenous over $\mathbb{Q}$ to $X_1(11)$.

4.3 Calculations of Euler characteristics

Recall that $\tau_p(E/F)$ is the integer defined by (20). Thus $\tau_p(E/F) = 0$ means that E has potential ordinary or potential multiplicative reduction at each prime v of F dividing p. If $p \geqslant 5$, Conjecture 1.12 asserts a necessary and sufficient condition for the Σ-Euler characteristic of $\mathcal{S}(E/F_\infty)$ to be finite. The necessity of this condition is easy and is contained in the following lemma.

Lemma 4.6. *Assume p is an odd prime. If $H^0(\Sigma, \mathcal{S}(E/F_\infty))$ is finite, then $\mathcal{S}(E/F)$ is finite and $\tau_p(E/F) = 0$.*

Proof. We use the fundamental diagram (43). We recall that, by Lemma 2.6, we have that $\mathrm{Ker}(\beta)$, $\mathrm{Coker}(\beta)$, and $\mathrm{Ker}(\alpha)$ are finite for all odd primes p. Now assume that $H^0(\Sigma, \mathcal{S}(E/F_\infty))$ is finite. It follows from (43) that both $\mathcal{S}(E/F)$ and $\mathrm{Coker}(\alpha)$ are finite. Since $\mathcal{S}(E/F)$ is finite, we deduce from Lemma 4.2 that $\mathrm{Coker}(\lambda_T(F))$ is finite. Using the fact that $\mathrm{Ker}(\beta)$, $\mathrm{Coker}(\lambda_T(F))$, and $\mathrm{Coker}(\alpha)$ are all finite, we conclude from (43) that $\mathrm{Ker}(\gamma) = \bigoplus \mathrm{Ker}(\gamma_v)$ is finite, where v runs over all places in T. But, if v is a place of F dividing p where E has potential supersingular reduction, then

$$\mathrm{Ker}(\gamma_v) = H^1(F_v, E)(p). \tag{104}$$

This is because, as we have remarked on several occasions, (101) holds when E has potential supersingular reduction at v. Since (104) is clearly infinite, we conclude from the finiteness of $\mathrm{Ker}(\gamma)$ that E does not have potential supersingular reduction at any v dividing p. This completes the proof of Lemma 4.6.

The remainder of this section will be devoted to the study of

$$\chi(\Sigma, \mathcal{S}(E/F_\infty))$$

under the hypotheses that $p \geqslant 5$, $\mathcal{S}(E/F)$ is finite, and E has good ordinary reduction at all primes v of F dividing p. Of course, this is a case where $\tau_p(E/F) = 0$, so that we certainly expect the Euler characteristic to be finite. Unfortunately, at present, we can only prove the finiteness of $H^i(\Sigma, \mathcal{S}(E/F_\infty))$ for $i = 0, 1$, without imposing further hypotheses. We expect that

$$H^i(\Sigma, \mathcal{S}(E/F_\infty)) = 0 \quad (i \geqslant 2), \tag{105}$$

but it is curious that we cannot even prove that the cohomology groups in (105) are finite. However, if we assume in addition that $\lambda_T(F_\infty)$ is surjective, then we can show that (105) holds and that our Conjecture 1.13 for the exact value of $\chi(\Sigma, \mathcal{S}(E/F_\infty))$ is indeed true.

We recall the fundamental diagram (43), and remind the reader that $\psi_T(F_\infty)$ denotes the map in the top right hand corner of the fundamental diagram.

Lemma 4.7. *Assume that* (i) $p \geqslant 5$, (ii) $\mathcal{S}(E/F)$ *is finite, and* (iii) E *has good ordinary reduction at all primes v of F dividing p. Then both* $H^0(\Sigma, \mathcal{S}(E/F_\infty))$ *and* $\mathrm{Coker}(\psi_T(F_\infty))$ *are finite. Moreover, the order of* $H^0(\Sigma, \mathcal{S}(E/F_\infty))$ *is equal to*

$$\xi_p(E/F) \cdot \#(H^3(\Sigma, E_{p^\infty})) \cdot \#(\mathrm{Coker}(\psi_T(F_\infty))), \tag{106}$$

where $\xi_p(E/F)$ *is given by* (33).

Proof. We simply compute orders using the fundamental diagram (43). We claim that

$$\begin{aligned} \mathrm{Ker}(\beta) &= H^1(\Sigma, E_{p^\infty}), \\ \mathrm{Coker}(\beta) &= H^2(\Sigma, E_{p^\infty}). \end{aligned} \tag{107}$$

This follows immediately from the inflation-restriction sequence, on noting that $H^2(G_T(F), E_{p^\infty}) = 0$ by Lemma 4.2, since $\mathcal{S}(E/F)$ is finite. As in Chapter 1, write $h_i(E)$ for the cardinality of $H^i(\Sigma, E_{p^\infty})$. Combining (107) with Serre's Theorem 1.2, we conclude that

$$\frac{\#(\mathrm{Ker}\,\beta)}{\#(\mathrm{Coker}\,\beta)} = \frac{h_1(E)}{h_2(E)} = \frac{h_0(E)}{h_3(E)}. \tag{108}$$

Next we analyse the map γ appearing in (43). Combining Propositions 3.9 and 3.11, we see that

$$\frac{\#(\operatorname{Ker}\gamma)}{\#(\operatorname{Coker}\gamma)} = \prod_{v|p} e_v^2 \times \left| \prod_{v\in\mathfrak{M}} c_v/L_v(E,1)\right|_p^{-1}, \tag{109}$$

where e_v denotes the order of the p-primary subgroup of $\widetilde{E}_v(k_v)$.

We now consider the following commutative diagram with exact rows, which is derived from the right side of (43), namely

$$\begin{array}{ccccccccc} 0 & \longrightarrow & \operatorname{Im}\lambda_T(F_\infty) & \longrightarrow & \bigoplus_{v\in T} J_v(F_\infty)^{\Sigma} & \longrightarrow & \operatorname{Coker}\lambda_T(F_\infty) & \longrightarrow & 0 \\ & & \uparrow{\scriptstyle\delta} & & \uparrow{\scriptstyle\gamma} & & \uparrow{\scriptstyle\varepsilon} & & \\ 0 & \longrightarrow & \operatorname{Im}\lambda_T(F) & \longrightarrow & \bigoplus_{v\in T} J_v(F) & \longrightarrow & \operatorname{Coker}\lambda_T(F) & \longrightarrow & 0. \end{array} \tag{110}$$

Here δ and ε are the obvious induced maps. We have already seen that γ has finite kernel and cokernel, and also Lemma 4.2 shows that $\operatorname{Coker}(\lambda_T(F))$ is finite of order $h_0(E)$. Applying the snake lemma to (110), we conclude that both δ and ε have finite kernels and cokernels, and that

$$\frac{\#(\operatorname{Ker}\delta)}{\#(\operatorname{Coker}\delta)} = \frac{\#(\operatorname{Ker}\gamma)}{\#(\operatorname{Coker}\gamma)} \times \frac{\#(\operatorname{Coker}\varepsilon)}{\#(\operatorname{Ker}\varepsilon)}. \tag{111}$$

It also follows that $\operatorname{Coker}(\lambda_T(F_\infty))$ is finite, and thus

$$\frac{\#(\operatorname{Coker}\varepsilon)}{\#(\operatorname{Ker}\varepsilon)} = \frac{\#(\operatorname{Coker}\lambda_T(F_\infty))}{h_0(E)}. \tag{112}$$

Finally, we also have the commutative diagram with exact rows given by

$$\begin{array}{ccccccccc} 0 & \to & \mathcal{S}(E/F_\infty)^{\Sigma} & \to & H^1(G_T(F_\infty), E_{p^\infty})^{\Sigma} & \to & \operatorname{Im}\lambda_T(F_\infty) & \to & 0 \\ & & \uparrow{\scriptstyle\alpha} & & \uparrow{\scriptstyle\beta} & & \uparrow{\scriptstyle\delta} & & \\ 0 & \to & \mathcal{S}(E/F) & \to & H^1(G_T(F), E_{p^\infty}) & \to & \operatorname{Im}\lambda_T(F) & \to & 0. \end{array} \tag{113}$$

It follows on applying the snake lemma to this diagram that

$$\frac{\#(H^0(\Sigma, \mathcal{S}(E/F_\infty)))}{\#(\mathcal{S}(E/F))} = \frac{\#(\operatorname{Coker}\beta)}{\#(\operatorname{Ker}\beta)} \times \frac{\#(\operatorname{Ker}\delta)}{\#(\operatorname{Coker}\delta)}. \tag{114}$$

Since $\mathcal{S}(E/F)$ is finite, we have $\mathcal{S}(E/F) = \text{Ш}(E/F)(p)$. Also, we recall the well known fact that $c_v \leqslant 4$ if v does not belong to the set $\mathfrak{M}$ of places v of F with $\operatorname{ord}_v(j_E) < 0$ (of course, $c_v = 1$ when E has good reduction at v). Combining (108), (109), (111), (112) and (114), we obtain the formula (106) for the order of $H^0(\Sigma, \mathcal{S}(E/F_\infty))$. This completes the proof of Lemma 4.7.

Lemma 4.8. *Assume that* (i) $p \geqslant 5$, (ii) $\mathcal{S}(E/F)$ *is finite, and* (iii) E *has good ordinary reduction at all primes* v *of* F *dividing* p. *Then* $H^1(\Sigma, \mathcal{S}(E/F_\infty))$ *is finite, and its order divides*

$$\#(H^3(\Sigma, E_{p^\infty})) \cdot \#(\mathrm{Coker}(\psi_T(F_\infty))). \tag{115}$$

Proof. From (42), we have the exact sequence

$$0 \longrightarrow \mathcal{S}(E/F_\infty) \longrightarrow H^1(G_T(F_\infty), E_{p^\infty}) \longrightarrow \mathrm{Im}(\lambda_T(F_\infty)) \longrightarrow 0. \tag{116}$$

Taking Σ-cohomology, and recalling (95), we obtain the exact sequence

$$H^1(G_T(F_\infty), E_{p^\infty})^\Sigma \xrightarrow{\theta} \mathrm{Im}(\lambda_T(F_\infty))^\Sigma \to H^1(\Sigma, \mathcal{S}(E/F_\infty)) \to H^3(\Sigma, E_{p^\infty}), \tag{117}$$

where θ is the obvious induced map. But clearly $\mathrm{Coker}(\theta)$ is finite, and its order divides the order of $\mathrm{Coker}(\psi_T(F_\infty))$. Lemma 4.8 is now plain from (115), and its proof is complete.

Lemma 4.9. *Assume that* (i) $p \geqslant 5$, (ii) $\mathcal{S}(E/F)$ *is finite,* (iii) E *has good ordinary reduction at all primes* v *of* F *dividing* p, *and* (iv) p *is unramified in* F. *Then the map* $\psi_T(F_\infty)$ *appearing in the fundamental diagram* (43) *is surjective.*

Proof. Let K_∞ denote the cyclotomic $\mathbb{Z}_p$-extension of K, and put $\Gamma = G(K_\infty/K)$. Now it is well known (see Theorem 1.4 of [13] or [7]) that hypotheses (ii) and (iii) of our lemma imply that $\mathcal{C}(E/K_\infty)$ is a torsion module over $\Lambda(\Gamma)$. As was already used crucially in the proof of Theorem 4.5, this in turn implies that the localization map $\lambda_T(K_\infty)$ is surjective (see [13], Lemma 4.6), so that we have the exact sequence

$$0 \longrightarrow \mathcal{S}(E/K_\infty) \longrightarrow H^1(G_T(K_\infty), E_{p^\infty}) \xrightarrow{\lambda_T(K_\infty)} \bigoplus_{v \in T} J_v(K_\infty) \longrightarrow 0. \tag{118}$$

In addition, it is well known (see [7], Proposition 4.15 or [13]) that our hypotheses that $\mathcal{S}(E/F)$ is finite and p is not ramified in F imply that

$$H^1(\Gamma, \mathcal{S}(E/K_\infty)) = 0. \tag{119}$$

Hence we obtain the exact sequence

$$0 \longrightarrow \mathcal{S}(E/K_\infty)^\Gamma \longrightarrow H^1(G_T(K_\infty), E_{p^\infty})^\Gamma \xrightarrow{\psi_T(K_\infty)} \bigoplus_{v \in T} J_v(K_\infty)^\Gamma \longrightarrow 0. \tag{120}$$

Now, by Lemma 3.16, the map γ appearing in the fundamental diagram (43) is surjective, because p is not ramified in F. Hence the vertical map κ in the commutative diagram

$$\begin{array}{ccc} H^1(G_T(F_\infty), E_{p^\infty})^\Sigma & \xrightarrow{\psi_T(F_\infty)} & \bigoplus_{v \in T} J_v(F_\infty)^\Sigma \\ \uparrow & & \uparrow \kappa \\ H^1(G_T(K_\infty), E_{p^\infty})^\Gamma & \xrightarrow{\psi_T(K_\infty)} & \bigoplus_{v \in T} J_v(K_\infty)^\Gamma \end{array}$$

is also surjective, because γ factors through κ. But then it is clear that the surjectivity of $\psi_T(K_\infty)$ implies the surjectivity of $\psi_T(F_\infty)$. This completes the proof of Lemma 4.9.

Now take $F = \mathbb{Q}$, and assume that $L(E,1) \neq 0$. Since $L(E,1) \neq 0$, Kolyvagin's theorem tells us that $\mathcal{S}(E/\mathbb{Q})$ is finite for every prime p. Thus Theorem 1.15 of Chapter 1 is an immediate consequence of Lemmas 4.7, 4.8 and 4.9.

Proposition 4.10. *Assume that* (i) $p \geqslant 5$, (ii) $\mathcal{S}(E/F)$ *is finite,* (iii) E *has good ordinary reduction at all primes* v *of* F *dividing* p, *and* (iv) $\lambda_T(F_\infty)$ *is surjective. Then Conjecture* 1.13 *holds for* E/F *and* p.

Proof. By virtue of (iv), we know from Proposition 4.3 that

$$H^i(\Sigma, \mathcal{S}(E/F_\infty)) = 0 \quad \text{for } i \geqslant 2.$$

We also have the exact sequence (100), and taking Σ-cohomology of it, we obtain the exact sequence

$$0 \longrightarrow \operatorname{Coker}(\psi_T(F_\infty)) \longrightarrow H^1(\Sigma, \mathcal{S}(E, F_\infty)) \longrightarrow H^3(\Sigma, E_{p^\infty}) \longrightarrow 0, \tag{121}$$

where the term on the right comes from (95). Thus we see that the exact order of $H^1(\Sigma, \mathcal{S}(E/F_\infty))$ is given by (115), and so we obtain from Lemma 4.7

$$\#(H^0(\Sigma, \mathcal{S}(E/F_\infty))) = \xi_p(E/F) \cdot \#(H^1(\Sigma, \mathcal{S}(E/F_\infty))). \tag{122}$$

Thus (27) is valid, and the proof of Proposition 4.10 is complete.

Example. Take $F = \mathbb{Q}$, and $p = 5$. Let E_0, E_1, E_2 denote, respectively, the elliptic curves (4), (5) and (103) of conductor 11. We have just shown that $\lambda_T(F_\infty)$ is surjective for all three curves, with $T = \{5, 11\}$. Hence Proposition 4.10 tells us that Conjecture 1.13 holds for $p = 5$ and all three curves. Thus we have

$$H^k(\Sigma, \mathcal{S}(E_i/F_\infty)) = 0 \quad (k \geqslant 2)$$

for $i = 0, 1, 2$. Moreover, as was explained in Chapter 1, we have

$$\chi(\Sigma, \mathcal{S}(E_0/F_\infty)) = 5^2, \quad \chi(\Sigma, \mathcal{S}(E_1/F_\infty)) = 5.$$

Similarly, one can show that

$$\chi(\Sigma, \mathcal{S}(E_2/F_\infty)) = 5^3,$$

noting that

$$Ш(E_2/\mathbb{Q})(5) = 0, \quad E_2(\mathbb{Q}) = 0.$$

Now take $F = \mathbb{Q}(\mu_5)$ and $p = 5$, and take E to be the elliptic curve E_1. We have $E_1(F)(5) = \mathbb{Z}/5\mathbb{Z}$, and, by a 5-descent (see [7], Chapter 4), we obtain $Ш(E_1/F)(5) = 0$. Now 11 splits completely in F, and $L_v(E, s) = (1-11^{-s})^{-1}$ for each of the four primes v of F dividing 11. Hence we conclude that

$$\chi(G(F_\infty/\mathbb{Q}(\mu_5)), \mathcal{S}(E_1/F_\infty)) = 5^4.$$

4.4 Rank calculations

In this last section, we only sketch the relationship between the surjectivity of $\lambda_T(F_\infty)$ and Conjecture 1.7. The basic idea is to compute $\Lambda(\Sigma)$-ranks (we recall that the notion of $\Lambda(\Sigma)$-rank is defined by (19)) along the dual of the exact sequence

$$0 \longrightarrow \mathcal{S}(E/F_\infty) \longrightarrow H^1(G_T(F_\infty), E_{p^\infty}) \xrightarrow{\lambda_T(F_\infty)} \bigoplus_{v \in T} J_v(F_\infty). \tag{123}$$

Let $t_p(E/F)$ denote the $\Lambda(\Sigma)$-rank of $C(E/F_\infty)$. It follows immediately from (123) and Corollary 2.11 that

$$t_p(E/F) = [F : \mathbb{Q}] - \Lambda(\Sigma)\text{-rank of dual of } \mathrm{Im}(\lambda_T(F_\infty)). \tag{124}$$

Thus the upper bound for $t_p(E/F)$ given in Theorem 1.8 is clear. To establish the lower bound for $t_p(E/F)$ in Theorem 1.8, we need to determine the $\Lambda(\Sigma)$-rank of the dual of $J_v(F_\infty)$ for all $v \in T$. We have already seen on several occasions that $J_v(F_\infty) = 0$ when v divides p, and E has potential supersingular reduction at v. The following result, which we do not prove here, is established in Susan Howson's Ph. D. thesis (see Proposition 6.8 and Theorem 6.9 of [18]).

Lemma 4.11. *Let r_v denote the $\Lambda(\Sigma)$-rank of the dual of $J_v(F_\infty)$. If p is any prime, and v does not divide p, then $r_v = 0$. If $p \geqslant 5$, v divides p, and E has potential ordinary or potential multiplicative reduction at v, then $r_v = [F_v : \mathbb{Q}_p]$.*

Now it is clear from (124) that

$$t_p(E/F) \geqslant [F:\mathbb{Q}] - \bigoplus_{v \in T} r_v.$$

Hence the lower bound for $t_p(E/F)$ asserted in Theorem 1.8 is clear from Lemma 4.11 and the remark made just before Lemma 4.11. This completes the proof of Theorem 1.8.

Theorem 4.12. *Assume $p \geqslant 5$. Then $\lambda_T(F_\infty)$ is surjective if and only if $C(E/F_\infty)$ has $\Lambda(\Sigma)$-rank equal to $\tau_p(E/F)$, where $\tau_p(E/F)$ is defined by* (20).

Proof. If $\lambda_T(F_\infty)$ is surjective, it is clear from (124) and the above determination of the $\Lambda(\Sigma)$-rank r_v of the dual of $J_v(F_\infty)$, that $C(E/F_\infty)$ has $\Lambda(\Sigma)$-rank equal to $\tau_p(E/F)$. Conversely, if $C(E/F_\infty)$ has $\Lambda(\Sigma)$-rank equal to $\tau_p(E/F)$, it follows from (124) that the dual of $\mathrm{Coker}(\lambda_T(F_\infty))$ has $\Lambda(\Sigma)$-rank equal to 0. This means the following. Let $\Phi = G(F_\infty/F_0)$, where $F_0 = F(E_p)$. Then the Iwasawa algebra $\Lambda(\Phi)$ has no divisors of zero. Thus the dual of $\mathrm{Coker}(\lambda_T(F_\infty))$ would be $\Lambda(\Phi)$-torsion. But a very well known argument using the Cassels-Poitou-Tate sequence shows that there is no non-zero $\Lambda(\Phi)$-torsion in the dual of $\mathrm{Coker}(\lambda_T(F_\infty))$ (see [18], Lemma 6.17 or [8], Proposition 11). Hence it follows that $\mathrm{Coker}(\lambda_T(F_\infty)) = 0$. This completes the proof of Theorem 4.12.

Finally, we remark that Theorem 1.14 is an immediate consequence of Theorem 4.12 and Proposition 4.10.

References

[1] P. Balister, S. Howson, *Note on Nakayama's lemma for compact Λ-modules*, Asian J. Math. **1** (1997), 214-219.

[2] K. Barré-Sirieix, G. Diaz, F. Gramain, G. Philibert, *Une preuve de la conjecture de Mahler-Manin*, Invent. Math. **124** (1996), 1-9.

[3] J. Coates, G. McConnell, *Iwasawa theory of modular elliptic curves of analytic rank at most* 1, J. London Math. Soc. **50** (1994), 243-264.

[4] J. Coates, R. Greenberg, *Kummer theory for abelian varieties over local fields*, Invent. Math. **124** (1996), 129-174.

[5] J. Coates, S. Howson, *Euler characteristics and elliptic curves*, Proc. Nat. Acad. Sci. USA **94** (1997), 11115-11117.

[6] J. Coates, S. Howson, *Euler characteristics and elliptic curves* II, in preparation.

[7] J. Coates, R. Sujatha, *Galois cohomology of elliptic curves*, Lecture Notes at the Tata Institute of Fundamental Research, Bombay (to appear).

[8] J. Coates, R. Sujatha, *Iwasawa theory of elliptic curves*, to appear in Proc. Number Theory Conference held at KIAS, Seoul, December 1997.

[9] J. Cremona, *Algorithms for modular elliptic curves*, 2nd edition, CUP (1997).

[10] J. Dixon, M. du Sautoy, A. Mann, D. Segal, *Analytic pro-p-groups*, LMS Lecture Notes **157**, CUP.

[11] K. Goodearl, R. Warfield, *An introduction to noncommutative Noetherian rings*, LMS Student Texts **16**, CUP.

[12] R. Greenberg, *Iwasawa theory for p-adic representations*, Advanced Studies in Pure Math. **17** (1989), 97-137.

[13] R. Greenberg, *Iwasawa theory for elliptic curves*, this volume.

[14] B. Gross, *Kolyvagin's work on modular elliptic curves*, LMS Lecture Notes **153**, CUP (1991), 235-256.

[15] Y. Hachimori, K. Matsuno, *An analogue of Kida's formula for the Selmer groups of elliptic curves*, to appear in J. of Algebraic Geometry.

[16] M. Harris, *p-adic representations arising from descent on abelian varieties*, Comp. Math. **39** (1979), 177-245.

[17] G. Hochschild, J.-P. Serre, *Cohomology of group extensions*, Trans. AMS **74** (1953), 110-134.

[18] S. Howson, *Iwasawa theory of elliptic curves for p-adic Lie extensions*, Ph. D. thesis, Cambridge 1998.

[19] K. Iwasawa, *On* $\mathbf{Z}_\ell$*-extensions of algebraic number fields*, Ann. of Math. **98** (1973), 246-326.

[20] H. Imai, *A remark on the rational points of abelian varieties with values in cyclotomic* $\mathbf{Z}_\ell$*-extensions*, Proc. Japan Acad. **51** (1975), 12-16.

[21] S. Lang, H. Trotter, *Frobenius distributions in* GL_2*-extensions*, Springer Lecture Notes **504** (1976), Springer.

[22] M. Lazard, *Groupes analytiques p-adiques*, Publ. Math. IHES **26** (1965), 389-603.

[23] W. McCallum, *Tate duality and wild ramification*, Math. Ann. **288** (1990), 553-558.

[24] Y. Ochi, Ph. D. thesis, Cambridge 1999.

[25] J.-P. Serre, *Cohomologie Galoisienne*, Springer Lecture Notes **5**, 5th edition (1994), Springer.

[26] J.-P. Serre, *Propriétés galoisiennes des points d'ordre fini des courbes elliptiques* Invent. Math. **15** (1972), 259-331.

[27] J.-P. Serre, *Abelian ℓ-adic representations*, 1968 Benjamin.

[28] J.-P. Serre, *Sur la dimension cohomologique des groupes profinis*, Topology **3** (1965), 413-420.

[29] J.-P. Serre, *Sur les groupes de congruence des variétés abéliennes* I, II, Izv. Akad. Nauk SSSR, **28** (1964), 3-20 and **35** (1971), 731-737.

[30] J.-P. Serre, *La distribution d'Euler-Poincaré d'un groupe profini*, to appear.

[31] J.-P. Serre, J. Tate, *Good reduction of abelian varieties*, Ann. of Math. **88** (1968), 492-517.

[32] J. Silverman, *The arithmetic of elliptic curves*, Graduate Texts in Math. **106** (1986), Springer.

[33] K. Wingberg, *On Poincaré groups*, J. London Math. Soc. **33** (1986), 271-278.

Emmanuel College
Cambridge CB2 3AP

Iwasawa Theory for Elliptic Curves

Ralph Greenberg

University of Washington

1. Introduction

The topics that we will discuss have their origin in Mazur's synthesis of the theory of elliptic curves and Iwasawa's theory of $\mathbb{Z}_p$-extensions in the early 1970s. We first recall some results from Iwasawa's theory. Suppose that F is a finite extension of $\mathbb{Q}$ and that F_∞ is a Galois extension of F such that $\mathrm{Gal}(F_\infty/F) \cong \mathbb{Z}_p$, the additive group of p-adic integers, where p is any prime. Equivalently, $F_\infty = \bigcup_{n\geq 0} F_n$, where, for $n \geq 0$, F_n is a cyclic extension of F of degree p^n and $F = F_0 \subset F_1 \subset \cdots \subset F_n \subset F_{n+1} \subset \cdots$. Let h_n denote the class number of F_n, p^{e_n} the exact power of p dividing h_n. Then Iwasawa proved the following result.

Theorem 1.1. *There exist integers λ, μ, and ν, which depend only on F_∞/F, such that $e_n = \lambda n + \mu p^n + \nu$ for $n \gg 0$.*

The idea behind the proof of this result is to consider the Galois group $X = \mathrm{Gal}(L_\infty/F_\infty)$, where L_∞ is the maximal abelian extension of F_∞ which is unramified at all primes of F_∞ and such that $\mathrm{Gal}(L_\infty/F_\infty)$ is a pro-p group. In fact, $L_\infty = \bigcup_{n\geq 0} L_n$, where L_n is the p-Hilbert class field of F_n for $n \geq 0$. Now L_∞/F is Galois and $\Gamma = \mathrm{Gal}(F_\infty/F)$ acts by inner automorphisms on the normal subgroup X of $\mathrm{Gal}(L_\infty/F)$. Thus, X is a $\mathbb{Z}_p$-module and Γ acts on X continuously and $\mathbb{Z}_p$-linearly. It is natural to regard X as a module over the group ring $\mathbb{Z}_p[\Gamma]$, but even better over the completed group ring

$$\Lambda = \mathbb{Z}_p[[\Gamma]] = \varprojlim \mathbb{Z}_p[\mathrm{Gal}(F_n/F)],$$

where the inverse limit is defined by the ring homomorphisms induced by the restriction maps $\mathrm{Gal}(F_m/F) \to \mathrm{Gal}(F_n/F)$ for $m \geq n \geq 0$. The ring Λ is sometimes called the "Iwasawa algebra" and has the advantage of being a complete local ring. More precisely, $\Lambda \cong \mathbb{Z}_p[[T]]$, where T is identified with $\gamma - 1 \in \Lambda$. Here $\gamma \in \Gamma$ is chosen so that $\gamma|_{F_1}$ is nontrivial, and 1 is the identity element of Γ (and of the ring Λ). Then γ generates a dense subgroup of Γ and the action of $T = \gamma - 1$ on X is "topologically nilpotent." This allows one to consider X as a Λ-module.

Iwasawa proves that X is a finitely generated, torsion Λ-module. There is a structure theorem for such Λ-modules which states that there exists a "pseudo-isomorphism"

$$X \sim \bigoplus_{i=1}^{t} \Lambda/(f_i(T)^{a_i}),$$

where each $f_i(T)$ is an irreducible element of Λ and the a_i's are positive integers. (We say that two finitely generated, torsion Λ-modules X and Y are pseudo-isomorphic when there exists a Λ-homomorphism from X to Y with finite kernel and cokernel. We then write $X \sim Y$.) It is natural to try to recover $\mathrm{Gal}(L_n/F_n)$ from $X = \mathrm{Gal}(L_\infty/F_\infty)$.

Suppose that F has only one prime lying over p and that this prime is totally ramified in F_∞/F. (Totally ramified in F_1/F suffices for this.) Then one can indeed recover $\mathrm{Gal}(L_n/F_n)$ from the Λ-module X. We have

$$\mathrm{Gal}(L_n/F_n) \cong X/(\gamma^{p^n} - 1)X.$$

The isomorphism is induced from the restriction map $X \to \mathrm{Gal}(L_n/F_n)$. Here is a brief sketch of the proof: $\mathrm{Gal}(F_\infty/F_n)$ is topologically generated by γ^{p^n}; one verifies that $(\gamma^{p^n} - 1)X$ is the commutator subgroup of $\mathrm{Gal}(L_\infty/F_n)$; and one proves that the maximal abelian extension of F_n contained in L_∞ is precisely $F_\infty L_n$. (This last step is where one uses the fact that there is only one prime of F_n lying over p.) Then one notices that $\mathrm{Gal}(L_n/F_n) \cong \mathrm{Gal}(F_\infty L_n/F_\infty)$. If F has more than one prime over p, one can still recover $\mathrm{Gal}(L_n/F_n)$ for $n \gg 0$, somehow taking into account the inertia subgroups of $\mathrm{Gal}(L_\infty/F_n)$ for primes over p. (Primes not lying over p are unramified.) One can find more details about the proof in [Wa2].

The invariants λ and μ can be obtained from X in the following way. Let $f(T)$ be a nonzero element of Λ: $f(T) = \sum_{i=0}^{\infty} c_i T^i$, where $c_i \in \mathbb{Z}_p$ for $i \geq 0$. Let $\mu(f) \geq 0$ be defined by: $p^{\mu(f)} | f(T)$, but $p^{\mu(f)+1} \nmid f(T)$ in Λ. Thus, $f(T)p^{-\mu(f)}$ is in Λ and has at least one coefficient in $\mathbb{Z}_p^\times$. Define $\lambda(f) \geq 0$ to be the smallest i such that $c_i p^{-\mu(f)} \in \mathbb{Z}_p^\times$. (Thus, $f(T) \in \Lambda^\times$ if and only if $\lambda(f) = \mu(f) = 0$.) Let $f(T) = \prod_{i=1}^{t} f_i(T)^{a_i}$. The ideal $(f(T))$ of Λ is called the "characteristic ideal" of X. Then it turns out that the λ and μ occurring in Iwasawa's theorem are given by $\lambda = \lambda(f)$, $\mu = \mu(f)$. For each i, there are two possibilities: either $f_i(T)$ is an associate of p, in which case $\mu(f_i) = 1$, $\lambda(f_i) = 0$, and $\Lambda/(f_i(T)^{a_i})$ is an infinite group of exponent p^{a_i}, or $f_i(T)$ is an associate of a monic polynomial of degree $\lambda(f_i)$, irreducible over $\mathbb{Q}_p$, and "distinguished" (which means that the nonleading coefficients are in $p\mathbb{Z}_p$), in which case $\mu(f_i) = 0$ and $\Lambda/(f_i(T)^{a_i})$ is isomorphic to $\mathbb{Z}_p^{\lambda(f_i)a_i}$ as a group. Then, $\lambda = \Sigma a_i \lambda(f_i)$, $\mu = \Sigma a_i \mu(f_i)$. The invariant λ can be described more simply as $\lambda = \mathrm{rank}_{\mathbb{Z}_p}(X/X_{\mathbb{Z}_p\text{-tors}})$, where $X_{\mathbb{Z}_p\text{-tors}}$ is the torsion subgroup of X. Equivalently, $\lambda = \dim_{\mathbb{Q}_p}(X \otimes_{\mathbb{Z}_p} \mathbb{Q}_p)$.

The invariants $\lambda = \lambda(F_\infty/F)$ and $\mu = \mu(F_\infty/F)$ are difficult to study. Iwasawa found examples of $\mathbb{Z}_p$-extensions F_∞/F where $\mu(F_\infty/F) > 0$. In his examples there are infinitely many primes of F which decompose completely in F_∞/F. In these lectures, we will concentrate on the "cyclotomic $\mathbb{Z}_p$-extension" of F which is defined as the unique subfield F_∞ of $F(\mu_{p^\infty})$ with $\Gamma = \mathrm{Gal}(F_\infty/F) \cong \mathbb{Z}_p$. Here μ_{p^∞} denotes the p-power roots of unity. It is easy to show that all nonarchimedean primes of F are finitely decomposed in F_∞/F. More precisely, if v is any such prime of F, then the corresponding decomposition subgroup $\Gamma(v)$ of Γ is of finite index. If $v \nmid p$, then the inertia subgroup is trivial, i.e., v is unramified. (This is true for any $\mathbb{Z}_p$-extension.) If $v|p$, then the corresponding inertia subgroup of Γ is of finite index. Iwasawa has conjectured that $\mu(F_\infty/F) = 0$ if F_∞/F is the cyclotomic $\mathbb{Z}_p$-extension. In the case where F is an abelian extension of $\mathbb{Q}$, this has been proved by Ferrero and Washington. (See [FeWa] or [Wa2].)

On the other hand, $\lambda(F_\infty/F)$ can be positive. The simplest example is perhaps the following. Let F be an imaginary quadratic field. Then all $\mathbb{Z}_p$-extensions of F are contained in a field $\widetilde{F}$ such that $\mathrm{Gal}(\widetilde{F}/F) \cong \mathbb{Z}_p^2$. (Thus, there are infinitely many $\mathbb{Z}_p$-extensions of F.) Letting F_∞/F still be the cyclotomic $\mathbb{Z}_p$-extension, one can verify that $\widetilde{F}/F_\infty$ is unramified if p is a prime that splits completely in $F/\mathbb{Q}$. Thus in this case, $F_\infty \subseteq \widetilde{F} \subseteq L_\infty$ and hence $X = \mathrm{Gal}(L_\infty/F_\infty)$ has a quotient $\mathrm{Gal}(\widetilde{F}/F_\infty) \cong \mathbb{Z}_p$. Therefore, $\lambda(F_\infty/F) \geq 1$ if p splits in $F/\mathbb{Q}$. Notice that, since $\widetilde{F}/F$ is abelian, the action of $T = \gamma - 1$ on $\mathrm{Gal}(\widetilde{F}/F_\infty)$ is trivial. Thus, X/TX is infinite. Now if one considers the Λ-module $Y = \Lambda/(f_i(T)^{a_i})$, where $f_i(T)$ is irreducible in Λ, then Y/TY is infinite if and only if $f_i(T)$ is an associate of T. Therefore, if F is an imaginary quadratic field in which p splits and if F_∞ is the cyclotomic $\mathbb{Z}_p$-extension of F, then $T|f(T)$, where $f(T)$ is a generator of the characteristic ideal of X. One can prove that $T^2 \nmid f(T)$. (This is an interesting exercise. It is easy to show that X/TX has $\mathbb{Z}_p$-rank 1. One must then show that X/T^2X also has $\mathbb{Z}_p$-rank 1. See [Gr1] for a more general "semi-simplicity" result.)

In contrast, suppose that F is again imaginary quadratic, but that p is inert in $F/\mathbb{Q}$. Then F has one prime over p, which is totally ramified in the cyclotomic $\mathbb{Z}_p$-extension F_∞/F. As we sketched earlier, it then turns out that X/TX is finite and isomorphic to the p-primary subgroup of the ideal class group of F. In particular, it follows that if p does not divide the class number of F, then $X = TX$. Nakayama's Lemma for Λ-modules then implies that $X = 0$ and hence $\lambda(F_\infty/F) = 0$ for any such prime p. In general, for arbitrary $n \geq 0$, the restriction map $X \to \mathrm{Gal}(L_n/F_n)$ induces an isomorphism

$$X/\theta_n X \xrightarrow{\sim} \mathrm{Gal}(L_n/F_n),$$

where $\theta_n = \gamma^{p^n} - 1 = (1+T)^{p^n} - 1$. We can think of $X/\theta_n X$ as X_{Γ_n}, the maximal quotient of X on which Γ_n acts trivially. Here $\Gamma_n = \mathrm{Gal}(F_\infty/F_n)$. It is interesting to consider the duals of these groups. Let

$$S_n = \mathrm{Hom}(\mathrm{Gal}(L_n/F_n), \mathbb{Q}_p/\mathbb{Z}_p), \qquad S_\infty = \mathrm{Hom}_{\mathrm{cont}}(X, \mathbb{Q}_p/\mathbb{Z}_p).$$

Then we can state that $S_n \cong S_\infty^{\Gamma_n}$, where the isomorphism is simply the dual of the map $X_{\Gamma_n} \xrightarrow{\sim} \mathrm{Gal}(L_n/F_n)$. Here $S_\infty^{\Gamma_n}$ denotes the subgroup of S_∞ consisting of elements fixed by Γ_n. The map $S_n \to S_\infty^{\Gamma_n}$ will be an isomorphism if F is any number field with just one prime lying over p, totally ramified in

F_∞/F. But returning to the case where F is imaginary quadratic and p splits in $F/\mathbb{Q}$, we have that S_∞^Γ is infinite. (It contains $\mathrm{Hom}(\mathrm{Gal}(\widetilde{F}/F_\infty), \mathbb{Q}_p/\mathbb{Z}_p)$ which is isomorphic to $\mathbb{Q}_p/\mathbb{Z}_p$.) Thus, $S_\infty^{\Gamma_n}$ is always infinite, but S_n is finite, for all $n \geq 0$. The groups S_n and S_∞ are examples of "Selmer groups," by which we mean that they are subgroups of Galois cohomology groups defined by imposing local restrictions. In fact, S_n is the group of cohomology classes in $H^1(G_{F_n}, \mathbb{Q}_p/\mathbb{Z}_p)$ which are unramified at all primes of F_n, and S_∞ is the similarly defined subgroup of $H^1(G_{F_\infty}, \mathbb{Q}_p/\mathbb{Z}_p)$. Here, for any field M, we let G_M denote the absolute Galois group of M. Also, the action of the Galois groups on $\mathbb{Q}_p/\mathbb{Z}_p$ is taken to be trivial. As is customary, we will denote the Galois cohomology group $H^i(G_M, *)$ by $H^i(M, *)$. We will denote $H^i(\mathrm{Gal}(K/M), *)$ by $H^i(K/M, *)$ for any Galois extension K/M. We always require cocycles to be continuous. Usually, the group indicated by $*$ will be a p-primary group which is given the discrete topology. We will also always understand $\mathrm{Hom}(\ ,\ )$ to refer to the set of continuous homomorphisms.

Now we come to Selmer groups for elliptic curves. Suppose that E is an elliptic curve defined over F. We will later recall the definition of the classical Selmer group $\mathrm{Sel}_E(M)$ for E over M, where M is any algebraic extension of F. Right now, we will just mention the exact sequence

$$0 \to E(M) \otimes (\mathbb{Q}/\mathbb{Z}) \to \mathrm{Sel}_E(M) \to \text{Ш}_E(M) \to 0,$$

where $E(M)$ denotes the group of M-rational points on E and $\text{Ш}_E(M)$ denotes the Shafarevich-Tate group for E over M. We denote the p-primary subgroups of $\mathrm{Sel}_E(M)$, $\text{Ш}_E(M)$ by $\mathrm{Sel}_E(M)_p$, $\text{Ш}_E(M)_p$. The p-primary subgroup of the first term above is $E(M) \otimes (\mathbb{Q}_p/\mathbb{Z}_p)$. Also, $\mathrm{Sel}_E(M)_p$ is a subgroup of $H^1(M, E[p^\infty])$, where $E[p^\infty]$ is the p-primary subgroup of $E(\overline{\mathbb{Q}})$. As a group, $E[p^\infty] \cong (\mathbb{Q}_p/\mathbb{Z}_p)^2$, but the action of G_F is quite nontrivial. Let F_∞/F denote the cyclotomic $\mathbb{Z}_p$-extension. We will now state a number of theorems and conjectures, which constitute part of what we call "Iwasawa Theory for E." Some of the theorems will be proved in these lectures. We always assume that F_∞ is the cyclotomic $\mathbb{Z}_p$-extension of F.

Theorem 1.2 (Mazur's Control Theorem). *Assume that E has good, ordinary reduction at all primes of F lying over p. Then the natural maps*

$$\mathrm{Sel}_E(F_n)_p \to \mathrm{Sel}_E(F_\infty)_p^{\Gamma_n}$$

have finite kernel and cokernel, of bounded order as n varies.

The natural maps referred to are those induced by the restriction maps $H^1(F_n, E[p^\infty]) \to H^1(F_\infty, E[p^\infty])$. One should compare this result with the remarks made above concerning S_n and $S_\infty^{\Gamma_n}$. We will discuss below the cases where E has either multiplicative or supersingular reduction at some primes of F lying over p. But first we state an important conjecture of Mazur.

Conjecture 1.3. *Assume that E has good, ordinary reduction at all primes of F lying over p. Then $\mathrm{Sel}_E(F_\infty)_p$ is Λ-cotorsion.*

Here $\Gamma = \mathrm{Gal}(F_\infty/F)$ acts naturally on the group $H^1(F_\infty, E[p^\infty])$, which is a torsion $\mathbb{Z}_p$-module, every element of which is killed by T^n for some n. Thus, $H^1(F_\infty, E[p^\infty])$ is a Λ-module. $\mathrm{Sel}_E(F_\infty)_p$ is invariant under the action of Γ and is thus a Λ-submodule. We say that $\mathrm{Sel}_E(F_\infty)_p$ is Λ-cotorsion if

$$X_E(F_\infty) = \mathrm{Hom}(\mathrm{Sel}_E(F_\infty)_p, \mathbb{Q}_p/\mathbb{Z}_p)$$

is Λ-torsion. Here $\mathrm{Sel}_E(F_\infty)_p$ is a p-primary group with the discrete topology. Its Pontryagin dual $X_E(F_\infty)$ is an abelian pro-p group, which we regard as a Λ-module. It is not hard to prove that $X_E(F_\infty)$ is finitely generated as a Λ-module (and so, $\mathrm{Sel}_E(F_\infty)_p$ is a "cofinitely generated" Λ-module). In the case where E has good, ordinary reduction at all primes of F over p, one can use theorem 1.2. For $X_E(F) = \mathrm{Hom}(\mathrm{Sel}_E(F)_p, \mathbb{Q}_p/\mathbb{Z}_p)$ is known to be finitely generated over $\mathbb{Z}_p$. (In fact, the weak Mordell-Weil theorem is proved by showing that $X_E(F)/pX_E(F)$ is finite.) Write $X = X_E(F_\infty)$ for brevity. Then, by theorem 1.2, X/TX is finitely generated over $\mathbb{Z}_p$. Hence, $X/\mathfrak{m}X$ is finite, where $\mathfrak{m} = (p, T)$ is the maximal ideal of Λ. By a version of Nakayama's Lemma (valid for profinite Λ-modules X), it follows that $X_E(F_\infty)$ is indeed finitely generated as a Λ-module. (This can actually be proved for any prime p, with no restriction on the reduction type of E.) Here is one important case where the above conjecture can be verified.

Theorem 1.4. *Assume that E has good, ordinary reduction at all primes of F lying over p. Assume also that $\mathrm{Sel}_E(F)_p$ is finite. Then $\mathrm{Sel}_E(F_\infty)_p$ is Λ-cotorsion.*

This theorem is an immediate corollary of theorem 1.2, using the following exercise: if X is a Λ-module such that X/TX is finite, then X is a torsion Λ-module. The hypothesis on $\mathrm{Sel}_E(F)_p$ is equivalent to assuming that both the Mordell-Weil group $E(F)$ and the p-Shafarevich-Tate group $\text{Ш}_E(F)_p$ are finite. A much deeper case where conjecture 1.3 is known is the following. The special case where E has complex multiplication had previously been settled by Rubin [Ru1].

Theorem 1.5 (Kato-Rohrlich). *Assume that E is defined over $\mathbb{Q}$ and is modular. Assume also that E has good, ordinary reduction or multiplicative reduction at p and that $F/\mathbb{Q}$ is abelian. Then $\mathrm{Sel}_E(F_\infty)_p$ is Λ-cotorsion.*

The case where E has multiplicative reduction at a prime v of F lying over p is somewhat analogous to the case where E has good, ordinary reduction at v. In both cases, the G_{F_v}-representation space $V_p(E) = T_p(E) \otimes \mathbb{Q}_p$ has an unramified 1-dimensional quotient. (Here $T_p(E)$ is the Tate-module for E; $V_p(E)$ is a 2-dimensional $\mathbb{Q}_p$-vector space on which the local Galois group G_{F_v} acts, where F_v is the v-adic completion of F.) It seems reasonable to believe that the analogue of Theorem 1.2 should hold. This was first suggested by Manin [Man] for the case $F = \mathbb{Q}$.

Conjecture 1.6. *Assume that E has good, ordinary reduction or multiplicative reduction at all primes of F lying over p. Then the natural maps*

$$\mathrm{Sel}_E(F_n)_p \to \mathrm{Sel}_E(F_\infty)_p^{\Gamma_n}$$

have finite kernel and cokernel, of bounded order as n varies.

For $F = \mathbb{Q}$, this is a theorem. In this case, Manin showed that it would suffice to prove that $\log_p(q_E) \neq 0$, where q_E denotes the Tate period for E, assuming that E has multiplicative reduction at p. But a recent theorem of Barré-Sirieix, Diaz, Gramain, and Philibert [B-D-G-P] shows that q_E is transcendental when the j-invariant j_E is algebraic. Since $j_E \in \mathbb{Q}$, it follows that $q_E p^{-\mathrm{ord}(q_E)}$ is not a root of unity and so $\log_p(q_E) \neq 0$. For arbitrary F, one would need to prove that $\log_p(N_{F_v/\mathbb{Q}_p}(q_E^{(v)})) \neq 0$ for all primes v of F lying over p where E has multiplicative reduction. Here F_v is the v-adic completion of F, $q_E^{(v)}$ the corresponding Tate period. This nonvanishing statement seems intractable at present.

If E has supersingular reduction at some prime v of F, then the "control theorem" undoubtedly fails. In fact, $\mathrm{Sel}_E(F_\infty)_p$ will not be Λ-cotorsion. More precisely, let

$$r(E,F) = \sum_{\mathrm{pss}} [F_v : \mathbb{Q}_p],$$

where the sum varies over the primes v of F where E has potentially supersingular reduction. Then one can prove the following result.

Theorem 1.7. *With the above notation, we have*

$$\mathrm{corank}_\Lambda(\mathrm{Sel}_E(F_\infty)_p) \geq r(E,F).$$

This result is due to P. Schneider. He conjectures that equality should hold here. (See [Sch2].) This would include for example a more general version of conjecture 1.3, where one assumes just that E has potentially ordinary or potentially multiplicative reduction at all primes of F lying over p. As a consequence of theorem 1.7, one finds that

$$\mathrm{corank}_{\mathbb{Z}_p}(\mathrm{Sel}_E(F_\infty)_p^{\Gamma_n}) \geq r(E,F)p^n$$

for $n \geq 0$. This follows from the fact that $\Lambda/\theta_n\Lambda \cong \mathbb{Z}_p^{p^n}$. (The ring $\Lambda/\theta_n\Lambda$ is just $\mathbb{Z}_p[\mathrm{Gal}(F_n/F)]$.) One uses the fact that there is a pseudo-isomorphism from $X_E(F_\infty)$ to $\Lambda^r \oplus Y$, where $r = \mathrm{rank}_\Lambda(X_E(F_\infty))$, which is the Λ-corank of $\mathrm{Sel}_E(F_\infty)_p$, and Y is the Λ-torsion submodule of $X_E(F_\infty)$. However, it is reasonable to make the following conjecture. We continue to assume that

F_∞/F is the cyclotomic $\mathbb{Z}_p$-extension, but make no assumptions on the reduction type for E at primes lying over p. The conjecture below follows from results of Kato and Rohrlich when F is abelian over $\mathbb{Q}$ and E is defined over $\mathbb{Q}$ and modular.

Conjecture 1.8. *The $\mathbb{Z}_p$-corank of* $\mathrm{Sel}_E(F_n)_p$ *is bounded as n varies.*

If this is so, then the map $\mathrm{Sel}_E(F_n)_p \to \mathrm{Sel}_E(F_\infty)^{\Gamma_n}$ must have infinite cokernel when n is sufficiently large, provided that we assume that E has potentially supersingular reduction at v for at least one prime v of F lying over p. Of course, assuming that the p-Shafarevich-Tate group is finite, the $\mathbb{Z}_p$-corank of $\mathrm{Sel}_E(F_n)_p$ is just the rank of the Mordell-Weil group $E(F_n)$. If one assumes that $E(F_n)$ does indeed have bounded rank as $n \to \infty$ then one can deduce the following nice consequence: *$E(F_\infty)$ is finitely generated. Hence, for some $n \geq 0$, $E(F_\infty) = E(F_n)$.* This is proved in Mazur's article [Maz1]. The crucial step is to show that $E(F_\infty)_{\mathrm{tors}}$ is finite. We refer the reader to Mazur (proposition 6.12) for a detailed proof of this helpful fact. (We will make use of it later. See also [Im] or [Ri].) Using this, one then argues as follows. Let $t = |E(F_\infty)_{\mathrm{tors}}|$. Choose m so that $\mathrm{rank}(E(F_m))$ is maximal. Then, for any $P \in E(F_\infty)$, we have $kP \in E(F_m)$ for some $k \geq 1$. Then $g(kP) = kP$ for all $g \in \mathrm{Gal}(F_\infty/F_m)$. That is, $g(P) - P$ is in $E(F_\infty)_{\mathrm{tors}}$ and hence $t(g(P) - P) = O_E$. This means that $tP \in E(F_m)$. Therefore, $tE(F_\infty) \subseteq E(F_m)$, from which it follows that $E(F_\infty)$ is finitely generated.

On the other hand, let us assume that E has good, ordinary reduction or multiplicative reduction at all primes v of F lying over p. Assume also that $\mathrm{Sel}_E(F_\infty)_p$ is Λ-cotorsion, as is conjectured. Then one can prove conjecture 1.8 very easily. Let λ_E denote the λ-invariant of the torsion Λ-module $X_E(F_\infty)$. That is, $\lambda_E = \mathrm{rank}_{\mathbb{Z}_p}(X_E(F_\infty)) = \mathrm{corank}_{\mathbb{Z}_p}(\mathrm{Sel}_E(F_\infty)_p)$. We get the following result.

Theorem 1.9. *Under the above assumptions, one has*

$$\mathrm{corank}_{\mathbb{Z}_p}(\mathrm{Sel}_E(F_n)_p) \leq \lambda_E.$$

In particular, the rank of the Mordell-Weil group $E(F_n)$ is bounded above by λ_E.

This result follows from the fact that the maps $\mathrm{Sel}_E(F_n)_p \to \mathrm{Sel}_E(F_\infty)_p$ have finite kernel. This turns out to be quite easy to prove, as we will see in section 3. Also, the rank of $E(F_n)$ is the $\mathbb{Z}_p$-corank of $E(F_n) \otimes (\mathbb{Q}_p/\mathbb{Z}_p)$, which is of course bounded above by $\mathrm{corank}_{\mathbb{Z}_p}(\mathrm{Sel}_E(F_n)_p)$. (Equality holds if $Ш_E(F_n)_p$ is finite.) Let $\lambda_E^{M\text{-}W}$ denote the maximum of $\mathrm{rank}(E(F_n))$ as n varies, which is just $\mathrm{rank}(E(F_\infty))$. Let $\lambda_E^{Ш} = \lambda_E - \lambda_E^{M\text{-}W}$. We let μ_E denote the μ-invariant of the Λ-module $X_E(F_\infty)$. If necessary to avoid confusion, we might write $\lambda_E = \lambda_E(F_\infty/F)$, $\mu_E = \mu_E(F_\infty/F)$, etc. Then we have the following analogue of Iwasawa's theorem.

Theorem 1.10. *Assume that E has good, ordinary reduction at all primes of F lying over p. Assume that $\mathrm{Sel}_E(F_\infty)_p$ is Λ-cotorsion and that $\text{Ш}_E(F_n)_p$ is finite for all $n \geq 0$. Then there exist λ, μ, and ν such that $|\text{Ш}_E(F_n)_p| = p^{e_n}$, where $e_n = \lambda n + \mu p^n + \nu$ for all $n \gg 0$. Here $\lambda = \lambda_E^{\text{Ш}}$ and $\mu = \mu_E$.*

As later examples will show, each of the invariants $\lambda_E^{M\text{-}W}$, $\lambda_E^{\text{Ш}}$, and μ_E can be positive. Mazur first pointed out the possibility that μ_E could be positive, giving the following example. Let $E = X_0(11)$, $p = 5$, $F = \mathbb{Q}$, and $F_\infty = \mathbb{Q}_\infty$ = the cyclotomic $\mathbb{Z}_5$-extension of $\mathbb{Q}$. Then $\mu_E = 1$. (In fact, $(f_E(T)) = (p)$.) There are three elliptic curves/$\mathbb{Q}$ of conductor 11, all isogenous. In addition to E, one of these elliptic curves has $\mu = 2$, another has $\mu = 0$. In general, suppose that $\phi : E_1 \to E_2$ is an F-isogeny, where E_1, E_2 are defined over F. Let $\Phi : \mathrm{Sel}_{E_1}(F_\infty)_p \to \mathrm{Sel}_{E_2}(F_\infty)_p$ denote the induced Λ-module homomorphism. It is not hard to show that the kernel and cokernel of Φ have finite exponent, dividing the exponent of $\ker(\phi)$. Thus, $\mathrm{Sel}_{E_1}(F_\infty)_p$ and $\mathrm{Sel}_{E_2}(F_\infty)_p$ have the same Λ-corank. If they are Λ-cotorsion, then the λ-invariants are the same. The characteristic ideals of $X_{E_1}(F_\infty)$ and $X_{E_2}(F_\infty)$ differ only by multiplication by a power of p. If $F = \mathbb{Q}$, then it seems reasonable to make the following conjecture. For arbitrary F, the situation seems more complicated. We had believed that this conjecture should continue to be valid, but counterexamples have recently been found by Michael Drinen.

Conjecture 1.11. *Let E be an elliptic curve defined over $\mathbb{Q}$. Assume that $\mathrm{Sel}_E(\mathbb{Q}_\infty)_p$ is Λ-cotorsion. Then there exists a $\mathbb{Q}$-isogenous elliptic curve E' such that $\mu_{E'} = 0$. In particular, if $E[p]$ is irreducible as a $(\mathbb{Z}/p\mathbb{Z})$-representation of $G_{\mathbb{Q}}$, then $\mu_E = 0$.*

Here $E[p] = \ker(E(\overline{\mathbb{Q}}) \xrightarrow{p} E(\overline{\mathbb{Q}}))$. P. Schneider has given a simple formula for the effect of an isogeny on the μ-invariant of $\mathrm{Sel}_E(F_\infty)_p$ for arbitrary F and for odd p. (See [Sch3] or [Pe2].) Thus, the above conjecture effectively predicts the value of μ_E for $F = \mathbb{Q}$.

Suppose that $\mathrm{Sel}_E(F_\infty)_p$ is Λ-cotorsion. Let $f_E(T)$ be a generator of the characteristic ideal of $X_E(F_\infty)$. Then $\lambda_E = \lambda(f_E)$ and $\mu_E = \mu(f_E)$. We have

$$X_E(F_\infty) \sim \prod_{i=1}^{t} \Lambda/(f_i(T)^{a_i})$$

where the $f_i(T)$'s are irreducible elements of Λ, and the a_i's are positive. If $(f_i(T)) = (p)$, then it is possible for $a_i > 1$. However, in contrast, it seems reasonable to make the following "semi-simplicity" conjecture.

Conjecture 1.12. *Let E be an elliptic curve defined over F. Assume that $\mathrm{Sel}_E(F_\infty)_p$ is Λ-cotorsion. The action of $\Gamma = \mathrm{Gal}(F_\infty/F)$ on $X_E(F_\infty) \otimes_{\mathbb{Z}_p} \mathbb{Q}_p$ is completely reducible. That is, $a_i = 1$ for all i's such that $f_i(T)$ is not an associate of p.*

Assume that E has good, ordinary reduction at all primes of F lying over p. Theorem 1.2 then holds. In particular, $\mathrm{corank}_{\mathbb{Z}_p}(\mathrm{Sel}_E(F)_p)$, which is equal to $\mathrm{rank}_{\mathbb{Z}_p}(X_E(F_\infty)/TX_E(F_\infty))$, would equal the power of T dividing $f_E(T)$, assuming the above conjecture. Also, the value of $\lambda_E^{M\text{-}W}$ would be equal to the number of roots of $f_E(T)$ of the form $\zeta - 1$, where ζ is a p-power root of unity, if we assume in addition the finiteness of $Ш_E(F_n)_p$ for all n. For conjecture 1.12 would imply that this number is equal to the $\mathbb{Z}_p$-rank of $X_E(F_\infty)/\theta_n X_E(F_\infty)$ for $n \gg 0$.

In section 4 we will introduce some theorems due to B. Perrin-Riou and to P. Schneider which give a precise relationship between $\mathrm{Sel}_E(F)_p$ and the behavior of $f_E(T)$ at $T = 0$. These theorems are important because they allow one to study the Birch and Swinnerton-Dyer conjecture by using the so-called "Main Conjecture" which states that one can choose the generator $f_E(T)$ so that it satisfies a certain interpolation property. We will give the statement of this conjecture for $F = \mathbb{Q}$, which was formulated by B. Mazur in the early 1970s (in the same paper [Maz1] where he proves theorem 1.2 and also in [M-SwD]).

Conjecture 1.13. *Assume that E is an elliptic curve defined over $\mathbb{Q}$ which has good, ordinary reduction at p. Then the characteristic ideal of $X_E(\mathbb{Q}_\infty)$ has a generator $f_E(T)$ with the properties:*

(i) $f_E(0) = (1 - \beta_p p^{-1})^2 L(E/\mathbb{Q}, 1)/\Omega_E$

(ii) $f_E(\phi(T)) = (\beta_p)^n L(E/\mathbb{Q}, \phi, 1)/\Omega_E \tau(\phi)$ *if ϕ is a finite order character of $\Gamma = \mathrm{Gal}(\mathbb{Q}_\infty/\mathbb{Q})$ of conductor $p^n > 1$.*

We must explain the notation. First of all, fix embeddings of $\overline{\mathbb{Q}}$ into $\mathbb{C}$ and into $\overline{\mathbb{Q}}_p$. $L(E/\mathbb{Q}, s)$ denotes the Hasse-Weil L-series for E over $\mathbb{Q}$. Ω_E denotes the real period for E, so that $L(E/\mathbb{Q}, 1)/\Omega_E$ is conjecturally in $\mathbb{Q}$. (If E is modular, then $L(E/\mathbb{Q}, s)$ has an analytic continuation to the complex plane, and, in fact, $L(E/\mathbb{Q}, 1)/\Omega_E \in \mathbb{Q}$.) Let $\widetilde{E}$ denote the reduction of E at p. The Euler factor for p in $L(E/\mathbb{Q}, s)$ is $((1 - \alpha_p p^{-s})(1 - \beta_p p^{-s}))^{-1}$, where α_p, $\beta_p \in \overline{\mathbb{Q}}$, $\alpha_p\beta_p = p$, $\alpha_p + \beta_p = 1 + p - |\widetilde{E}(\mathbb{F}_p)|$. Choose α_p to be the p-adic unit under the fixed embedding $\overline{\mathbb{Q}} \to \overline{\mathbb{Q}}_p$. Thus, $\beta_p p^{-1} = \alpha_p^{-1}$. For every complex-valued, finite order Dirichlet character ϕ, $L(E/\mathbb{Q}, \phi, s)$ denotes the twisted Hasse-Weil L-series. In the above interpolation property, ϕ is a Dirichlet character whose associated Artin character factors through Γ. Using the fixed embeddings chosen above, we can consider ϕ as a continuous homomorphism $\phi : \Gamma \to \overline{\mathbb{Q}}_p^\times$ of finite order, i.e., $\phi(\gamma) = \zeta$, where ζ is a p-power root of unity in $\overline{\mathbb{Q}}_p$. Then $\phi(T) = \phi(\gamma - 1) = \zeta - 1$, which is in the maximal ideal of $\overline{\mathbb{Z}}_p$. Hence $f_E(\phi(T)) = f_E(\zeta - 1)$ converges in $\overline{\mathbb{Q}}_p$. The complex number $L(E/\mathbb{Q}, \phi, 1)/\Omega_E$ should be algebraic. In (ii), we regard it as an element of $\overline{\mathbb{Q}}_p$, as well as the Gaussian sum $\tau(\phi)$. For $p > 2$, conjecture 1.13 has been proven by Rubin when E has complex multiplication. (See [Ru2].) If E is a modular elliptic curve with good, ordinary reduction at p, then the existence of some power series satisfying the stated interpolation property (i) and (ii) was proven by Mazur and Swinnerton-Dyer in the early 1970s. We will denote it by $f_E^{\text{anal}}(T)$. (See [M-SwD] or [M-T-T].) Conjecturally, this power series should be in Λ. This is proven in [St] if $E[p]$ is irreducible as a $G_{\mathbb{Q}}$-module. In general, it is only known to be $\Lambda \otimes_{\mathbb{Z}_p} \mathbb{Q}_p$. That is, $p^t f_E^{\text{anal}}(T) \in \Lambda$ for some $t \geq 0$. Kato then proves that the characteristic ideal at least contains $p^m f_E^{\text{anal}}(T)$ for some $m \geq 0$. Rohrlich proves that $L(E/\mathbb{Q}, \phi, 1) \neq 0$ for all but finitely many characters ϕ of Γ, which is equivalent to the statement $f_E^{\text{anal}}(T) \neq 0$ as an element of $\Lambda \otimes_{\mathbb{Z}_p} \mathbb{Q}_p$. One can use Kato's theorem to prove conjecture 1.13 when E admits a cyclic $\mathbb{Q}$-isogeny of degree p, where p is odd and the kernel of the isogeny satisfies a certain condition (namely, the hypotheses in proposition 5.10 in these notes). This will be discussed in [GrVa].

Continuing to assume that $E/\mathbb{Q}$ is modular and that p is a prime where E has good, ordinary reduction, the so-called p-adic L-function $L_p(E/\mathbb{Q}, s)$ can be defined in terms of $f_E^{\text{anal}}(T)$. We first define a canonical character

$$\kappa : \Gamma \to 1 + 2p\mathbb{Z}_p$$

induced by the cyclotomic character $\chi : \text{Gal}(\mathbb{Q}(\mu_{p^\infty})/\mathbb{Q}) \xrightarrow{\sim} \mathbb{Z}_p^\times$ composed with the projection map to the second factor in the canonical decomposition $\mathbb{Z}_p^\times = \mu_{p-1} \times (1 + p\mathbb{Z}_p)$ for odd p, or $\mathbb{Z}_2^\times = \{\pm 1\} \times (1 + 4\mathbb{Z}_2)$ for $p = 2$. Thus, κ is an isomorphism. For $s \in \mathbb{Z}_p$, define $L_p(E/\mathbb{Q}, s)$ by

$$L_p(E/\mathbb{Q}, s) = f_E^{\text{anal}}(\kappa(\gamma)^{s-1} - 1).$$

The power series converges since $\kappa(\gamma)^{s-1} - 1 \in p\mathbb{Z}_p$. (Note: Let $t \in \mathbb{Z}_p$. The continuous group homomorphism $\kappa^t : \Gamma \to 1 + p\mathbb{Z}_p$ can be extended uniquely to a continuous $\mathbb{Z}_p$-linear ring homomorphism $\kappa^t : \Lambda \to \mathbb{Z}_p$. We have $\kappa^t(T) = \kappa(\gamma)^t - 1$ and $\kappa^t(f(T)) = f(\kappa(\gamma)^t - 1)$ for any $f(T) \in \Lambda$. Thus, $L_p(E/\mathbb{Q}, s)$ is $\kappa^{s-1}(f_E^{\text{anal}}(T))$.) The functional equations for the Hasse-Weil L-series give a simple relation between the values $L(E/\mathbb{Q}, \phi, 1)$ and $L(E/\mathbb{Q}, \phi^{-1}, 1)$ occurring in the interpolation property for $f_E^{\text{anal}}(T)$. Since $f_E^{\text{anal}}(T)$ is determined by its interpolation property, one can deduce a simple relation between $f_E^{\text{anal}}(T)$ and $f_E^{\text{anal}}((1+T)^{-1} - 1)$. Omitting the details, one obtains a functional equation for $L_p(E/\mathbb{Q}, s)$:

$$L_p(E/\mathbb{Q}, 2 - s) = w_E \langle N_E \rangle^{s-1} L_p(E/\mathbb{Q}, s)$$

for all $s \in \mathbb{Z}_p$. Here w_E is the sign which occurs in the functional equation for the Hasse-Weil L-series $L(E/\mathbb{Q}, s)$, N_E is the conductor of E, and $\langle N_E \rangle$ is the projection of N_E to $1 + 2p\mathbb{Z}_p$ as above.

The final theorem we will state is motivated by conjecture 1.13 and the above functional equation for the p-adic L-function $L_p(E/\mathbb{Q}, s)$. The functional equation is in fact equivalent to the relation between $f_E^{\text{anal}}(T)$ and $f_E^{\text{anal}}((1+T)^{-1} - 1)$ mentioned above. In particular, $f_E^{\text{anal}}(T^\iota)/f_E^{\text{anal}}(T)$ should be in $\Lambda^\times$, where $T^\iota = (1+T)^{-1} - 1$. The analogue of this statement is true for $f_E(T)$. More generally (for any F), we have:

Theorem 1.14. *Assume that E is an elliptic curve defined over F with good, ordinary reduction or multiplicative reduction at all primes of F lying over p. Assume that $\mathrm{Sel}_E(F_\infty)_p$ is Λ-cotorsion. Then the characteristic ideal of $X_E(F_\infty)$ is fixed by the involution ι of Λ induced by $\iota(\gamma) = \gamma^{-1}$ for all $\gamma \in \Gamma$.*

A proof of this result can be found in [Gr2] using the Duality Theorems of Poitou and Tate. There it is dealt with in a much more general context—that of Selmer groups attached to "ordinary" p-adic representations.

We will prove theorem 1.2 completely in the following two sections. Our approach is quite different than the approach in Mazur's article and in Manin's more elementary expository article. We first prove that, when E has good, ordinary or multiplicative reduction at primes over p, the p-primary subgroups of $\mathrm{Sel}_E(F_n)$ and of $\mathrm{Sel}_E(F_\infty)$ have a very simple and elegant description. This is the main content of section 2. Once we have this, it is quite straightforward to prove theorem 1.2 and also a conditional result concerning conjecture 1.6 which we do in section 3. In this approach we avoid completely the need to study the norm map for formal groups over local fields, which is crucial in the approach in [Maz1] and [Man]. We also can use our description of the p-Selmer group to determine the p-adic valuation of $f_E(0)$, under the assumption that E has good, ordinary reduction at primes over p and that $\mathrm{Sel}_E(F)_p$ is finite. Section 4 is devoted to this comparatively easy special case of results of B. Perrin-Riou and P. Schneider found in [Pe1], [Sch1]. Their results give an expression involving a p-adic height determinant for the p-adic valuation of $(f_E(T)/T^r)|_{T=0}$, where $r = \mathrm{rank}(E(F))$, under suitable hypotheses. Finally, in section 5, (which is by far the longest section of this article) we will discuss a variety of examples to illustrate the results of sections 3 and 4 and also how our description of the p-Selmer group can be used for calculation. We also include in section 5 a number of remarks taken from [Maz1] (some of which are explained quite differently here) as well as various results which don't seem to be in the existing literature. Throughout this article, we have tried to include $p = 2$ in all of the main results. Perhaps surprisingly, this turns out not to be so complicated.

We will have very little to say about the case where E has supersingular reduction at some primes over p. In recent years, this has become a very lively aspect of Iwasawa theory. We just refer the reader to [Pe4] as an introduction. In [Pe4], one finds the following concrete application of the theory described there: *Suppose that $E/\mathbb{Q}$ has supersingular reduction at p and that* $\mathrm{Sel}_E(\mathbb{Q})_p$ *is finite. Then* $\mathrm{Sel}_E(\mathbb{Q}_n)_p$ *has bounded $\mathbb{Z}_p$-corank as n varies.* This is, of course, a special case of conjecture 1.8. In the case where E has good, ordinary reduction over p, theorem 1.4 gives the same conclusion. Another topic that we will not pursue is the behavior of the p-Selmer group in other $\mathbb{Z}_p$-extensions—for example, the anti-cyclotomic $\mathbb{Z}_p$-extension of an imaginary quadratic field. The analogues of conjectures 1.3 and 1.8 can in fact be false. We refer the reader to [Be], [BeDa1, 2], and [Maz4] for a discussion of this topic. We also will not pursue the analytic side of Iwasawa theory—

questions involving the properties of p-adic L-functions and the p-adic version of a Birch and Swinnerton-Dyer conjecture. For this, one can learn something from the articles [M-SwD], [B-G-S], and [M-T-T]. Many of the ideas we discuss here can be extended to a far more general context. For an introduction to this, we refer the reader to [CoSc] and to [Gr2,3].

The author is grateful to the Fondazione Centro Internazionale Matematico Estivo and to Carlo Viola for the invitation to give lectures in Cetraro. This article is an extensively expanded version of those lectures, based considerably on research which was partially supported by the National Science Foundation. The author is also grateful for the support and hospitality of the American Institute of Mathematics during the Winter of 1998, when many of the results and examples described in section 5 were obtained. We want to thank Karl Rubin for many valuable discussions and for his help in the details of several examples, Ted McCabe for carrying out numerous calculations of p-adic L-functions which allowed us to verify the main conjecture in many cases, and Ken Kramer for explaining his results about elliptic curves with 2-power isogenies. We are also grateful to John Coates for many helpful remarks and to Y. Hachimori, K. Matsuno and T. Ochiai for finding a number of mistakes in the text.

2. Kummer Theory for E

Let E be an elliptic curve defined over a number field F. If M is any algebraic extension of F, Kummer theory for E over M leads quite naturally to the classical definition of the Selmer group $\mathrm{Sel}_E(M)$. The main objective of this section is to give a simplified description of its p-primary subgroup $\mathrm{Sel}_E(M)_p$ under the hypothesis that E has either good, ordinary reduction or multiplicative reduction at all primes of F lying over p. We will assume that M is either a finite extension or a $\mathbb{Z}_p$-extension of F.

Kummer theory for the multiplicative group $M^\times$ is quite familiar. Regarding M as a subfield of $\overline{F}$, a fixed algebraic closure of F (or $\mathbb{Q}$), we can define the Kummer homomorphism

$$k: M^\times \otimes (\mathbb{Q}/\mathbb{Z}) \to H^1(M, \overline{F}^\times_{\mathrm{tors}})$$

as follows. Let $a \in M^\times$. Let $\alpha = a \otimes (m/n + \mathbb{Z}) \in M^\times \otimes (\mathbb{Q}/\mathbb{Z})$. Choose $b \in \overline{F}^\times$ such that $b^n = a^m$, using the fact that $\overline{F}^\times$ is a divisible group. Then one defines $k(\alpha)$ to be the class of the 1-cocycle ϕ_α given by $\phi_\alpha(g) = g(b)/b$ for all $g \in G_M = \mathrm{Gal}(\overline{F}/M)$. The values of ϕ_α are in $\overline{F}^\times_{\mathrm{tors}}$, the group of roots of unity in $\overline{F}$. The Kummer homomorphism is an isomorphism. Injectivity is easy to verify. Surjectivity is a consequence of Hilbert's Theorem 90, which asserts that $H^1(M, \overline{F}^\times) = 0$.

Since $E(\overline{F})$ is divisible, one can imitate the above definition, obtaining an exact sequence

$$0 \to E(M) \otimes (\mathbb{Q}/\mathbb{Z}) \xrightarrow{k} H^1(M, E(\overline{F})_{\mathrm{tors}}) \to H^1(M, E(\overline{F})) \to 0.$$

If $\alpha = a\otimes(m/n+\mathbb{Z}) \in E(M)\otimes(\mathbb{Q}/\mathbb{Z})$, then $k(\alpha)$ is the class of the 1-cocycle ϕ_α given by $\phi_\alpha(g) = g(b)-b$ for all $g \in G_M$. Here $b \in E(\overline{F})$ satisfies $nb = ma$ on $E(\overline{F})$. However, in general, $H^1(M, E(\overline{F}))$ is nonzero. We will fix a prime p and concentrate on the p-primary subgroups of the above groups. We let $\kappa = \kappa_M$ denote the corresponding Kummer homomorphism:

$$\kappa : E(M)\otimes(\mathbb{Q}_p/\mathbb{Z}_p) \to H^1(M, E[p^\infty]).$$

If η is any prime of M, we define M_η to be the union of the η-adic completions of all finite extensions of F contained in M. Thus, if η lies over the prime v of F, then M_η is an algebraic extension of F_v. By fixing an embedding $\overline{F} \to \overline{F}_v$ extending the embedding $M \to M_\eta$, one can identify G_{M_η} with a subgroup of G_M, which of course is just the decomposition subgroup for some prime of $\overline{F}$ lying over η. We will let κ_η denote the Kummer homomorphism for E over M_η:

$$\kappa_\eta : E(M_\eta)\otimes(\mathbb{Q}_p/\mathbb{Z}_p) \to H^1(M_\eta, E[p^\infty]).$$

This is defined exactly as above. Now we can give the classical definition of the p-primary subgroup of the Selmer group for E over M:

$$\mathrm{Sel}_E(M)_p = \ker\big(H^1(M, E[p^\infty]) \to \prod_\eta H^1(M_\eta, E[p^\infty])/\mathrm{Im}(\kappa_\eta)\big),$$

where η runs over all primes of M and the map is induced by $\phi \to (\phi|_{G_\eta})_\eta$ for any 1-cocycle ϕ. We will denote the class of a 1-cocycle ϕ by $[\phi]$. Thus $[\phi]$ is in $\mathrm{Sel}_E(M)_p$ if and only if $[\phi|_{G_{M_\eta}}] \in \mathrm{Im}(\kappa_\eta)$ for all η. Obviously, $\mathrm{Im}(\kappa) \subseteq \mathrm{Sel}_E(M)_p$. The corresponding quotient $\mathrm{Sel}_E(M)_p/\mathrm{Im}(\kappa)$ is, by definition, $Ш_E(M)_p$.

Faltings has proved that E is determined up to F-isogeny by the G_F-representation space $V_p(E) = T_p(E)\otimes\mathbb{Q}_p$, where $T_p(E)$ denotes the p-adic Tate module for E. More precisely, the G_F-module $E[p^\infty] \cong V_p(E)/T_p(E)$ determines E up to an F-isogeny of degree prime to p. Now $\mathrm{Sel}_E(M)_p$ is not changed by such F-isogenies, and hence one might hope to define it in a way which involves only the G_F-module $E[p^\infty]$. To do this, it suffices to give such a description of the subgroup $\mathrm{Im}(\kappa_\eta)$ of $H^1(M_\eta, E[p^\infty])$ for all primes η of M. We will now proceed to do this under the assumption that E has good, ordinary or multiplicative reduction at all primes of F over p.

Assume at first that M is a finite extension of F. Then $\eta|v$ for some prime v of F, and $\eta|l$ for some prime l of $\mathbb{Q}$ (possible $l=\infty$). If l is a finite prime, then we have a theorem of Lutz: $E(M_\eta) \cong \mathbb{Z}_l^{[M_\eta:\mathbb{Q}_l]} \times U$ as a group, where $U = E(M_\eta)_{\mathrm{tors}}$ is finite. Now $\mathbb{Z}_l\otimes(\mathbb{Q}_p/\mathbb{Z}_p) = 0$ if $l \neq p$, whereas $\mathbb{Z}_p\otimes(\mathbb{Q}_p/\mathbb{Z}_p) \cong \mathbb{Q}_p/\mathbb{Z}_p$. Also, $U\otimes(\mathbb{Q}_p/\mathbb{Z}_p) = 0$. If $l=\infty$, then $M_\eta \cong \mathbb{R}$ or $\mathbb{C}$. In this case, $E(M_\eta) \cong T^{[M_\eta:\mathbb{R}]}\times U$, where $T = \mathbb{R}/\mathbb{Z}$ and $|U| \le 2$. Since T is divisible, we have $T\otimes(\mathbb{Q}_p/\mathbb{Z}_p) = 0$. We then obtain the following result.

Proposition 2.1. *If $\eta \nmid p$, then* $\mathrm{Im}(\kappa_\eta) = 0$. *If $\eta | p$, then*

$$\mathrm{Im}(\kappa_\eta) \cong (\mathbb{Q}_p/\mathbb{Z}_p)^{[M_\eta:\mathbb{Q}_p]}.$$

The first assertion can also be explained by using the fact that, for $\eta \nmid p$, $H^1(M_\eta, E[p^\infty])$ is a finite group. But $E(M_\eta) \otimes (\mathbb{Q}_p/\mathbb{Z}_p)$, and hence $\mathrm{Im}(\kappa_\eta)$ are divisible groups. Even if M_η is an infinite extension of F_v, it is clear from the above that $\mathrm{Im}(\kappa_\eta) = 0$ if $\eta \nmid p$.

Assume that E has good, ordinary reduction at v, where v is a prime of F lying over p. Then, considering $E[p^\infty]$ as a subgroup of $E(\overline{F}_v)$, we have the reduction map $E[p^\infty] \to \widetilde{E}[p^\infty]$, where $\widetilde{E}$ is the reduction of E modulo v. Define C_v by

$$C_v = \ker\left(E[p^\infty] \to \widetilde{E}[p^\infty]\right).$$

Now $E[p^\infty] \cong (\mathbb{Q}_p/\mathbb{Z}_p)^2$, $\widetilde{E}[p^\infty] \cong \mathbb{Q}_p/\mathbb{Z}_p$ as groups. It is easy to see that $C_v \cong \mathbb{Q}_p/\mathbb{Z}_p$. (In fact, $C_v = \mathcal{F}(\overline{\mathfrak{m}})[p^\infty]$, where $\mathcal{F}$ is the formal group of height 1 for E and $\overline{\mathfrak{m}}$ is the maximal ideal of the integers of $\overline{F}_v$.) A characterization in terms of $E[p^\infty]$ is that C_v is G_{F_v}-invariant and $E[p^\infty]/C_v$ is the maximal unramified quotient of $E[p^\infty]$. Let M be a finite extension of F. If η is a prime of M lying above v, then we can consider M_η as a subfield of $\overline{F}_v$ containing F_v. (The identification will not matter.) We then have a natural map

$$\lambda_\eta : H^1(M_\eta, C_v) \to H^1(M_\eta, E[p^\infty]).$$

Here is a description of $\mathrm{Im}(\kappa_\eta)$.

Proposition 2.2. $\mathrm{Im}(\kappa_\eta) = \mathrm{Im}(\lambda_\eta)_{\mathrm{div}}$.

Proof. The idea is quite simple. We know that $\mathrm{Im}(\kappa_\eta)$ and $\mathrm{Im}(\lambda_\eta)$ are p-primary groups, that $\mathrm{Im}(\kappa_\eta)$ is divisible, and has $\mathbb{Z}_p$-corank $[M_\eta : \mathbb{Q}_p]$. It suffices to prove two things: (i) $\mathrm{Im}(\kappa_\eta) \subseteq \mathrm{Im}(\lambda_\eta)$ and (ii) $\mathrm{Im}(\lambda_\eta)$ has $\mathbb{Z}_p$-corank equal to $[M_\eta : \mathbb{Q}_p]$. To prove (i), let $c \in \mathrm{Im}(\kappa_\eta)$. We show that $c \in \ker(H^1(M_\eta, E[p^\infty]) \to H^1(M_\eta, \widetilde{E}[p^\infty]))$, which coincides with $\mathrm{Im}(\lambda_\eta)$. Let f_v denote the residue field of F_v, $\overline{f}_v$ its algebraic closure—the residue field of $\overline{F}_v$. If $b \in E(\overline{F}_v)$, we let $\widetilde{b} \in \widetilde{E}(\overline{f}_v)$ denote its reduction. Let ϕ be a cocycle representing c. Then $\phi(g) = g(b) - b$ for all $g \in G_{M_\eta}$, where $b \in E(\overline{F}_v)$. The 1-cocycle induced by $E[p^\infty] \to \widetilde{E}[p^\infty]$ is $\widetilde{\phi}$, given by $\widetilde{\phi}(g) = g(\widetilde{b}) - \widetilde{b}$ for all $g \in G_{M_\eta}$. But $\widetilde{\phi}$ represents a class $\widetilde{c}$ in $H^1(M_\eta, \widetilde{E}[p^\infty])$ which becomes trivial in $H^1(M_\eta, \widetilde{E}(\overline{f}_v))$, i.e. $\widetilde{\phi}$ is a 1-coboundary. Finally, the key point is that $\widetilde{E}(\overline{f}_v)$ is a torsion group, $\widetilde{E}[p^\infty]$ is its p-primary subgroup, and hence the map $H^1(M_\eta, \widetilde{E}[p^\infty]) \to H^1(M_\eta, \widetilde{E}(\overline{f}_v))$ must be injective. Thus, $\widetilde{c}$ is trivial, and therefore $c \in \mathrm{Im}(\lambda_\eta)$.

Now we calculate the $\mathbb{Z}_p$-corank of $\mathrm{Im}(\lambda_\eta)$. We have the exact sequence

$$E[p^\infty]^{G_{M_\eta}} \to \widetilde{E}[p^\infty]^{G_{M_\eta}} \to H^1(M_\eta, C_v) \xrightarrow{\lambda_\eta} H^1(M_\eta, E[p^\infty]).$$

If m_η denotes the residue field of M_η, then $\widetilde{E}[p^\infty]^{G_{M_\eta}}$ is just the p-primary subgroup of $\widetilde{E}(m_\eta)$, a finite group. Thus, $\ker(\lambda_\eta)$ is finite. The following lemma then suffices to prove (ii). If $\psi : G_{F_v} \to \mathbb{Z}_p^\times$ is a continuous homomorphism, we will let $(\mathbb{Q}_p/\mathbb{Z}_p)(\psi)$ denote the group $\mathbb{Q}_p/\mathbb{Z}_p$ together with the action of G_{F_v} given by ψ.

Lemma 2.3. *$H^1(M_\eta, (\mathbb{Q}_p/\mathbb{Z}_p)(\psi))$ has $\mathbb{Z}_p$-corank equal to $[M_\eta : \mathbb{Q}_p] + \delta$, where $\delta = 1$ if $\psi|_{G_{M_\eta}}$ is either the trivial character or the cyclotomic character of G_{M_η} and $\delta = 0$ otherwise.*

Remark. Because of the importance of this lemma, we will give a fairly self-contained proof using local class field theory and techniques of Iwasawa Theory. But we then show how to obtain the same result as a simple application of the Duality theorems of Poitou and Tate.

Proof. The case where ψ is trivial follows from local class field theory. Then $H^1(M_\eta, (\mathbb{Q}_p/\mathbb{Z}_p)(\psi)) = \mathrm{Hom}(\mathrm{Gal}(M_\eta^{ab}/M_\eta), \mathbb{Q}_p/\mathbb{Z}_p)$. The well-known structure of $M_\eta^\times$ implies that $\mathrm{Gal}(M_\eta^{ab}/M_\eta) \cong \mathbb{Z}_p^{[M_\eta:\mathbb{Q}_p]} \times \widehat{\mathbb{Z}} \times (M_\eta^\times)_{\mathrm{tors}}$, where $\widehat{\mathbb{Z}}$ is the profinite completion of $\mathbb{Z}$. The lemma is clear in this case. If $\psi|_{G_{M_\eta}}$ is the cyclotomic character, then $(\mathbb{Q}_p/\mathbb{Z}_p)(\psi) \cong \mu_{p^\infty}$ as G_{M_η}-modules. Then $H^1(M_\eta, \mu_{p^\infty}) \cong (M_\eta^\times) \otimes (\mathbb{Q}_p/\mathbb{Z}_p)$, which indeed has the stated $\mathbb{Z}_p$-corank.

Now suppose we are not in one of the above two cases. For brevity, we will write M for M_η. Let M_∞ be the extension of M cut out by $\psi|_{G_M}$. Thus, $G = \mathrm{Gal}(M_\infty/M) \cong \mathrm{Im}(\psi|_{G_M})$. If ψ has finite order, one can reduce to studying the action of G on $\mathrm{Gal}(M_\infty^{ab}/M_\infty)$ since M_∞ would just be a finite extension of $\mathbb{Q}_p$. We will do something similar if ψ has infinite order. Then, $G \cong \Delta \times H$, where Δ is finite and $H \cong \mathbb{Z}_p$. If p is odd, $|\Delta|$ divides $p-1$. If $p = 2$, $|\Delta| = 1$ or 2. Let $C = (\mathbb{Q}_p/\mathbb{Z}_p)(\psi)$. The inflation-restriction sequence gives

$$0 \to H^1(G, C) \to H^1(M, C) \to H^1(M_\infty, C)^G \to H^2(G, C).$$

Now let h be a topological generator of H. Then $H^1(H, C) = C/(h-1)C = 0$ because, considering $h-1$ as an endomorphism of C, $\ker(h-1)$ is finite and $\mathrm{Im}(h-1)$ is divisible. Thus, $H^1(G, C) = 0$ if p is odd, and has order ≤ 2 if $p = 2$. On the other hand, $H^2(H, C) = 0$ since H has p-cohomological dimension 1. Then $H^2(G, C) = 0$ if p is odd, and again has order ≤ 2 if $p = 2$. Thus, it is enough to study

$$H^1(M_\infty, C)^G = \mathrm{Hom}_G(\mathrm{Gal}(M_\infty^{ab}/M_\infty), C).$$

Let $X = \mathrm{Gal}(L_\infty/M_\infty)$, where L_∞ is the maximal abelian pro-p extension of M_∞. We will prove the rest of lemma 2.3 by studying the structure of X as a module for $\mathbb{Z}_p[[\Delta \times H]] = \Lambda[\Delta]$, where $\Lambda = \mathbb{Z}_p[[H]] \cong \mathbb{Z}_p[[T]]$, with $T = h - 1$. The results are due to Iwasawa.

For any $n \geq 0$, let $H_n = H^{p^n}$. Let $M_n = M_\infty^{H_n}$. The commutator subgroup of $\mathrm{Gal}(L_\infty/M_n)$ is $(h^{p^n}-1)X$ and so, if L_n is the maximal abelian extension of M_n contained in L_∞, then $\mathrm{Gal}(L_n/M_n) \cong H_n \times (X/(h^{p^n}-1)X)$. But L_n is the maximal abelian pro-p extension of M_n and, by local class field theory, this Galois group is isomorphic to $\mathbb{Z}_p^{[M_n:\mathbb{Q}_p]+1} \times W_n$, where W_n denotes the group of p-power roots of unity contained in M_n. Consequently, if we put $t = [M_0 : \mathbb{Q}_p] = |\Delta| \cdot [M : \mathbb{Q}_p]$, we have

$$X/(h^{p^n}-1)X \cong \mathbb{Z}_p^{tp^n} \times W_n.$$

Now, the structure theory for Λ-modules states that $X/X_{\Lambda\text{-tors}}$ is isomorphic to a submodule of Λ^r, with finite index, where $r = \mathrm{rank}_\Lambda(X)$. Also, we have $\Lambda/(h^{p^n}-1)\Lambda \cong \mathbb{Z}_p^{p^n}$ for $n \geq 0$. It follows that $r = t$. One can also see that $X_{\Lambda\text{-tors}} \cong \varprojlim W_n$, where this inverse limit is defined by the norm maps $M_m^\times \to M_n^\times$ for $m \geq n$. If W_n has bounded order (i.e., if $\mu_{p^\infty} \not\subseteq M_\infty$), then $X_{\Lambda\text{-tors}} = 0$. Thus, $X \subseteq \Lambda^t$. To get more precise information about the structure of X, choose n large enough so that $h^{p^n}-1$ annihilates Λ^t/X. We then have

$$(h^{p^n}-1)X \subseteq (h^{p^n}-1)\Lambda^t \subseteq X \subseteq \Lambda^t.$$

We can see easily from this that Λ^t/X is isomorphic to the torsion subgroup of $X/(h^{p^n}-1)X$. That is, $\Lambda^t/X \cong W$, where $W = M_\infty^\times \cap \mu_{p^\infty}$. On the other hand, if $\mu_{p^\infty} \subseteq M_\infty$, then $X_{\Lambda\text{-tors}} \cong \mathbb{Z}_p(1)$, the Tate module for μ_{p^∞}. In this case, $X/X_{\Lambda\text{-tors}}$ is free and hence $X \cong \Lambda^t \times \mathbb{Z}_p(1)$.

In the preceding discussion, the Λ-module Λ^t is in fact canonical. It is the reflexive hull of $X/X_{\Lambda\text{-tors}}$. Thus, the action of Δ on X gives an action on Λ^t. Examining the above arguments more carefully, one finds that, for p odd, Λ^t is isomorphic to $\Lambda[\Delta]^{[M:\mathbb{Q}_p]}$. (One just studies the Λ-module X^ϕ for each character ϕ of Δ. Recall that $|\Delta|$ divides $p-1$ and hence each character ϕ has values in $\mathbb{Z}_p^\times$.) For $p = 2$, we can at least make such an identification up to a group of exponent 2. For the proof of lemma 2.3, it suffices to point out that $\mathrm{Hom}_{\Delta\times H}(\Lambda[\Delta], C)$ is isomorphic to $\mathbb{Q}_p/\mathbb{Z}_p$ and that $\mathrm{Hom}_{\Delta\times H}(\mathbb{Z}_p(1), C)$ is finite. (We are assuming now that $C \not\cong \mu_{p^\infty}$ as G_M-modules.) This completes the proof of lemma 2.3 and consequently proposition 2.2, since one sees easily that $\delta = 0$ when $C = C_v$. ■

The above discussion of the $\Lambda[\Delta]$-module structure of X gives a more precise result concerning $H^1(M_\eta, (\mathbb{Q}_p/\mathbb{Z}_p)(\psi))$. Assume that p is odd and that ψ has infinite order. If the extension of M_η cut out by the character ψ of G_{M_η} contains μ_{p^∞}, then we see that

$$H^1(M_\eta, C) \cong (\mathbb{Q}_p/\mathbb{Z}_p)^{[M_\eta:\mathbb{Q}_p]} \times \mathrm{Hom}_{G_{M_\eta}}(\mathbb{Z}_p(1), C), \tag{1}$$

where as above $C = (\mathbb{Q}_p/\mathbb{Z}_p)(\psi)$. The factor $\mathrm{Hom}_{G_{M_\eta}}(\mathbb{Z}_p(1), C)$ is just $H^0(M_\eta, C \otimes \chi^{-1})$, where χ denotes the cyclotomic character. Even if W is finite, we can prove (1). For if g_0 is a topological generator of $\Delta \times H$, then the torsion subgroup of $X/(g_0 - \psi(g_0))X$ is isomorphic to the kernel of $g_0 - \psi(g_0)$ acting on $\Lambda^t/X \cong W$. (It is seen to be $((g_0 - \psi(g_0))\Lambda^t \cap X)/(g_0 - \psi(g_0))X$.) But this in turn is isomorphic to $W/(g_0 - \psi(g_0))W$, whose dual is easily identified with $\mathrm{Hom}_{G_{M_\eta}}(\mathbb{Z}_p(1), (\mathbb{Q}_p/\mathbb{Z}_p)(\psi))$.

We have attempted to give a rather self-contained "Iwasawa-theoretic" approach to studying the above local Galois cohomology group. This suffices for the proof of proposition 2.2. But using results of Poitou and Tate is often easier and more effective. We will illustrate this. Let $C = (\mathbb{Q}_p/\mathbb{Z}_p)(\psi)$. Let T denote its Tate module and $V = T \otimes_{\mathbb{Z}_p} \mathbb{Q}_p$. The $\mathbb{Z}_p$-corank of $H^1(G_{M_\eta}, C)$ is just $\dim_{\mathbb{Q}_p}(H^1(M_\eta, V))$. (Cocycles are required to be continuous. V has its $\mathbb{Q}_p$-vector space topology. Similarly, T has its natural topology and is compact.) Letting h_i denote $\dim_{\mathbb{Q}_p}(H^i(M_\eta, V))$, then the Euler characteristic for V over M_η is given by

$$h_0 - h_1 + h_2 = -[M_\eta : \mathbb{Q}_p] \dim_{\mathbb{Q}_p}(V)$$

for any G_{M_η}-representation space V. We have $\dim_{\mathbb{Q}_p}(V) = 1$ and so the $\mathbb{Z}_p$-corank of $H^1(M_\eta, (\mathbb{Q}_p/\mathbb{Z}_p)(\psi))$ is $[M_\eta : \mathbb{Q}_p] + h_0 + h_2$. Poitou-Tate Duality implies that $H^2(M_\eta, V)$ is dual to $H^0(M_\eta, V^*)$, where $V^* = \mathrm{Hom}(V, \mathbb{Q}_p(1))$. It is easy to see from this that $\delta = h_0 + h_2$, proving lemma 2.3 again.

The exact sequence $0 \to T \to V \to C \to 0$ induces the exact sequence

$$H^1(M_\eta, V) \stackrel{\alpha}{\to} H^1(M_\eta, C) \stackrel{\beta}{\to} H^2(M_\eta, T) \stackrel{\gamma}{\to} H^2(M_\eta, V).$$

The image of α is the maximal divisible subgroup of $H^1(G_{M_\eta}, C)$. The kernel of γ is the torsion subgroup of $H^2(M_\eta, T)$. Of course, $\mathrm{coker}(\alpha) \cong \mathrm{Im}(\beta) \cong \ker(\gamma)$. Poitou-Tate Duality implies that $H^2(M_\eta, T)$ is dual to $H^0(M_\eta, \mathrm{Hom}(T, \mu_{p^\infty})) = \mathrm{Hom}_{G_{M_\eta}}(T, \mu_{p^\infty})$. The action of G_{M_η} on T is by ψ; the action on μ_{p^∞} is by χ. Thus, $\mathrm{Hom}_{G_{M_\eta}}(T, \mu_{p^\infty})$ can be identified with the dual of $H^0(M_\eta, (\mathbb{Q}_p/\mathbb{Z}_p)(\chi\psi^{-1}))$. If $\psi|_{G_{M_\eta}} = \chi|_{G_{M_\eta}}$, then we find that $H^2(M_\eta, T) \cong \mathbb{Z}_p$, $\mathrm{Im}(\beta) = 0$, and therefore $H^1(M_\eta, C)$ is divisible. Otherwise, we find that $H^2(M_\eta, T)$ is finite and that

$$H^1(M_\eta, C)/H^1(M_\eta, C)_{\mathrm{div}} \cong (\mathbb{Q}_p/\mathbb{Z}_p)(\psi\chi^{-1})^{G_{M_\eta}}, \tag{2}$$

which is a finite cyclic group, indeed isomorphic to $\mathrm{Hom}_{G_{M_\eta}}(\mathbb{Z}_p(1), C)$. This argument works even for $p = 2$.

We want to mention here one useful consequence of the above discussion. Again we let $C = (\mathbb{Q}_p/\mathbb{Z}_p)(\psi)$, where $\psi : G_{F_v} \to \mathbb{Z}_p^\times$ is a continuous homomorphism, v is any prime of F lying over p. If η is a prime of F_∞ lying over v, then $(F_\infty)_\eta$ is the cyclotomic $\mathbb{Z}_p$-extension of F_v. By lemma 2.3, the $\mathbb{Z}_p$-corank of $H^1((F_n)_\eta, C)$ differs from $[(F_n)_\eta : F_v]$ by at most 1. Thus, if we let $\Gamma_v = \mathrm{Gal}((F_\infty)_\eta/F_v)$, then it follows that as $n \to \infty$

$$\mathrm{corank}_{\mathbb{Z}_p}(H^1((F_\infty)_\eta, C)^{\Gamma_v^{p^n}}) = p^n[F_v : \mathbb{Q}_p] + O(1).$$

The structure theory of Λ-modules then implies that $H^1((F_\infty)_\eta, C)$ has corank equal to $[F_v : \mathbb{Q}_p]$ as a $\mathbb{Z}_p[[\Gamma_v]]$-module. Assume that ψ is unramified and that the maximal unramified extension of F_v contains no p-th roots of unity. (If the ramification index e_v for v over p is $\leq p-2$, then this will be true. If $F = \mathbb{Q}$, this is true for all $p \geq 3$.) Then by (2) we see that $H^1(F_v, C)$ is divisible. The $\mathbb{Z}_p$-corank of $H^1(F_v, C)$ is $[F_v : \mathbb{Q}_p] + \delta$, where $\delta = 0$ if ψ is nontrivial, $\delta = 1$ if ψ is trivial. By the inflation-restriction sequence we see that $H^1((F_\infty)_\eta, C)^{\Gamma_v} \cong (\mathbb{Q}_p/\mathbb{Z}_p)^{[F_v:\mathbb{Q}_p]}$. It follows that $H^1((F_\infty)_\eta, C)$ is $\mathbb{Z}_p[[\Gamma_v]]$-cofree of corank $[F_v : \mathbb{Q}_p]$, under the hypotheses that ψ is unramified and $e_v \leq p-2$. These remarks are a special case of results proved in [Gr2].

Now we return to the case where $C_v = \ker(E[p^\infty] \to \widetilde{E}[p^\infty])$. The action of G_{F_v} on C_v is by a character ψ, the action on $\widetilde{E}[p^\infty]$ is by a character ϕ, and we have $\psi\phi = \chi$ since the Weil pairing $T_p(E) \wedge T_p(E) \cong \mathbb{Z}_p(1)$ means that χ is the determinant of the representation of G_{F_v} on $T_p(E)$. Note that ϕ has infinite order. The same is true for ψ since ψ and χ become equal after restriction to the inertia subgroup $G_{F_v^{\mathrm{unr}}}$. This explains why $\delta = 0$ for $\psi|_{G_{M_\eta}}$, as used to prove proposition 2.2. In this case, $\chi\psi^{-1} = \phi$ and hence $H^0(G_{M_\eta}, (\mathbb{Q}_p/\mathbb{Z}_p)(\chi\psi^{-1}))$ is isomorphic to $\widetilde{E}(m_\eta)_p$, where m_η is the residue field for M_η. These facts lead to a version of proposition 2.2 for some infinite extensions of F_v.

Proposition 2.4. *Assume that K is a Galois extension of F_v, that* $\mathrm{Gal}(K/F_v)$ *contains an infinite pro-p subgroup, and that the inertia subgroup of* $\mathrm{Gal}(K/F_v)$ *is of finite index. Then* $\mathrm{Im}(\kappa_K) = \mathrm{Im}(\lambda_K)$*, where κ_K is the Kummer homomorphism for E over K and λ_K is the canonical homomorphism*

$$H^1(K, C_v) \to H^1(K, E[p^\infty]).$$

Proof. Let M run over all finite extensions of F_v contained in K. Then $\mathrm{Im}(\kappa_K) = \varinjlim \mathrm{Im}(\kappa_M)$, $\mathrm{Im}(\lambda_K) = \varinjlim \mathrm{Im}(\lambda_M)$, and $\mathrm{Im}(\kappa_M) = \mathrm{Im}(\lambda_M)_{\mathrm{div}}$ by proposition 2.2. But $\mathrm{Im}(\lambda_M)/\mathrm{Im}(\lambda_M)_{\mathrm{div}}$ has order bounded by $|\widetilde{E}(m)_p|$, where m is the residue field of M. Now $|m|$ is bounded by assumption. Hence it follows that $\mathrm{Im}(\lambda_K)/\mathrm{Im}(\kappa_K)$ is a finite group. On the other hand, G_K has p-cohomological dimension 1 because of the hypothesis that $\mathrm{Gal}(K/F_v)$ contains an infinite pro-p subgroup. (See Serre, Cohomologie Galoisienne, Chapitre II, §3.) Thus if C is a divisible, p-primary G_K-module, then the exact sequence $0 \to C[p] \to C \xrightarrow{p} C \to 0$ induces the cohomology exact sequence $H^1(K, C) \xrightarrow{p} H^1(K, C) \to H^2(K, C[p])$. The last group is zero and hence $H^1(K, C)$ is divisible. Applying this to $C = C_v$, we see that $\mathrm{Im}(\lambda_K)$ is divisible and so $\mathrm{Im}(\kappa_K) = \mathrm{Im}(\lambda_K)$. ∎

If F_∞ denotes the cyclotomic $\mathbb{Z}_p$-extension of F, then every prime v of F lying over p is ramified in F_∞/F. If η is a prime of F_∞ over v, then

$K = (F_\infty)_\eta$ satisfies the hypothesis of proposition 2.4 since the inertia subgroup of $\Gamma = \mathrm{Gal}(F_\infty/F)$ for η is infinite, pro-p, and has finite index in Γ. Propositions 2.1, 2.2, and 2.4 will allow us to give a fairly straightforward proof of theorem 1.2, which we will do in section 3. However, in section 4 it will be useful to have more precise information about $\mathrm{Im}(\lambda_\eta)/\mathrm{Im}(\kappa_\eta)$, where η is a prime for a finite extension M of F lying over p. What we will need is the following.

Proposition 2.5. *Let M_η be a finite extension of F_v, where $v|p$. Let m_η be the residue field for M_η. Then*

$$\mathrm{Im}(\lambda_\eta)/\mathrm{Im}(\kappa_\eta) \cong \widetilde{E}(m_\eta)_p.$$

Proof. The proof comes out of the following diagram:

$$\begin{array}{ccccccccc} 0 \longrightarrow & \mathcal{F}(\mathfrak{m}) \otimes (\mathbb{Q}_p/\mathbb{Z}_p) & \xrightarrow{\kappa_{\mathcal{F}}} & H^1(M_\eta, C_v) & \longrightarrow & H^1(M_\eta, \mathcal{F}(\overline{\mathfrak{m}}))_p & \longrightarrow 0 \\ & \downarrow & & \downarrow \lambda_\eta & & \downarrow \epsilon & \\ 0 \longrightarrow & E(M_\eta) \otimes (\mathbb{Q}_p/\mathbb{Z}_p) & \xrightarrow{\kappa_\eta} & H^1(M_\eta, E[p^\infty]) & \longrightarrow & H^1(M_\eta, E(\overline{F}_v))_p & \longrightarrow 0 \end{array}$$

Here $\mathcal{F}$ is the formal group for E (which has height 1), $\mathfrak{m}$ is the maximal ideal of M_η. The upper row is the Kummer sequence for $\mathcal{F}(\mathfrak{m})$, based on the fact that $\mathcal{F}(\overline{\mathfrak{m}})$ is divisible. The first vertical arrow is surjective since $\mathcal{F}(\mathfrak{m})$ has finite index in $E(M_\eta)$. Comparing $\mathbb{Z}_p$-coranks, one sees that $\mathrm{Im}(\kappa_{\mathcal{F}}) = H^1(G_{M_\eta}, C_v)_{\mathrm{div}}$. A simple diagram chase shows that the map

$$H^1(M_\eta, C_v)/H^1(M_\eta, C_v)_{\mathrm{div}} \longrightarrow \mathrm{Im}(\lambda_\eta)/\mathrm{Im}(\lambda_\eta)_{\mathrm{div}} \qquad (3)$$

is surjective and has kernel isomorphic to $\ker(\epsilon)$. The exact sequence

$$0 \to \mathcal{F}(\overline{\mathfrak{m}}) \to E(\overline{F}_v) \to \widetilde{E}(\overline{f}_v) \to 0,$$

together with the fact that the reduction map $E(M_\eta) \to \widetilde{E}(m_\eta)$ is surjective implies that ϵ is injective. (For the surjectivity of the reduction map, see proposition 2.1 of [Si].) Therefore, the map (3) is an isomorphism. Combining this with the observation preceding proposition 2.4, we get the stated conclusion. ∎

Assume now that E has split, multiplicative reduction at v. Then one has an exact sequence

$$0 \to C_v \to E[p^\infty] \to \mathbb{Q}_p/\mathbb{Z}_p \to 0$$

where $C_v \cong \mu_{p^\infty}$. The proof of proposition 2.2 can be made to work and gives the following result. *For any algebraic extension K of F_v, we have* $\mathrm{Im}(\kappa_K) = \mathrm{Im}(\lambda_K)$. It is enough to prove this when $[K:F_v] < \infty$. Then

$\mathrm{Im}(\kappa_K)$ is divisible and has $\mathbb{Z}_p$-corank $[K:\mathbb{Q}_p]$. $H^1(K,C_v)$ is divisible and has $\mathbb{Z}_p$-corank $[K:\mathbb{Q}_p]+1$. But the kernel of $\lambda_K: H^1(K,C_v) \to H^1(K,E[p^\infty])$ is isomorphic to $\mathbb{Q}_p/\mathbb{Z}_p$. Thus, $\mathrm{Im}(\lambda_K)$ and $\mathrm{Im}(\kappa_K)$ are both divisible and have the same $\mathbb{Z}_p$-corank. The inclusion $\mathrm{Im}(\kappa_K) \subseteq \mathrm{Im}(\lambda_K)$ can be seen by noting that in defining κ_K, one can assume that $\alpha \in E(K) \otimes (\mathbb{Q}_p/\mathbb{Z}_p)$ has been written as $\alpha = a \otimes (1/p^t)$, where $a \in \mathcal{F}(\mathfrak{m})$. Here $\mathcal{F}$ is the formal group for E, $\mathfrak{m}$ is the maximal ideal for K. Then, since $\mathcal{F}(\overline{\mathfrak{m}})$ is divisible, one can choose $b \in \mathcal{F}(\overline{\mathfrak{m}})$ so that $p^t b = a$. The 1-cocycle ϕ_α then has values in $C_v = \mathcal{F}(\overline{\mathfrak{m}})[p^\infty]$. Alternatively, the equality $\mathrm{Im}(\kappa_K) = \mathrm{Im}(\lambda_K)$ can be verified quite directly by using the Tate parametrization for E.

If E has nonsplit, multiplicative reduction, then the above assertion still holds for p odd. That is, $\mathrm{Im}(\kappa_K) = \mathrm{Im}(\lambda_K)$ for every algebraic extension K of F_v. We can again assume that $[K:F_v] < \infty$. If E becomes split over K, then the argument in the preceding paragraph applies. If not, then lemma 2.3 and (2) imply that $H^1(K,C_v)$ is divisible and has $\mathbb{Z}_p$-corank $[K:\mathbb{Q}_p]$. Just as in the case of good, ordinary reduction, we see that $\mathrm{Im}(\kappa_K) = \mathrm{Im}(\lambda_K)$. (It is analogous to the case where $\widetilde{E}(k)_p = 0$, where k is the residue field of K.) Now assume that $p = 2$. If $[K:F_v] < \infty$ and E is nonsplit over K, then we have $H^1(K,C_v)/H^1(K,C_v)_{\mathrm{div}} \cong \mathbb{Z}/2\mathbb{Z}$ by (2), since $\psi\chi^{-1}$ will be the unramified character of G_K of order 2. Thus, we obtain that $\mathrm{Im}(\kappa_K) = \mathrm{Im}(\lambda_K)_{\mathrm{div}}$ and that $[\mathrm{Im}(\lambda_K) : \mathrm{Im}(\kappa_K)] \le 2$. Using the same argument as in the proof of proposition 2.5, we find that this index is equal to the Tamagawa factor $[E(K) : \mathcal{F}(\mathfrak{m}_K)]$ for E over K. This equals 1 or 2 depending on whether $\mathrm{ord}_K(j_E)$ is odd or even. Finally, we remark that proposition 2.4 holds when E has multiplicative reduction. The proof given there works because the index $[\mathrm{Im}(\lambda_M) : \mathrm{Im}(\kappa_M)]$ is bounded.

For completeness, we will state a result of Bloch and Kato describing $\mathrm{Im}(\kappa_K)$ when E has good, supersingular reduction and $[K : F_v] < \infty$. It involves the ring B_{cris} of Fontaine. Define

$$H^1_f(K, V_p(E)) = \ker\left(H^1(K, V_p(E)) \to H^1(K, V_p(E) \otimes B_{\mathrm{cris}})\right).$$

The result is that $\mathrm{Im}(\kappa_K)$ is the image of $H^1_f(K, V_p(E))$ under the canonical map $H^1(K, V_p(E)) \to H^1(K, V_p(E)/T_p(E))$, noting that $V_p(E)/T_p(E)$ is isomorphic to $E[p^\infty]$. This description is also correct if E has good, ordinary reduction.

If E has supersingular reduction at v, where $v|p$, and if K is any ramified $\mathbb{Z}_p$-extension of F_v, then the analogue of proposition 2.4 is true. In this case, $C_v = E[p^\infty]$ since $\widetilde{E}[p^\infty] = 0$. Thus, the result is that $\mathrm{Im}(\kappa_K) = H^1(K, E[p^\infty])$. Perhaps the easiest way to prove this is to use the analogue of Hilbert's theorem 90 for formal groups proved in [CoGr]. If $\mathcal{F}$ denotes the formal group (of height 2) associated to E, then $H^1(K, \mathcal{F}(\overline{\mathfrak{m}})) = 0$. (This is a special case of Corollary 3.2 in [CoGr].) Just as in the case of Kummer theory for the multiplicative group, we then obtain an isomorphism

$$\kappa_K^{\mathcal{F}} : \mathcal{F}(\mathfrak{m}_K) \otimes (\mathbb{Q}_p/\mathbb{Z}_p) \xrightarrow{\sim} H^1(K, C_v)$$

because $C_v = \mathcal{F}(\overline{\mathfrak{m}})[p^\infty]$. We get the result stated above immediately, since

$$E(K) \otimes (\mathbb{Q}_p/\mathbb{Z}_p) = \mathcal{F}(\mathfrak{m}_K) \otimes (\mathbb{Q}_p/\mathbb{Z}_p).$$

The assertion that $\mathrm{Im}(\kappa_K) = H^1(K, E[p^\infty])$ is proved in [CoGr] under the hypotheses that E has potentially supersingular reduction at v and that K/F_v is a "deeply ramified extension" (which means that K/F_v has infinite conductor, i.e., $K \not\subset F_v^{(t)}$ for any $t \geq 1$, where $F_v^{(t)}$ denotes the fixed field for the t-th ramification subgroup of $\mathrm{Gal}(\overline{F}_v/F_v)$). A ramified $\mathbb{Z}_p$-extension K of F_v is the simplest example of a deeply ramified extension. As an illustration of how this result affects the structure of Selmer groups, consider the definition of $\mathrm{Sel}_E(M)_p$ given near the beginning of this section. If E has potentially supersingular reduction at a prime v of F lying over p and if M_η/F_v is deeply ramified for all $\eta|v$, then the groups $H^1(M_\eta, E[p^\infty])/\mathrm{Im}(\kappa_\eta)$ occurring in the definition of $\mathrm{Sel}_E(M)_p$ are simply zero. In particular, if $M = F_\infty$, the cyclotomic $\mathbb{Z}_p$-extension of F, then the primes η of F_∞ lying over primes of F where E has potentially supersingular reduction can be omitted in the local conditions defining $\mathrm{Sel}_E(F_\infty)_p$. This is the key to proving theorem 1.7.

One extremely important consequence of the fact that the Selmer group for an elliptic curve E has a description involving just the Galois representations attached to the torsion points on E is that one can then attempt to introduce analogously-defined "Selmer groups" and to study all the natural questions associated to such objects in a far more general context. We will illustrate this idea by considering Δ, the normalized cusp form of level 1, weight 12. Its q-expansion is $\Delta = \sum_{n=1}^{\infty} \tau(n)q^n$, where $\tau(n)$ is Ramanujan's tau function. Deligne attached to Δ a compatible system $\{V_l(\Delta)\}$ of l-adic representations of $G_{\mathbb{Q}}$. Consider a prime p such that $p \nmid \tau(p)$. For such a prime p, Mazur and Wiles have proved that the action of $G_{\mathbb{Q}_p}$ on $V_p(\Delta)$ is reducible (where one fixes an embedding $\overline{\mathbb{Q}} \to \overline{\mathbb{Q}}_p$, identifying $G_{\mathbb{Q}_p}$ with a subgroup of $G_{\mathbb{Q}}$). More precisely, there is an exact sequence

$$0 \to W_p(\Delta) \to V_p(\Delta) \to U_p(\Delta) \to 0$$

where $W_p(\Delta)$ is 1-dimensional and $G_{\mathbb{Q}_p}$-invariant, the action of $G_{\mathbb{Q}_p}$ on $U_p(\Delta)$ is unramified, and the action of Frob_p on $U_p(\Delta)$ is multiplication by α_p (where α_p is the p-adic unit root of $t^2 - \tau(p)t + p^{11}$). Let $T_p(\Delta)$ be any $G_{\mathbb{Q}}$-invariant $\mathbb{Z}_p$-lattice in $V_p(\Delta)$. (It turns out to be unique up to homothety for $p \nmid \tau(p)$, except for $p = 691$, when there are two possible choices up to homothety.) Let $A = V_p(\Delta)/T_p(\Delta)$. As a group, $A \cong (\mathbb{Q}_p/\mathbb{Z}_p)^2$. Let C denote the image of $W_p(\Delta)$ in A. Then $C \cong \mathbb{Q}_p/\mathbb{Z}_p$ as a group. Here then is a definition of the p-Selmer group $S_A(\mathbb{Q})_p$ for A over $\mathbb{Q}$:

$$S_A(\mathbb{Q})_p = \ker\big(H^1(\mathbb{Q}, A) \to \prod_v H^1(\mathbb{Q}_v, A)/L_v\big),$$

where v runs over all primes of $\mathbb{Q}$. Here we take $L_v = 0$ for $v \neq p$, analogously to the elliptic curve case. One defines $L_p = \mathrm{Im}(\lambda_p)_{\mathrm{div}}$, where

$$\lambda_p : H^1(\mathbb{Q}_p, C) \to H^1(\mathbb{Q}_p, A)$$

is the natural map. In [Gr3], one can find a calculation of $S_A(\mathbb{Q})_p$, and also $S_A(\mathbb{Q}_\infty)_p$, for $p = 11, 23$, and 691. One can make similar definitions whenever one has a p-adic Galois representation with suitable properties.

3. Control Theorems

We will now give a proof of theorem 1.2. It is based on the description of the images of the local Kummer homomorphisms presented in section 2, specifically propositions 2.1, 2.2, and 2.4. We will also prove a special case of conjecture 1.6. Let E be any elliptic curve defined over F. Let M be an algebraic extension of F. For every prime η of M, we let

$$\mathcal{H}_E(M_\eta) = H^1(M_\eta, E[p^\infty])/\mathrm{Im}(\kappa_\eta).$$

Let $\mathcal{P}_E(M) = \prod_\eta \mathcal{H}_E(M_\eta)$, where η runs over all primes of M. Thus,

$$\mathrm{Sel}_E(M)_p = \ker\left(H^1(M, E[p^\infty]) \to \mathcal{P}_E(M)\right),$$

where the map is induced by restricting cocycles to decomposition groups. Also, we put

$$\mathcal{G}_E(M) = \mathrm{Im}\left(H^1(M, E[p^\infty]) \to \mathcal{P}_E(M)\right).$$

Let $F_\infty = \bigcup_n F_n$ be the cyclotomic $\mathbb{Z}_p$-extension. Consider the following commutative diagram with exact rows:

$$\begin{array}{ccccccccc}
0 & \longrightarrow & \mathrm{Sel}_E(F_n)_p & \longrightarrow & H^1(F_n, E[p^\infty]) & \longrightarrow & \mathcal{G}_E(F_n) & \longrightarrow & 0 \\
 & & \downarrow{\scriptstyle s_n} & & \downarrow{\scriptstyle h_n} & & \downarrow{\scriptstyle g_n} & & \\
0 & \longrightarrow & \mathrm{Sel}_E(F_\infty)_p^{\Gamma_n} & \longrightarrow & H^1(F_\infty, E[p^\infty])^{\Gamma_n} & \longrightarrow & \mathcal{G}_E(F_\infty)^{\Gamma_n}. & &
\end{array}$$

Here $\Gamma_n = \mathrm{Gal}(F_\infty/F_n) = \Gamma^{p^n}$. The maps s_n, h_n, and g_n are the natural restriction maps. The snake lemma then gives the exact sequence

$$0 \to \ker(s_n) \to \ker(h_n) \to \ker(g_n) \to \mathrm{coker}(s_n) \to \mathrm{coker}(h_n).$$

Therefore, we must study $\ker(h_n)$, $\mathrm{coker}(h_n)$, and $\ker(g_n)$, which we do in a sequence of lemmas.

Lemma 3.1. *The kernel of h_n is finite and has bounded order as n varies.*

Proof. By the inflation-restriction sequence, $\ker(h_n) \cong H^1(\Gamma_n, B)$, where B is the p-primary subgroup of $E(F_\infty)$. This group B is in fact finite and hence $H^1(\Gamma_n, B) = \mathrm{Hom}(\Gamma_n, B)$ for $n \gg 0$. Lemma 3.1 follows immediately. But it is not necessary to know the finiteness of B. If γ denotes a topological generator of Γ, then $H^1(\Gamma_n, B) = B/(\gamma^{p^n} - 1)B$. Since $E(F_n)$ is finitely generated, the kernel of $\gamma^{p^n} - 1$ acting on B is finite. Now B_{div} has finite $\mathbb{Z}_p$-corank. It is clear that

$$B_{\mathrm{div}} \subseteq (\gamma^{p^n} - 1)B \subseteq B.$$

Thus, $H^1(\Gamma_n, B)$ has order bounded by $[B : B_{\mathrm{div}}]$, which is independent of n. If we use the fact that B is finite, then $\ker(h_n)$ has the same order as $H^0(\Gamma_n, B)$, namely $|E(F_n)_p|$. ■

Lemma 3.2. $\mathrm{Coker}(h_n) = 0$.

Proof. The sequence $H^1(F_n, E[p^\infty]) \to H^1(F_\infty, E[p^\infty])^{\Gamma_n} \to H^2(\Gamma_n, B)$ is exact, where $B = H^0(F_\infty, E[p^\infty])$ again. But $\Gamma_n \cong \mathbb{Z}_p$ is a free pro-p group. Hence $H^2(\Gamma_n, B) = 0$. Thus, h_n is surjective as claimed. ■

Let v be any prime of F. We will let v_n denote any prime of F_n lying over v. To study $\ker(g_n)$, we focus on each factor in $\mathcal{P}_E(F_n)$ by considering

$$r_{v_n} : \mathcal{H}_E((F_n)_{v_n}) \to \mathcal{H}_E((F_\infty)_\eta)$$

where η is any prime of F_∞ lying above v_n. ($\mathcal{P}_E(F_\infty)$ has a factor for all such η's, but the kernels will be the same.) If v is archimedean, then v splits completely in F_∞/F, i.e., $F_v = K_\eta$. Thus, $\ker(r_{v_n}) = 0$. For nonarchimedean v, we consider separately $v \nmid p$ and $v \mid p$.

Lemma 3.3. *Suppose v is a nonarchimedean prime not dividing p. Then* $\ker(r_{v_n})$ *is finite and has bounded order as n varies. If E has good reduction at v, then* $\ker(r_{v_n}) = 0$ *for all n.*

Proof. By proposition 2.1, $\mathcal{H}_E(M_\eta) = H^1(M_\eta, E[p^\infty])$ for every algebraic extension M_η of F_v. Let $B_v = H^0(K, E[p^\infty])$, where $K = (F_\infty)_\eta$. Since v is unramified and finitely decomposed in F_∞/F, K is the unramified $\mathbb{Z}_p$-extension of F_v (in fact, the only $\mathbb{Z}_p$-extension of F_v). The group B_v is isomorphic to $(\mathbb{Q}_p/\mathbb{Z}_p)^e \times$ (a finite group), where $0 \le e \le 2$. Let $\Gamma_{v_n} = \mathrm{Gal}(K/(F_n)_{v_n})$, which is isomorphic to $\mathbb{Z}_p$, topologically generated by γ_{v_n}, say. Then

$$\ker(r_{v_n}) \cong H^1(\Gamma_{v_n}, B_v) \cong B_v/(\gamma_{v_n} - 1)B_v.$$

Since $E((F_n)_{v_n})$ has a finite p-primary subgroup, it is clear that $(\gamma_{v_n} - 1)B_v$ contains $(B_v)_{\mathrm{div}}$ (just as in the proof of lemma 3.1) and hence

$$|\ker(r_{v_n})| \le |B_v/(B_v)_{\mathrm{div}}|. \tag{4}$$

This bound is independent of n and v_n. We have equality if $n \gg 0$. Now assume that E has good reduction at v. Then, since $v \nmid p$, $F_v(E[p^\infty])/F_v$ is unramified. It is clear that $K \subseteq F_v(E[p^\infty])$ and that $\Delta = \mathrm{Gal}(F_v(E[p^\infty])/K)$ is a finite, cyclic group of order prime to p. It then follows that $B_v = E[p^\infty]^\Delta$ is divisible. Therefore, $\ker(r_{v_n}) = 0$ as stated. ∎

One can determine the precise order of $\ker(r_{v_n})$, where $v_n | v$ and v is any nonarchimedean prime of F not dividing p where E has bad reduction. This will be especially useful in section 4, where we will need $|\ker(r_v)|$. The result is: $|\ker(r_v)| = c_v^{(p)}$, *where* $c_v^{(p)}$ *is the highest power of* p *dividing the Tamagawa factor* c_v *for* E *at* v. Recall that $c_v = [E(F_v) : E_0(F_v)]$, where $E_0(F_v)$ is the subgroup of local points which have nonsingular reduction at v. First we consider the case where E has additive reduction at v. Then $H^0(I_v, E[p^\infty])$ is finite, where I_v denotes the inertia subgroup of G_{F_v}. Hence B_v is finite because $I_v \subseteq G_K$. Also, $E_0(F_v)$ is a pro-l group, where l is the characteristic of the residue field for v, i.e., $v | l$. (Note: Using the notation in [Si], chapter 5, we have $|\widetilde{E}_{ns}(f_v)| = |f_v| =$ a power of l and $E_1(F_v)$ is pro-l.) Since $l \neq p$, we have $c_v^{(p)} = |E(F_v)_p|$, which in turn equals $|B_v/(\gamma_v - 1)B_v|$. Hence $|\ker(r_v)| = c_v^{(p)}$ when E has additive reduction at v. (It is known that $c_v \leq 4$ when E has additive reduction at v. Thus, for such v, $\ker(r_v) = 0$ if $p \geq 5$.) Now assume that E has split, multiplicative reduction at v. Then $c_v = \mathrm{ord}_v(q_E^{(v)}) = -\mathrm{ord}_v(j_E)$, where $q_E^{(v)}$ denotes the Tate period for E at v. Thus, $q_E^{(v)} = \pi_v^{c_v} \cdot u$, where u is a unit of F_v and π_v is a uniformizing parameter. One can verify easily that the group of units in K is divisible by p. By using the Tate parametrization one can show that $B_v/(B_v)_{\mathrm{div}}$ is cyclic of order $c_v^{(p)}$ and that Γ_v acts trivially on this group. Thus, $|\ker(r_{v_n})| = c_v^{(p)}$ for all $n \geq 0$. B_v might be infinite. In fact, $(B_v)_{\mathrm{div}} = \mu_{p^\infty}$ if $\mu_p \subseteq F_v$; $(B_v)_{\mathrm{div}} = 0$ if $\mu_p \not\subseteq F_v$. Finally, assume that E has nonsplit, multiplicative reduction at v. Then $c_v = 1$ or 2, depending on whether $\mathrm{ord}_v(j_E)$ is odd or even. Using the Tate parametrization, one can see that B_v is divisible when p is odd (and then $\ker(r_v) = 0$). If $p = 2$, E will have split, multiplicative reduction over K and so again $B_v/(B_v)_{\mathrm{div}}$ has order related to $\mathrm{ord}_v(q_E^{(v)})$. But γ_v acts by -1 on this quotient. Hence $H^1(\Gamma_v, B_v)$ has order 1 or 2, depending on the parity of $\mathrm{ord}_v(q_E^{(v)})$. Hence, in all cases, $|\ker(r_v)| = c_v^{(p)}$.

Now assume that $v | p$. For each n, we let f_{v_n} denote the residue field for $(F_n)_{v_n}$. It doesn't depend on the choice of v_n. Also, since v_n is totally ramified in F_∞/F_n for $n \gg 0$, the finite field f_{v_n} stabilizes to f_η, the residue field of $(F_\infty)_\eta$. We let $\widetilde{E}$ denote the reduction of E at v. Then we have

Lemma 3.4. *Assume that* E *has good, ordinary reduction at* v. *Then*

$$|\ker(r_{v_n})| = |\widetilde{E}(f_{v_n})_p|^2.$$

It is finite and has bounded order as n *varies.*

Proof. Let $C_v = \ker(E[p^\infty] \to \widetilde{E}[p^\infty])$, where we regard $E[p^\infty]$ as a subgroup of $E(\overline{F}_v)$. Considering $(F_n)_{v_n}$ as a subfield of $\overline{F}_v$, we have $\mathrm{Im}(\kappa_{v_n}) = \mathrm{Im}(\lambda_{v_n})_{\mathrm{div}}$ by proposition 2.2. By proposition 2.4, we have $\mathrm{Im}(\kappa_\eta) = \mathrm{Im}(\lambda_\eta)$, since the inertia subgroup of $\mathrm{Gal}(F_\infty/F)$ for v has finite index. Thus, we can factor r_{v_n} as follows.

$$\begin{array}{ccc} H^1((F_n)_{v_n}, E[p^\infty])/\mathrm{Im}(\lambda_{v_n})_{\mathrm{div}} & \xrightarrow{a_{v_n}} & H^1((F_n)_{v_n}, E[p^\infty])/\mathrm{Im}(\lambda_{v_n}) \\ & \searrow^{r_{v_n}} & \downarrow{\scriptstyle b_{v_n}} \\ & & H^1((F_\infty)_\eta, E[p^\infty])/\mathrm{Im}(\lambda_\eta) \end{array}$$

Now a_{v_n} is clearly surjective. Hence $|\ker(r_{v_n})| = |\ker(a_{v_n})| \cdot |\ker(b_{v_n})|$. By proposition 2.5, we have $|\ker(a_{v_n})| = |\widetilde{E}(f_{v_n})_p|$. For the proof of proposition 1.2, just the boundedness of $|\ker(a_{v_n})|$ (and of $|\ker(b_{v_n})|$) suffices. To study $\ker(b_{v_n})$ we use the following commutative diagram.

$$\begin{array}{ccccccc} H^1((F_n)_{v_n}, C_v) & \xrightarrow{\lambda_{v_n}} & H^1((F_n)_{v_n}, E[p^\infty]) & \xrightarrow{\pi_{v_n}} & H^1((F_n)_{v_n}, \widetilde{E}[p^\infty]) & \to & 0 \\ \downarrow & & \downarrow & & \downarrow{\scriptstyle d_{v_n}} & & \\ H^1((F_\infty)_\eta, C_v) & \xrightarrow{\lambda_\eta} & H^1((F_\infty)_\eta, E[p^\infty]) & \xrightarrow{\pi_\eta} & H^1((F_\infty)_\eta, \widetilde{E}[p^\infty]) & \to & 0 \end{array} \tag{5}$$

The surjectivity of the first row follows from Poitou-Tate Duality, which gives $H^2(M, C_v) = 0$ for any finite extension M of F_v. (Note that $C_v \not\cong \mu_{p^\infty}$ for the action of G_M.) Thus, $\ker(b_{v_n}) \cong \ker(d_{v_n})$. But

$$\ker(d_{v_n}) \cong H^1((F_\infty)_\eta/(F_n)_{v_n}, \widetilde{E}(f_\eta)_p) \cong \widetilde{E}(f_\eta)_p/(\gamma_{v_n} - 1)\widetilde{E}(f_\eta)_p$$

where γ_{v_n} is a topological generator of $\mathrm{Gal}((F_\infty)_\eta/(F_n)_{v_n})$. Now $\widetilde{E}(f_\eta)_p$ is finite and the kernel and cokernel of $\gamma_{v_n} - 1$ have the same order, namely $|\widetilde{E}(f_{v_n})_p|$. This is the order of $\ker(d_{v_n})$. Lemma 3.4. follows. ∎

Let Σ_0 denote the finite set of nonarchimedean primes of F which either lie over p or where E has bad reduction. If $v \notin \Sigma_0$ and v_n is a prime of F_n lying over v, then $\ker(r_{v_n}) = 0$. For each $v \in \Sigma_0$, lemmas 3.3 and 3.4 show that $|\ker(r_{v_n})|$ is bounded as n varies. The number of primes v_n of F_n lying over any nonarchimedean prime v is also bounded. Consequently, we have proved the following lemma.

Lemma 3.5. *The order of* $\ker(g_n)$ *is bounded as* n *varies.*

Lemma 3.1 implies that $\ker(s_n)$ is finite and has bounded order no matter what type of reduction E has at $v|p$. Lemmas 3.2 and 3.5 show that $\mathrm{coker}(s_n)$ is finite and of bounded order, assuming that E has good, ordinary reduction at all $v|p$. Thus, theorem 1.2 is proved.

It is possible for s_n to be injective for all n. A simple sufficient condition for this is: *$E(F)$ has no element of order p*. For then $E(F_\infty)$ will have no p-torsion, since $\Gamma = \mathrm{Gal}(F_\infty/F)$ is a pro-p group. Thus $\ker(h_n)$ and hence $\ker(s_n)$ would be trivial for all n. A somewhat more subtle result will be proved later, in proposition 3.9.

It is also possible for s_n to be surjective for all n. Still assuming that E has good, ordinary reduction at all primes of F lying over v, here is a sufficient condition for this: *For each $v|p$, $\widetilde{E}_v(f_v)$ has no element of order p and, for each v where E has bad reduction, $E[p^\infty]^{I_v}$ is divisible.* The first part of this condition implies that $\widetilde{E}_v(f_{v_n})_p = 0$ for all $v|p$ and all n, again using the fact that Γ is pro-p. Thus, $\ker(r_{v_n}) = 0$ by lemma 3.4. In the second part of this condition, I_v denotes the inertia subgroup of G_{F_v}. Note that $v \nmid p$. It is easy to see that if $E[p^\infty]^{I_v}$ is divisible, the same is true of $B_v = H^0((F_\infty)_\eta, E[p^\infty])$ for $\eta|v$. Thus, $\ker(r_{v_n}) = 0$ for $v_n|v$, because of (4). The second part of this condition is equivalent to $p \nmid c_v$.

We want to now discuss the case where E has multiplicative reduction at some $v|p$. In this case, one can attempt to imitate the proof of lemma 3.4, taking $C_v = \mathcal{F}(\overline{\mathfrak{m}})[p^\infty]$. We first assume that E has split, multiplicative reduction. Then $C_v \cong \mu_{p^\infty}$ and we have an exact sequence

$$0 \to \mu_{p^\infty} \to E[p^\infty] \to \mathbb{Q}_p/\mathbb{Z}_p \to 0$$

of G_{F_v}-modules, where the action on $\mathbb{Q}_p/\mathbb{Z}_p$ is trivial. Then $H^1((F_n)_{v_n}, \mu_{p^\infty})$ and hence $\mathrm{Im}(\lambda_{v_n})$ are divisible. We have $\mathrm{Im}(\kappa_{v_n}) = \mathrm{Im}(\lambda_{v_n})$ as well as $\mathrm{Im}(\kappa_\eta) = \mathrm{Im}(\lambda_\eta)$. Thus, $\ker(r_{v_n}) = \ker(b_{v_n})$, where b_{v_n} is the map

$$b_{v_n} : H^1((F_n)_{v_n}, E[p^\infty])/\mathrm{Im}(\lambda_{v_n}) \to H^1((F_\infty)_\eta, E[p^\infty])/\mathrm{Im}(\lambda_\eta).$$

For any algebraic extension M of F_v, we have an exact sequence

$$H^1(M, \mu_{p^\infty}) \xrightarrow{\lambda_M} H^1(M, E[p^\infty]) \xrightarrow{\pi_M} H^1(M, \mathbb{Q}_p/\mathbb{Z}_p) \xrightarrow{\delta_M} H^2(M, \mu_{p^\infty}) \to 0.$$

If $[M{:}F_v] < \infty$, then Poitou-Tate Duality shows that $H^2(M, \mu_{p^\infty}) \cong \mathbb{Q}_p/\mathbb{Z}_p$, whereas $H^2(M, E[p^\infty]) = 0$, which gives the surjectivity of δ_M. Thus, π_M is not surjective in contrast to the case where E has good, ordinary reduction at v. We let $\pi_{v_n} = \pi_{(F_n)_{v_n}}$, $\pi_\eta = \pi_{(F_\infty)_\eta}$. Thus, $\ker(b_{v_n})$ can be identified with $\mathrm{Im}(\pi_{v_n}) \cap \ker(d_{v_n})$, where d_{v_n} is the map

$$d_{v_n} : H^1((F_n)_{v_n}, \mathbb{Q}_p/\mathbb{Z}_p) \to H^1((F_\infty)_\eta, \mathbb{Q}_p/\mathbb{Z}_p).$$

The kernel of d_{v_n} is quite easy to describe. We have

$$\ker(d_{v_n}) = \mathrm{Hom}(\mathrm{Gal}((F_\infty)_\eta/(F_n)_{v_n}), \mathbb{Q}_p/\mathbb{Z}_p)$$

which is isomorphic to $\mathbb{Q}_p/\mathbb{Z}_p$ as a group. The image of π_{v_n} is more interesting to describe. It depends on the Tate period q_E for E, which is defined by the equation $j(q_E) = j_E$, solving this equation for $q_E \in F_v^\times$. Here $j(q) =$

$q^{-1} + 744 + 196884q + \cdots$ for $|q|_v < 1$ and j_E is the j-invariant for E. Since $j_E \in F$ is algebraic, the theorem of [B-D-G-P] referred to in section 1 implies that q_E is transcendental. Also, we have $|q_E|_v = |j_E|_v^{-1}$. Let

$$\mathrm{rec}_M : M^\times \to \mathrm{Gal}(M^{ab}/M)$$

denote the reciprocity map of local class field theory. We will prove the following result.

Proposition 3.6. *Let M be a finite extension of F_v. Then*

$$\mathrm{Im}(\pi_M) = \{\psi \in \mathrm{Hom}(\mathrm{Gal}(M^{ab}/M), \mathbb{Q}_p/\mathbb{Z}_p) \mid \psi(\mathrm{rec}_M(q_E)) = 0\}.$$

If M is a $\mathbb{Z}_p$-extension of F_v, then π_M is surjective.

Proof. The last statement is clear since G_M has p-cohomological dimension 1 if M/F_v has profinite degree divisible by p^∞. For the first statement, the exact sequence

$$0 \to \mu_{p^n} \to E[p^n] \to \mathbb{Z}/p^n\mathbb{Z} \to 0$$

induces a map $\pi_M^{(n)}: H^1(M, E[p^n]) \to H^1(M, \mathbb{Z}/p^n\mathbb{Z})$ for every $n \geq 1$. Because of the Weil pairing, we have $\mathrm{Hom}(E[p^n], \mu_{p^n}) \cong E[p^n]$. Thus, by Poitou-Tate Duality, $\pi_M^{(n)}$ is adjoint to the natural map

$$H^1(M, \mu_{p^n}) \to H^1(M, E[p^n])$$

whose kernel is easy to describe. It is generated by the class of the 1-cocycle $\phi: G_M \to \mu_{p^n}$ given by $\phi(g) = g(\sqrt[p^n]{q_E})/\sqrt[p^n]{q_E}$ for all $g \in G_M$. The pairing

$$H^1(M, \mu_{p^n}) \times H^1(M, \mathbb{Z}/p^n\mathbb{Z}) \to \mathbb{Z}/p^n\mathbb{Z}$$

is just $(\phi_q, \psi) \to \psi(\mathrm{rec}_M(q))$ for $q \in M^\times$, where ϕ_q is the 1-cocycle associated to q as above, i.e., the image of q under the Kummer homomorphism $M^\times/(M^\times)^{p^n} \to H^1(M, \mu_{p^n})$. This implies that

$$\mathrm{Im}(\pi_M^{(n)}) = \{\psi \in \mathrm{Hom}(\mathrm{Gal}(M^{ab}/M), \mathbb{Z}/p^n\mathbb{Z}) \mid \psi(\mathrm{rec}_M(q_E)) = 0\},$$

from which the first part of proposition 3.6 follows by just taking a direct limit. ∎

Still assuming that E has split, multiplicative reduction at v, the statement that $\ker(r_{v_n})$ is finite is equivalent to the assertion that $\ker(d_{v_n}) \not\subseteq \mathrm{Im}(\pi_{v_n})$. In this case, we show that $|\ker(r_{v_n})|$ is bounded as n varies. For let $\sigma = \mathrm{rec}_{F_v}(q_E)|_{(F_\infty)_\eta} \in \mathrm{Gal}((F_\infty)_\eta/F_v)$. Let $e_n = [(F_n)_{v_n}:F_v]$. Then we have $\mathrm{rec}_{(F_n)_{v_n}}(q_E)|_{(F_\infty)_\eta} = \sigma^{e_n}$. It is clear that

$$\ker(r_{v_n}) = \{\psi \in \mathrm{Hom}(\mathrm{Gal}((F_\infty)_\eta/(F_n)_{v_n}), \mathbb{Q}_p/\mathbb{Z}_p) \mid \psi(\sigma^{e_n}) = 0\}$$

has order equal to $[\mathrm{Gal}((F_\infty)_\eta/(F_n)_{v_n}):\overline{\langle\sigma^{e_n}\rangle}]$. But $\mathrm{Gal}((F_\infty)_\eta/F_v)\cong\mathbb{Z}_p$. This index is constant for $n\ge 0$. Thus, $\ker(r_{v_n})$ is finite and of constant order as n varies provided that $\sigma\ne\mathrm{id}$. Let $\mathbb{Q}_p^{\mathrm{cyc}}$ denote the cyclotomic $\mathbb{Z}_p$-extension of $\mathbb{Q}_p$. Then $(F_\infty)_\eta=F_v\mathbb{Q}_p^{\mathrm{cyc}}$. We have the following diagram

$$\begin{array}{ccc} F_v^\times & \xrightarrow{\mathrm{rec}} & \mathrm{Gal}((F_\infty)_\eta/F_v) \\ \downarrow{\scriptstyle N_{F_v/\mathbb{Q}_p}} & & \downarrow{\scriptstyle \mathrm{rest}} \\ \mathbb{Q}_p^\times & \xrightarrow{\mathrm{rec}} & \mathrm{Gal}(\mathbb{Q}_p^{\mathrm{cyc}}/\mathbb{Q}_p) \end{array}$$

where the horizontal arrows are the reciprocity maps. It is known that the group of universal norms for $\mathbb{Q}_p^{\mathrm{cyc}}/\mathbb{Q}_p$ is precisely $\mu\cdot\langle p\rangle$, where μ denotes the roots of unity in $\mathbb{Q}_p$. This of course coincides with the kernel of the reciprocity map $\mathbb{Q}_p^\times\to\mathrm{Gal}(\mathbb{Q}_p^{\mathrm{cyc}}/\mathbb{Q}_p)$ and also coincides with the kernel of $\log_p$ (where we take Iwasawa's normalization $\log_p(p)=0$.) Also, it is clear that $\sigma\ne\mathrm{id}\Leftrightarrow\sigma|_{\mathbb{Q}_p^{\mathrm{cyc}}}\ne\mathrm{id}$. Thus we have shown that $\ker(r_{v_n})$ is finite if and only if $\log_p(N_{F_v/\mathbb{Q}_p}(q_E))\ne 0$. The order will then be constant and is determined by the projection of $N_{F_v/\mathbb{Q}_p}(q_E)$ to $\mathbb{Z}_p^\times$ in the decomposition $\mathbb{Q}_p^\times=\langle p\rangle\times\mathbb{Z}_p^\times$. One finds that

$$|\ker(r_{v_n})|\sim\log_p(N_{F_v/\mathbb{Q}_p}(q_E))/2p[F_v\cap\mathbb{Q}_p^{\mathrm{cyc}}:\mathbb{Q}_p]$$

where $\sim$ indicates that the two sides have the same p-adic valuation.

Assume now that p is odd and that E has nonsplit, multiplicative reduction. We then show that $\ker(r_{v_n})=0$. We have an exact sequence

$$0\to\mu_{p^\infty}\otimes\phi\to E[p^\infty]\to(\mathbb{Q}_p/\mathbb{Z}_p)\otimes\phi\to 0$$

where ϕ is the unramified character of G_{F_v} of order 2. As discussed in section 2, we have $\mathrm{Im}(\kappa_{v_n})=\mathrm{Im}(\lambda_{v_n})$. Also π_{v_n} is surjective. We can identify $\ker(r_{v_n})$ with $\ker(d_{v_n})$, where d_{v_n} is the map

$$H^1((F_n)_{v_n},(\mathbb{Q}_p/\mathbb{Z}_p)(\phi))\to H^1((F_\infty)_\eta,(\mathbb{Q}_p/\mathbb{Z}_p)(\phi))$$

whose kernel is clearly zero. Thus, as stated, $\ker(r_{v_n})=0$. (The value of $N_{F_v/\mathbb{Q}_p}(q_E)$ is not relevant in this case.) If $p=2$, then $|\mathrm{Im}(\lambda_{v_n})/\mathrm{Im}(\lambda_{v_n})_{\mathrm{div}}|$ is easily seen to be at most 2. Hence, if E has nonsplit, multiplicative reduction over $(F_n)_{v_n}$, we have $|\ker(r_{v_n})|\le 2$. (Note: It can happen that $(F_\infty)_\eta$ contains the unramified quadratic extension of F_v. Thus E can become split over $(F_n)_{v_n}$ for $n>0$.) We will give the order of $\ker(r_v)$ when E has nonsplit, multiplicative reduction at $v|2$. The kernel of a_v has order $[\mathrm{Im}(\lambda_v):\mathrm{Im}(\kappa_v)]$, which is just the Tamagawa factor for E at v. (See the discussion following the proof of proposition 2.5.) On the other hand, $\ker(b_v)\cong\ker(d_v)$ and this group has order 2. Thus, $|\ker(r_v)|\sim 2c_v$, where c_v denotes the Tamagawa factor for E at v.

The above observations together with lemmas 3.1–3.3 provide a proof of the following result in the direction of conjecture 1.6.

Proposition 3.7. *Assume that E is an elliptic curve defined over F which has good, ordinary reduction or multiplicative reduction at all primes v of F lying over p. Assume also that $\log_p(N_{F_v/\mathbb{Q}_p}(q_E^{(v)})) \neq 0$ for all v where E has multiplicative reduction. Then the maps*

$$s_n \colon \mathrm{Sel}_E(F_n)_p \to \mathrm{Sel}_E(F_\infty)_p^{\Gamma_n}$$

have finite kernel and cokernel, of bounded order as n varies.

In the above result, $q_E^{(v)}$ denotes the Tate period for E over F_v. If $j_E \in \mathbb{Q}_p$, then so is $q_E^{(v)}$. Thus, $N_{F_v/\mathbb{Q}_p}(q_E^{(v)}) = (q_E^{(v)})^{[F_v:\mathbb{Q}_p]}$ is transcendental according to the theorem of Barré-Sirieix, Diaz, Gramain, and Philibert. Perhaps, it is reasonable to conjecture in general that $N_{F_v/\mathbb{Q}_p}(q_E^{(v)})$ is transcendental whenever $j_E \in F_v \cap \overline{\mathbb{Q}}$. Then the hypothesis $\log_p(N_{F_v/\mathbb{Q}_p}(q_E^{(v)})) \neq 0$ obviously holds. This hypothesis is unnecessary in proposition 3.7, if p is odd, for those v's where E has nonsplit, multiplicative reduction. (For $p = 2$, one needs the hypothesis when E has split reduction over $(F_\infty)_\eta$.)

Let X be a profinite Λ-module, where $\Lambda = \mathbb{Z}_p[[T]]$, $T = \gamma - 1$, as in section 1. Here are some facts which are easily proved or can be found in [Wa2].

(1) $X = TX \Rightarrow X = 0$.
(2) *X/TX finite $\Rightarrow X$ is a finitely generated, torsion Λ-module.*
(3) *X/TX finitely generated over $\mathbb{Z}_p \Rightarrow X$ is finitely generated over Λ.*
(4) *Assume that X is a finitely generated, torsion Λ-module. Let θ_n denote $\gamma^{p^n} - 1 \in \Lambda$ for $n \geq 0$. Then there exist integers a, b, and c such that the $\mathbb{Z}_p$-torsion subgroup of $X/\theta_n X$ has order p^{e_n}, where $e_n = an + bp^n + c$ for $n \gg 0$.*

We sketch an argument for (4). Let $f(T)$ be a generator for the characteristic ideal of X, assuming that X is finitely generated and torsion over Λ. If we have $f(\zeta - 1) \neq 0$ for all p-power roots of unity, then $X/\theta_n X$ is finite for all $n \geq 0$ and one estimates its order by studying $\prod f(\zeta - 1)$, where ζ runs over the p^n-th roots of unity. One then could take $a = \lambda(f)$, $b = \mu(f)$ in (4). Suppose $X = \Lambda/(h(T)^e)$, where $h(T)$ is an irreducible element of Λ. If $h(T) \nmid \theta_n$ for all n, then we are in the case just discussed. This is true for $(h(T)) = p\Lambda$ for example. If $h(T)|\theta_{n_0}$ for some $n_0 \geq 0$, then write $\theta_n = h(T)\phi_n$, for $n \geq n_0$, where $\phi_n \in \Lambda$. Since $\theta_n = (1+T)^{p^n} - 1$ has no multiple factors, we have $h(T) \nmid \phi_n$. Then we get an exact sequence

$$0 \to Y/\phi_n Y \to X/\theta_n X \to \Lambda/h(T)\Lambda \to 0$$

for $n \geq n_0$. Here $Y = (h(T))/(h(T)^e) \cong \Lambda/(h(T)^{e-1})$. Then $Y/\phi_n Y$ is finite and one estimates its growth essentially as mentioned above. Now $\Lambda/h(T)\Lambda$ is a free $\mathbb{Z}_p$-module of rank $= \lambda(h)$. Thus the $\mathbb{Z}_p$-torsion subgroup of $X/\theta_n X$ is $Y/\phi_n Y$ whose order is given by a formula as above. In general, X is pseudo-isomorphic to a direct sum of Λ-modules of the form $\lambda/(h(T)^e)$ and one can reduce to that case. One sees that $b = \mu(f)$, where $f = f(T)$ generates the characteristic ideal of X. Also, $a = \lambda(f) - \lambda_0$, where $\lambda_0 = \max(\mathrm{rank}_{\mathbb{Z}_p}(X/\theta_n X))$. The $\mathbb{Z}_p$-rank of $X/\theta_n X$ clearly stabilizes, equal to λ_0 for $n \gg 0$.

These facts together with the results of this section have some immediate consequences, some of which we state here without trying to be as general as possible. For simplicity, we take $\mathbb{Q}$ as the base field.

Proposition 3.8. *Let E be an elliptic curve with good, ordinary reduction at p. We make the following assumptions:*

(i) *p does not divide $|\widetilde{E}(\mathbb{F}_p)|$, where $\widetilde{E}$ denotes the reduction of E at p.*
(ii) *If E has split, multiplicative reduction at l, where $l \neq p$, then $p \nmid \mathrm{ord}_l(j_E)$. If E has nonsplit, multiplicative reduction at l, then either p is odd or $\mathrm{ord}_l(j_E)$ is odd.*
(iii) *If E has additive reduction at l, then $E(\mathbb{Q}_l)$ has no point of order p.*

Then the map $\mathrm{Sel}_E(\mathbb{Q})_p \to \mathrm{Sel}_E(\mathbb{Q}_\infty)_p^\Gamma$ is surjective. If $\mathrm{Sel}_E(\mathbb{Q})_p = 0$, then $\mathrm{Sel}_E(\mathbb{Q}_\infty)_p = 0$ also.

Remark. The comments in the paragraph following the proof of lemma 3.3 allow us to restate hypotheses (ii) and (iii) in the following way: $p \nmid c_l$ *for all* $l \neq p$. Here c_l is the Tamagawa factor for E at l. If E has good reduction at l, then $c_l = 1$. If E has additive reduction at l, then $c_l \leq 4$. Thus, hypothesis (iii) is automatically satisfied for any $p \geq 5$. If E has nonsplit, multiplicative reduction at l, then hypothesis (ii) holds for any $p \geq 3$. On the other hand, if E has split, multiplicative reduction at l, then there is no restriction on the primes which could possibly divide c_l. Hypothesis (i) is equivalent to $a_p \not\equiv 1 \pmod{p}$, where $a_p = 1 + p - |\widetilde{E}(\mathbb{F}_p)|$.

Proof. We refer back to the sequence at the beginning of this section. We have $\mathrm{coker}(h_n) = 0$ by lemma 3.2. The surjectivity of the map s_0 would follow from the assertion $\ker(g_0) = 0$. But the above assumptions simply guarantee that the map $\mathcal{P}_E(\mathbb{Q}) \to \mathcal{P}_E(\mathbb{Q}_\infty)$ is injective and hence that $\ker(g_0) = 0$. For by lemma 3.4, (i) implies that $\ker(r_p) = 0$. If E has multiplicative reduction at $l \neq p$ then (ii) implies that $\mathrm{ord}_l(q_E^{(l)})$ is not divisible by p. This means $\mathbb{Q}_l(\sqrt[p]{q_E^{(l)}})/\mathbb{Q}_l$ is ramified. Thus $H^0(L, E[p^\infty])$ is a divisible group, where L denotes the maximal unramified extension of $\mathbb{Q}_l$. Now $\mathrm{Gal}(L/\mathbb{Q}_l) \cong \hat{\mathbb{Z}}$. The cyclotomic $\mathbb{Z}_p$-extension of $\mathbb{Q}_l$ is $(\mathbb{Q}_\infty)_\eta$, where $\eta | l$. Thus, $(\mathbb{Q}_\infty)_\eta \subseteq L$. Let $H = \mathrm{Gal}(L/(\mathbb{Q}_\infty)_\eta)$. Then H acts on $H^0(L, E[p^\infty])$ through a finite cyclic group of order prime to p. Thus, it is easy to see that $H^0((\mathbb{Q}_\infty)_\eta, E[p^\infty])$ is divisible and hence, from (4), we have $\ker(r_l) = 0$. Assume now that E has additive reduction at l (where, of course, $l \neq p$). Then $E[p^\infty]^{I_l}$ is finite, where I_l denotes G_L, the inertia subgroup of $G_{\mathbb{Q}_l}$. We know that if E has potentially good reduction at l, then I_l acts on $E[p^\infty]$ through a quotient of order $2^a 3^b$. Thus $E[p^\infty]^{I_l} = 0$ if $p \geq 5$, and (iii) is then not important. If $p = 2$ or 3, (iii) suffices to conclude that $H^0((\mathbb{Q}_\infty)_\eta, E[p^\infty]) = 0$ since $\mathrm{Gal}((\mathbb{Q}_\infty)_\eta/\mathbb{Q}_l)$ is pro-p. Thus, again, $\ker(r_l) = 0$. (We are essentially repeating some previous observations.) Finally, if $\mathrm{Sel}_E(\mathbb{Q})_p$ is trivial, then so is $\mathrm{Sel}_E(\mathbb{Q}_\infty)^\Gamma$. Let $X = X_E(\mathbb{Q}_\infty)$. Then $X/TX = 0$, which implies that $X = 0$. Hence $\mathrm{Sel}_E(\mathbb{Q}_\infty) = 0$ as stated. ∎

If we continue to take $F = \mathbb{Q}$, then we now know that the restriction map $\mathrm{Sel}_E(\mathbb{Q})_p \to \mathrm{Sel}_E(\mathbb{Q}_\infty)_p^\Gamma$ has finite cokernel if E has good, ordinary or multiplicative reduction at p. (In fact, potentially ordinary or potentially multiplicative reduction would suffice.) Thus, if $\mathrm{Sel}_E(\mathbb{Q})_p$ is finite, then so is $\mathrm{Sel}_E(\mathbb{Q}_\infty)_p^\Gamma$. Hence, for $X = X_E(\mathbb{Q}_\infty)$, we would have that X/TX is finite. Thus, X would be a Λ-torsion module. In addition, we would have $T \nmid f_E(T)$.

Assume that E has good, ordinary reduction at p. If p is odd, then the map $\mathrm{Sel}_E(\mathbb{Q}_n)_p \to \mathrm{Sel}_E(\mathbb{Q}_\infty)^{\Gamma_n}$ is actually injective for all $n \geq 0$. To see this, let $B = H^0(\mathbb{Q}_\infty, E[p^\infty])$. Then $\ker(h_n) = H^1(\Gamma_n, B)$. The inertia subgroup I_p of $G_{\mathbb{Q}_p}$ acts on $\ker(E[p] \to \widetilde{E}[p])$ by the Teichmüller character ω. That is,

$$\ker(E[p] \to \widetilde{E}[p]) \cong \mu_p$$

for the action of I_p. On the other hand, I_p acts on B through $\mathrm{Gal}((\mathbb{Q}_\infty)_\eta/\mathbb{Q}_p)$, where η denotes the unique prime of $\mathbb{Q}_\infty$ lying over p. This Galois group is pro-p, being isomorphic to $\mathbb{Z}_p$. Since $p > 2$, ω has nontrivial order and this order is relatively prime to p. It follows that

$$B \cap \ker(E[p^\infty] \to \widetilde{E}[p^\infty]) = \{O_E\}$$

and therefore B maps injectively into $\widetilde{E}[p^\infty]$. Thus, I_p acts trivially on B Since p is totally ramified in $\mathbb{Q}_\infty/\mathbb{Q}$, it is clear that $\Gamma = \mathrm{Gal}(\mathbb{Q}_\infty/\mathbb{Q})$ also acts trivially on B. That is,

$$B = E(\mathbb{Q}_\infty)_p = E(\mathbb{Q})_p.$$

Hence $\ker(h_n) = \mathrm{Hom}(\Gamma_n, B)$ for all $n \geq 0$. Now suppose that ϕ is a nontrivial element of $\mathrm{Hom}(\Gamma_n, B)$. Let $I_p^{(n)}$ denote the inertia subgroup of $G_{(\mathbb{Q}_n)_\eta}$. Then ϕ clearly remains nontrivial when restricted to

$$H^1(I_p^{(n)}, \widetilde{E}[p^\infty]) = \mathrm{Hom}(I_p^{(n)}, \widetilde{E}[p^\infty]).$$

But this implies that $[\phi] \notin \mathrm{Sel}_E(\mathbb{Q}_n)_p$. Hence $\ker(s_n) = \ker(h_n) \cap \mathrm{Sel}_E(\mathbb{Q}_n)_p$ is trivial as claimed. This argument also applies if E has multiplicative reduction at p. More generally, the argument gives the following result. We let F be any number field. For any prime v of F lying over p, we let $e(v/p)$ denote the ramification index for $F_v/\mathbb{Q}_p$.

Proposition 3.9. *Let E be an elliptic curve defined over F. Assume that there is at least one prime v of F lying over p with the following properties:*

(i) *E has good, ordinary reduction or multiplicative reduction at v,*
(ii) $e(v/p) \le p-2$.

Then the map $\mathrm{Sel}_E(F_n)_p \to \mathrm{Sel}_E(F_\infty)_p$ *is injective for all* $n \ge 0$.

Theorem 1.10 is also an application of the results described in this section. One applies the general fact (4) about torsion Λ-modules to $X = X_E(F_\infty)$. Then, $X/\theta_n X$ is the Pontryagin dual of $\mathrm{Sel}_E(F_\infty)_p^{\Gamma_n}$. The torsion subgroup of $X/\theta_n X$ is then dual to $\mathrm{Sel}_E(F_\infty)_p^{\Gamma_n}/(\mathrm{Sel}_E(F_\infty)_p^{\Gamma_n})_{\mathrm{div}}$. One compares this to $\mathrm{Sel}_E(F_n)_p/(\mathrm{Sel}_E(F_n)_p)_{\mathrm{div}}$, which is precisely $\text{Ш}_E(F_n)_p$ under the assumption of finiteness. One must show that the orders of the relevant kernels and cokernels stabilize, which we leave for the reader. One then obtains the formula for the growth of $|\text{Ш}_E(F_n)_p|$, with the stated λ and μ.

We want to mention one other useful result. It plays a role in Li Guo's proof of a parity conjecture for elliptic curves with complex multiplication. (See [Gu2].)

Proposition 3.10. *Assume that E is an elliptic curve/F and that* $\mathrm{Sel}_E(F_\infty)_p$ *is* Λ*-cotorsion. Let* $\lambda_E = \mathrm{corank}_{\mathbb{Z}_p}(\mathrm{Sel}_E(F_\infty)_p)$. *Assume also that p is odd. Then*

$$\mathrm{corank}_{\mathbb{Z}_p}(\mathrm{Sel}_E(F)_p) \equiv \lambda_E \pmod 2.$$

Proof. The maps $H^1(F_n, E[p^\infty]) \to H^1(F_\infty, E[p^\infty])$ have finite kernels of bounded order as n varies, by lemma 3.1. Thus, $\mathrm{corank}_{\mathbb{Z}_p}(\mathrm{Sel}_E(F_n)_p)$ is bounded above by λ_E. Let λ'_E denote the maximum of these $\mathbb{Z}_p$-coranks. Then $\mathrm{corank}_{\mathbb{Z}_p}(\mathrm{Sel}_E(F_n)_p) = \lambda'_E$ for all $n \ge n_0$, say. For brevity, we let $S_n = \mathrm{Sel}_E(F_n)_p$, $T_n = (S_n)_{\mathrm{div}}$, and $U_n = S_n/T_n$, which is finite. The restriction map $S_0 \to S_n^{\mathrm{Gal}(F_n/F)}$, and hence the map $T_0 \to T_n^{\mathrm{Gal}(F_n/F)}$, have finite kernel and cokernel. Since the nontrivial $\mathbb{Q}_p$-irreducible representations of $\mathrm{Gal}(F_n/F)$ have degree divisible by $p-1$, it follows easily that $\mathrm{corank}_{\mathbb{Z}_p}(T_n) \equiv \mathrm{corank}_{\mathbb{Z}_p}(T_0) \pmod{p-1}$. Hence

$$\mathrm{corank}_{\mathbb{Z}_p}(\mathrm{Sel}_E(F)_p) \equiv \lambda'_E \pmod{p-1}.$$

Since p is odd, this gives a congruence modulo 2. Let $S_\infty = \mathrm{Sel}_E(F_\infty)_p$ and let $T_\infty = \varinjlim T_n$, which is a Λ-submodule of S_∞. Also, $T_\infty \cong (\mathbb{Q}_p/\mathbb{Z}_p)^{\lambda'_E}$. Let $U_\infty = S_\infty/T_\infty = \varinjlim U_n$. The map $T_n \to T_\infty$ is obviously surjective for all $n \ge n_0$ (since the kernel is finite). This implies that

$$|\ker(U_n \to U_\infty)| \le |\ker(S_n \to S_\infty)|$$

for $n \ge n_0$, which is of bounded order as n varies. Now a well-known theorem of Cassels states that there exists a nondegenerate, skew-symmetric pairing

$$U_n \times U_n \to \mathbb{Q}_p/\mathbb{Z}_p.$$

This forces $|U_n|$ to be a perfect square. More precisely, if the abelian group U_n is decomposed as a direct product of cyclic groups of orders $p^{e_n^{(i)}}$, $1 \le i \le g_n$, say, then g_n is even and one can arrange the terms so that $e_n^{(1)} = e_n^{(2)} \ge \cdots \ge e_n^{(g_n-1)} = e_n^{(g_n)}$. We refer to [Gul] for a proof of this elementary result. (See lemma 3, page 157 there.) Since the kernels of the maps $U_n \to U_\infty$ have bounded order, the $\mathbb{Z}_p$-corank u of U_∞ can be determined from the behavior of the $e_n^{(i)}$'s as $n \to \infty$, namely, the first u of the $e_n^{(i)}$'s will be unbounded, the rest bounded as $n \to \infty$. Thus u is even. Since $u = \lambda_E - \lambda'_E$, it follows that

$$\lambda_E \equiv \lambda'_E \ (\mathrm{mod}\ 2).$$

Combining that with the previous congruence, we get proposition 3.10. ■

Appendix to Section 3. We would like to give a different and rather novel proof of a slightly weaker form of proposition 3.6, which is in fact adequate for proving proposition 3.7. We let $\widetilde{M}$ denote the composition of all $\mathbb{Z}_p$-extensions of M. For any $q \in M^\times$, we let $\widetilde{M}_q$ denote the composition of all $\mathbb{Z}_p$-extensions M_∞ of M such that $\mathrm{rec}_M(q)|_{M_\infty}$ is trivial, i.e., the image of q under the reciprocity map $M^\times \to \mathrm{Gal}(M_\infty/M)$ is trivial. This means that $q \in N_{M_n/M}(M_n^\times)$ for all $n \ge 0$, where M_n denotes the n-th layer in M_∞/M. We then say that q is a universal norm for the $\mathbb{Z}_p$-extension M_∞/M. We will show that

$$\mathrm{Im}(\pi_M)_{\mathrm{div}} = \mathrm{Hom}(\mathrm{Gal}(\widetilde{M}_{q_E}/M), \mathbb{Q}_p/\mathbb{Z}_p). \tag{6}$$

The proof is based on the following observation:

Proposition 3.11. *Assume that $q \in M^\times$ is a universal norm for the $\mathbb{Z}_p$-extension M_∞/M. Then the image of $\langle q\rangle \otimes (\mathbb{Q}_p/\mathbb{Z}_p)$ under the composite map*

$$\langle q\rangle \otimes (\mathbb{Q}_p/\mathbb{Z}_p) \to M^\times \otimes (\mathbb{Q}_p/\mathbb{Z}_p) \xrightarrow{\sim} H^1(M, \mu_{p^\infty}) \to H^1(M_\infty, \mu_{p^\infty})$$

is contained in $H^1(M_\infty, \mu_{p^\infty})_{\Lambda\text{-div}}$, where $\Lambda = \mathbb{Z}_p[[\mathrm{Gal}(M_\infty/M)]]$.

Proof. To justify this, note that the inflation-restriction sequence shows that the natural map

$$H^1(M_n, \mu_{p^\infty}) \to H^1(M_\infty, \mu_{p^\infty})^{\Gamma_n}$$

is surjective and has finite kernel. Here $\Gamma = \mathrm{Gal}(M_\infty/M)$, $\Gamma_n = \Gamma^{p^n} = \mathrm{Gal}(M_\infty/M_n)$. But $H^1(M_n, \mu_{p^n})$ is isomorphic to $(\mathbb{Q}_p/\mathbb{Z}_p)^{tp^n+1}$ as a group, where $t = [M:\mathbb{Q}_p]$. Thus, $H^1(M_\infty, \mu_{p^\infty})^{\Gamma_n}$ is divisible and has $\mathbb{Z}_p$-corank tp^n+1. If $X = H^1(M_\infty, \mu_{p^\infty})\hat{}$, then X is a finitely generated Λ-module with the property that $X/\theta_n X \cong \mathbb{Z}_p^{tp^n+1}$ for all $n \geq 0$, where $\theta_n = \gamma^{p^n} - 1 \in \Lambda$, and γ is some topological generator of Γ. It is not hard to deduce from this that $X \cong \Lambda^t \times \mathbb{Z}_p$, where $\mathbb{Z}_p = X_{\Lambda\text{-tors}}$ is just $\Lambda/\theta_0\Lambda$. Letting $\widehat{\Lambda}$ denote the Pontryagin dual of Λ, regarded as a discrete Λ-module, we have

$$H^1(M_\infty, \mu_{p^\infty}) \cong \widehat{\Lambda}^t \times (\mathbb{Q}_p/\mathbb{Z}_p),$$

where the action of Γ on $\mathbb{Q}_p/\mathbb{Z}_p$ is trivial. Thus, $H^1(M_\infty, \mu_{p^\infty})_{\Lambda\text{-div}} \cong \widehat{\Lambda}^t$, noting that the Pontryagin dual of a torsion-free Λ-module is Λ-divisible. Hence $(H^1(M_\infty, \mu_{p^\infty})_{\Lambda\text{-div}})^\Gamma$ has $\mathbb{Z}_p$-corank t. The maximal divisible subgroup of its inverse image in $M^\times \otimes (\mathbb{Q}_p/\mathbb{Z}_p)$ is isomorphic to $(\mathbb{Q}_p/\mathbb{Z}_p)^t$. We must show that this "canonical subgroup" of $M^\times \otimes (\mathbb{Q}_p/\mathbb{Z}_p)$, which the $\mathbb{Z}_p$-extension M_∞/M determines, contains $\langle q\rangle \otimes (\mathbb{Q}_p/\mathbb{Z}_p)$ whenever q is a universal norm for M_∞/M. Since $\mathrm{Gal}(M_\infty/M)$ is torsion-free, we may assume that $q \notin (M^\times)^p$. For every $n \geq 0$, choose $q_n \in M_n^\times$ so that $N_{M_n/M}(q_n) = q$. Fix $m \geq 1$. Consider $\alpha = q \otimes (1/p^m)$. In $M_n^\times \otimes (\mathbb{Q}_p/\mathbb{Z}_p)$, we have $N_{M_n/M}(\alpha_n) = \alpha$, where $\alpha_n = q_n \otimes (1/p^m)$. Let $\widetilde{\alpha}$, $\widetilde{\alpha}_n$ denote the images of α, α_n in $M_\infty^\times \otimes (\mathbb{Q}_p/\mathbb{Z}_p)/(M_\infty^\times \otimes (\mathbb{Q}_p/\mathbb{Z}_p))_{\Lambda\text{-div}}$. The action of Γ on this group is trivial. Hence $p^n\widetilde{\alpha}_n = \widetilde{\alpha}$. But $\widetilde{\alpha}_n$ has order dividing p^m. Since n is arbitrary, we have $\widetilde{\alpha} = 0$, which of course means that the image of $q \otimes (1/p^m)$ is in $H^1(M_\infty, \mu_{p^\infty})_{\Lambda\text{-div}}$. This is true for any $m \geq 1$, as claimed. ∎

We now will prove (6). We know that $H^1(M, \mathbb{Q}_p/\mathbb{Z}_p)$ has $\mathbb{Z}_p$-corank $t+1$. Thus, $\mathrm{Im}(\pi_M)$ has $\mathbb{Z}_p$-corank t, which is also the $\mathbb{Z}_p$-corank of $\mathrm{Gal}(\widetilde{M}_{q_E}/M)$. To justify (6), it therefore suffices to prove that $\mathrm{Hom}(\mathrm{Gal}(M_\infty/M), \mathbb{Q}_p/\mathbb{Z}_p)$ is contained in $\mathrm{Im}(\pi_M)$ for all $\mathbb{Z}_p$-extensions M_∞ of M contained in $\widetilde{M}_{q_E}$. We do this by studying the following diagram

$$\begin{array}{ccccccc}
0 \longrightarrow & H^1(M, \mu_{p^\infty})/B & \longrightarrow & H^1(M, E[p^\infty]) & \xrightarrow{\pi_M} & H^1(M, \mathbb{Q}_p/\mathbb{Z}_p) \\
 & \downarrow{\scriptstyle a} & & \downarrow{\scriptstyle b} & & \downarrow{\scriptstyle c} \\
0 \longrightarrow & (H^1(M_\infty, \mu_{p^\infty})/B_\infty)^\Gamma & \longrightarrow & H^1(M_\infty, E[p^\infty])^\Gamma & \xrightarrow{e} & H^1(M_\infty, \mathbb{Q}_p/\mathbb{Z}_p)^\Gamma
\end{array}$$

where B is the image of $\langle q_E\rangle \otimes (\mathbb{Q}_p/\mathbb{Z}_p)$ in $H^1(M, \mu_{p^\infty})$, which is the kernel of the map $H^1(M, \mu_{p^\infty}) \to H^1(M, E[p^\infty])$. Thus the first row is exact. We define B_∞ as the image of B under the restriction map. The exactness of the second row follows similarly, noting that B_∞ is the image of $\langle q_E\rangle \otimes (\mathbb{Q}_p/\mathbb{Z}_p)$ in $H^1(M_\infty, \mu_{p^\infty})$. Now $\ker(c) = \mathrm{Hom}(\mathrm{Gal}(M_\infty/M), \mathbb{Q}_p/\mathbb{Z}_p)$ is isomorphic to $\mathbb{Q}_p/\mathbb{Z}_p$. We prove that $\ker(c) \subseteq \mathrm{Im}(\pi_M)$ by showing that $\mathrm{Im}(c \circ \pi_M) = \mathrm{Im}(e \circ b)$ has $\mathbb{Z}_p$-corank $t-1$. The first row shows that $H^1(M, E[p^\infty])$ has $\mathbb{Z}_p$-corank $2t$. Since b is surjective and has finite kernel, the $\mathbb{Z}_p$-corank of $H^1(M_\infty, E[p^\infty])^\Gamma$ is also $2t$. But $H^1(M_\infty, \mu_{p^\infty}) \cong \widehat{\Lambda}^t \times (\mathbb{Q}_p/\mathbb{Z}_p)$ and B_∞ is contained in the Λ-divisible submodule corresponding to $\widehat{\Lambda}^t$ by proposition

3.11. One can see from this that $H^1(M_\infty, \mu_{p^\infty})/B_\infty$ is also isomorphic to $\hat{\Lambda}^t \times (\mathbb{Q}_p/\mathbb{Z}_p)$. (This is an exercise on Λ-modules: If X is a free Λ-module of finite rank and Y is a Λ-submodule such that X/Y has no $\mathbb{Z}_p$-torsion, then Y is a free Λ-module too.) It now follows that $(H^1(M_\infty, \mu_{p^\infty})/B_\infty)^\Gamma$ has $\mathbb{Z}_p$-corank $t+1$. Therefore, $\mathrm{Im}(e)$ indeed has $\mathbb{Z}_p$-corank $t-1$.

4. Calculation of an Euler Characteristic

This section will concern the evaluation of $f_E(0)$. We will assume that E has good, ordinary reduction at all primes of F lying over p. We will also assume that $\mathrm{Sel}_E(F)_p$ is finite. By theorem 1.4, $\mathrm{Sel}_E(F_\infty)_p$ is then Λ-cotorsion. By definition, $f_E(T)$ is a generator of the characteristic ideal of the Λ-module $X_E(F_\infty) = \mathrm{Hom}(\mathrm{Sel}_E(F_\infty)_p, \mathbb{Q}_p/\mathbb{Z}_p)$. Since $\mathrm{Sel}_E(F_\infty)_p^\Gamma$ is finite by theorem 1.2, it follows that $X_E(F_\infty)/TX_E(F_\infty)$ is finite. Hence $T \nmid f_E(T)$ and so $f_E(0) \neq 0$. The following theorem is a special case of a result of B. Perrin-Riou (if E has complex multiplication) and of P. Schneider (in general). (See [Pe1] and [Sch1].) For every prime v of F lying over p, we let $\tilde{E}_v$ denote the reduction of E modulo v, which is defined over the residue field f_v. For primes v where E has bad reduction, we let $c_v = [E(F_v){:}E_0(F_v)]$ as before, where $E_0(F_v)$ denotes the subgroup of points with nonsingular reduction modulo v. The highest power of p dividing c_v is denoted by $c_v^{(p)}$. Also, if $a, b \in \mathbb{Q}_p^\times$, we write $a \sim b$ to indicate that a and b have the same p-adic valuation.

Theorem 4.1. *Assume that E is an elliptic curve defined over F with good, ordinary reduction at all primes of F lying over p. Assume also that* $\mathrm{Sel}_E(F)_p$ *is finite. Then*

$$f_E(0) \sim \Big(\prod_{v \text{ bad}} c_v^{(p)}\Big)\Big(\prod_{v|p} |\tilde{E}_v(f_v)_p|^2\Big)|\mathrm{Sel}_E(F)_p|/|E(F)_p|^2.$$

Note that under the above hypotheses, $\mathrm{Sel}_E(F)_p = Ш_E(F)_p$. Also, we have $|\tilde{E}_v(f_v)| = (1-\alpha_v)(1-\beta_v)$, where $\alpha_v\beta_v = N(v)$, $\alpha_v + \beta_v = a_v \in \mathbb{Z}$, and $p \nmid a_v$. It follows that $\alpha_v, \beta_v \in \mathbb{Q}_p$. We can assume that $\alpha_v \in \mathbb{Z}_p^\times$. Hence $p \mid |\tilde{E}_v(f_v)|$ if and only if $a_v \equiv 1 \pmod{p}$. We say in this case that v is an anomalous prime for E, a terminology introduced by Mazur who first pointed out the interest of such primes for the Iwasawa theory of E. In [Maz1], one finds an extensive discussion of them.

We will prove theorem 4.1 by a series of lemmas. We begin with a general fact about Λ-modules.

Lemma 4.2. *Assume that S is a cofinitely generated, cotorsion Λ-module. Let $f(T)$ be a generator of the characteristic ideal of $X = \mathrm{Hom}(S, \mathbb{Q}_p/\mathbb{Z}_p)$. Assume that S^Γ is finite. Then S_Γ is finite, $f(0) \neq 0$, and $f(0) \sim |S^\Gamma|/|S_\Gamma|$.*

Remark. Note that $H^i(\Gamma, S) = 0$ for $i > 1$. Hence the quantity $|S^\Gamma|/|S_\Gamma|$ is the Euler characteristic $|H^0(\Gamma, S)|/|H^1(\Gamma, S)|$. Also, the assumption that S^Γ is finite in fact implies that S is cofinitely generated and cotorsion as a Λ-module.

Proof of lemma 4.2. By assumption, we have that X/TX is finite. Now X is pseudo-isomorphic to a direct sum of Λ-modules of the form $Y = \Lambda/(g(T))$. For each such Y, we have $Y/TY = \Lambda/(T, g(T)) = \mathbb{Z}_p/(g(0))$. Thus, Y/TY is finite if and only if $g(0) \neq 0$. In this case, we have $\ker(T{:}Y \to Y) = 0$. From this, one sees that X/TX is finite if and only if $f(0) \neq 0$, and then obviously $\ker(T{:}X \to X)$ would be finite. Thus, S_Γ is finite. Since both Euler characteristics and the characteristic power series of Λ-modules behave multiplicatively in exact sequences, it is enough to verify the final statement when S is finite and when $\mathrm{Hom}(S, \mathbb{Q}_p/\mathbb{Z}_p) = \Lambda/(g(T))$. In the first case, the Euler characteristic is 1 and the characteristic ideal is Λ. The second case is clear from the above remarks about Y. ■

Referring to the diagram at the beginning of section 3, we will denote s_0, h_0, and g_0 simply by s, h, and g.

Lemma 4.3. *Under the assumptions of theorem 4.1, we have*

$$|\mathrm{Sel}_E(F_\infty)_p^\Gamma| = |\mathrm{Sel}_E(F)_p||\ker(g)|/|E(F)_p|.$$

Proof. We have $|(\mathrm{Sel}_E(F_\infty)_p^\Gamma|/|\mathrm{Sel}_E(F)_p| = |\mathrm{coker}(s)|/|\ker(s)|$, where all the groups occurring are finite. By lemma 3.2, $\mathrm{coker}(h) = 0$. Thus, we have an exact sequence: $0 \to \ker(s) \to \ker(h) \to \ker(g) \to \mathrm{coker}(s) \to 0$. It follows that $|\mathrm{coker}(s)|/|\ker(s)| = |\ker(g)|/|\ker(h)|$. Now we use the fact that $E(F_\infty)_p$ is finite. Then

$$\ker(h) = H^1(\Gamma, E(F_\infty)_p) = (E(F_\infty)_p)_\Gamma$$

has the same order as $H^0(\Gamma, E(F_\infty)_p) = E(F)_p$. These facts give the formula in lemma 4.3. ■

The proof of theorem 4.1 clearly rests now on studying $|\ker(g)|$. The results of section 3 allow us to study $\ker(r)$, factor by factor, where r is the natural map

$$r : \mathcal{P}_E(F) \to \mathcal{P}_E(F_\infty).$$

It will be necessary for us to replace $\mathcal{P}_E(*)$ by a much smaller group. Let Σ denote the set of primes of F where E has bad reduction or which divide p or ∞. By lemma 3.3, we have $\ker(r_v) = 0$ if $v \notin \Sigma$. Let $\mathcal{P}_E^\Sigma(F) = \prod_v \mathcal{H}_E(F_v)$, where the product is over all primes of F in Σ. We consider $\mathcal{P}_E^\Sigma(F)$ as a subgroup of $\mathcal{P}_E(F)$. Clearly, $\ker(r) \subseteq \mathcal{P}_E^\Sigma(F)$. Thus $|\ker(r)| = \prod_v |\ker(r_v)|$, where v again varies over all primes in Σ. For $v|p$, the order of $\ker(r_v)$ is given in lemma 3.4. For $v \nmid p$, the remarks after the proof of lemma 3.3 show that $|\ker(r_v)| \sim c_v^{(p)}$. We then obtain the following result.

Lemma 4.4. *Assume that E/F has good, ordinary reduction at all $v|p$. Then* $|\ker(r)| \sim (\prod_{v \text{ bad}} c_v^{(p)})(\prod_{v|p} |\tilde{E}_v(f_v)_p|^2)$.

Now let $\mathcal{G}_E^{\Sigma}(F) = \mathrm{Im}\left(H^1(F_\Sigma/F, E[p^\infty]) \to \mathcal{P}_E^{\Sigma}(F)\right)$, where F_Σ denotes the maximal extension of F unramified outside of Σ. Then

$$\ker(g) = \ker(r) \cap \mathcal{G}_E^{\Sigma}(F).$$

We now recall a theorem of Cassels which states that $\mathcal{P}_E^{\Sigma}(F)/\mathcal{G}_E^{\Sigma}(F) \cong E(F)_p$. (We will sketch a proof of this later, using the Duality Theorem of Poitou and Tate.) It is interesting to consider theorem 4.1 in the case where $E(F)_p = 0$, which is of course true for all but finitely many primes p. Then, by Cassels' theorem, $\ker(g) = \ker(r)$. Lemmas 4.3, 4.4 then show that the right side of $\sim$ in theorem 4.1 is precisely $|\mathrm{Sel}_E(F_\infty)_p^{\Gamma}|$. Therefore, in this special case, by lemma 4.2, theorem 4.1 is equivalent to asserting that $(\mathrm{Sel}_E(F_\infty)_p)_\Gamma = 0$. It is an easy exercise to see that this in turn is equivalent to asserting that the Λ-module $X_E(F_\infty)$ has no finite, nonzero Λ-submodules. In section 5 we will give an example where $X_E(F_\infty)$ does have a finite, nonzero Λ-submodule. All the hypotheses of this section will hold, but of course $E(F)$ will have an element of order p.

The following general fact will be useful in the rest of the proof of theorem 4.1. We will assume that G is a profinite group and that A is a discrete, p-primary abelian group on which G acts continuously.

Lemma 4.5. *Assume that G has p-cohomological dimension $n \geq 1$ and that A is a divisible group. Then $H^n(G, A)$ is a divisible group.*

Proof. Consider the exact sequence $0 \to A[p] \to A \xrightarrow{p} A \to 0$, where the map $A \xrightarrow{p} A$ is of course multiplication by p. This induces an exact sequence

$$H^n(G, A) \xrightarrow{p} H^n(G, A) \to H^{n+1}(G, A).$$

Since the last group is zero, $H^n(G, A)$ is divisible by p. The lemma follows because $H^n(G, A)$ is a p-primary group. ■

We have actually already applied this lemma once, namely in the proof of proposition 2.4. We will apply it to some other cases. A good reference for the facts we use is [Se2]. Let v be a nonarchimedean prime of F, η a prime of F_∞ lying above v. Then $\mathrm{Gal}((F_\infty)_\eta/F_v) \cong \mathbb{Z}_p$, as mentioned earlier. Thus, $G_{(F_\infty)_\eta}$ has p-cohomological dimension 1. Hence $H^1((F_\infty)_\eta, E[p^\infty])$ must be divisible, and consequently the same is true for $\mathcal{H}_E((F_\infty)_\eta)$. As another example, $\mathrm{Gal}(F_\Sigma/F)$ has p-cohomological dimension 2 if p is any odd prime. Let $A_s = E[p^\infty] \otimes (\kappa^s)$, where $\kappa : \Gamma \to 1 + 2p\mathbb{Z}_p$ is an isomorphism and $s \in \mathbb{Z}$. (A_s is something like a Tate twist of the G_F-module $E[p^\infty]$. One could even take $s \in \mathbb{Z}_p$.) It then follows that $H^2(F_\Sigma/F, A_s)$ is a divisible group.

Lemma 4.6. *Assume that* $\mathrm{Sel}_E(F_\infty)_p$ *is* Λ*-cotorsion. Then the map*

$$H^1(F_\Sigma/F_\infty, E[p^\infty]) \to \mathcal{P}_E^\Sigma(F_\infty)$$

is surjective.

Remark. We must define $\mathcal{P}_E^\Sigma(F_\infty)$ carefully. For any prime v in Σ, we define

$$\mathcal{P}_E^{(v)}(F_\infty) = \varinjlim_n \mathcal{P}_E^{(v)}(F_n)$$

where $\mathcal{P}_E^{(v)}(F_n) = \prod_{v_n|v} \mathcal{H}_E((F_n)_{v_n})$ and $\mathcal{H}_E(*)$ is as defined at the beginning of section 3. The maps $\mathcal{P}_E^{(v)}(F_n) \to \mathcal{P}_E^{(v)}(F_{n+1})$ are easily defined, considering separately the case where v_n is inert or ramified in F_{n+1}/F_n (where one uses a restriction map) or where v_n splits completely in F_{n+1}/F_n (where one uses a "diagonal" map). If v is nonarchimedean, then v is finitely decomposed in F_∞/F and one can more simply define $\mathcal{P}_E^{(v)}(F_\infty) = \prod_{\eta|v} \mathcal{H}_E((F_\infty)_\eta)$, where η runs over the finite set of primes of F_∞ lying over v. If v is archimedean, then v splits completely in F_∞/F. We know that $\mathrm{Im}(\kappa_{v_n}) = 0$ for $v_n|v$. Thus, $\mathcal{H}_E((F_n)_{v_n}) = \mathcal{H}_E(F_v) = H^1(F_v, E[p^\infty])$. Usually, this group is zero. But it can be nonzero if $p = 2$ and $F_v = \mathbb{R}$. In fact,

$$H^1(F_v, E[2^\infty]) \cong E(F_v)/E(F_v)_{\mathrm{con}},$$

where $E(F_v)_{\mathrm{con}}$ denotes the connected component of the identity of $E(F_v)$. Therefore, obviously $H^1(F_v, E[2^\infty])$ has order 1 or 2. The order is 2 if $E[2]$ is contained in $E(F_v)$. We have

$$\mathcal{P}_E^{(v)}(F_n) \cong H^1(F_v, E[2^\infty]) \otimes \mathbb{Z}_2[\mathrm{Gal}(F_n/F)],$$

which is either zero or isomorphic to $(\mathbb{Z}/2\mathbb{Z})[\mathrm{Gal}(F_n/F)]$. In each of the above cases, $\mathcal{P}_E^{(v)}(F_\infty)$ can be regarded naturally as a Λ-module. If v is nonarchimedean then the remarks following lemma 4.5 show that, as a group, $\mathcal{P}_E^{(v)}(F_\infty)$ is divisible. If v is archimedean, then usually $\mathcal{P}_E^{(v)}(F_\infty) = 0$. But, if $p = 2$, $F_v = \mathbb{R}$, and $E[2]$ is contained in $E(F_v)$, then one sees that $\mathcal{P}_E^{(v)}(F_\infty) \cong \mathrm{Hom}(\Lambda/2\Lambda, \mathbb{Z}/2\mathbb{Z})$ as a Λ-module. (One uses the fact that $\mathcal{P}_E^{(v)}(F_\infty)^{\Gamma_n} \cong \mathcal{P}_E^{(v)}(F_n)$ for all $n \geq 0$ and the structure of $\mathcal{P}_E^{(v)}(F_n)$ mentioned above.) Finally, we define $\mathcal{P}_E^\Sigma(F_\infty) = \prod_{v\in\Sigma} \mathcal{P}_E^{(v)}(F_\infty)$.

Proof of Lemma 4.6. We can regard $\mathcal{P}_E^\Sigma(F_\infty)$ as a Λ-module. The idea of the proof is to show that the image of the above map is a Λ-submodule of $\mathcal{P}_E^\Sigma(F_\infty)$ with finite index and that any such Λ-submodule must be $\mathcal{P}_E^\Sigma(F_\infty)$. We will explain the last point first. If p is odd, the remarks above show that each factor in $\mathcal{P}_E^\Sigma(F_\infty)$ is divisible. Hence $\mathcal{P}_E^\Sigma(F_\infty)$ is divisible and therefore has no proper subgroups of finite index. If $p = 2$, one has to observe that the factor $\mathcal{P}_E^{(v)}(F_\infty)$ of $\mathcal{P}_E^\Sigma(F_\infty)$ coming from an archimedean prime v of F is a Λ-module whose Pontryagin dual is either zero or isomorphic to $(\Lambda/2\Lambda)$. Since $\Lambda/2\Lambda$ has no nonzero, finite Λ-submodules, we see that $\mathcal{P}_E^{(v)}(F_\infty)$ has no proper Λ-submodules of finite index. Since the factors $\mathcal{P}_E^{(v)}(F_\infty)$ for nonarchimedean v are still divisible, it follows again that $\mathcal{P}_E^\Sigma(F_\infty)$ has no proper Λ-submodules of finite index.

Now we will prove that the image of the map in the lemma has finite index. (It is clearly a Λ-submodule.) To give the idea of the proof, assume first that $\mathrm{Sel}_E(F_n)_p$ is finite for all $n \geq 0$. Then the cokernel of the map $H^1(F_\Sigma/F_n, E[p^\infty]) \to \mathcal{P}_E^\Sigma(F_n)$ is isomorphic to $E(F_n)_p$ by a theorem of Cassels. But $|E(F_n)_p|$ is bounded since it is known that $E(F_\infty)_p$ is finite. It clearly follows that the cokernel of the corresponding map over F_∞ is also finite. To give the proof in general, we use a trick of twisting the Galois module $E[p^\infty]$. We let A_s be defined as above, where $s \in \mathbb{Z}$. As G_{F_∞}-modules, $A_s = E[p^\infty]$. Thus, $H^1(F_\infty, A_s) = H^1(F_\infty, E[p^\infty])$. But the action of Γ changes in a simple way, namely $H^1(F_\infty, A_s) = H^1(F_\infty, E[p^\infty]) \otimes (\kappa^s)$. Now we can define Selmer groups for A_s as suggested at the end of section 2. One just imitates the description of the p-Selmer group for E. For the local condition at v dividing p, one uses $C_v \otimes (\kappa^s)$. For v not dividing p, we require 1-cocycles to be locally trivial. We let $S_{A_s}(F_n)$, $S_{A_s}(F_\infty)$ denote the Selmer groups defined in this way. Then $S_{A_s}(F_\infty) = \mathrm{Sel}_E(F_\infty)_p \otimes (\kappa^s)$ as Λ-modules. Now we are assuming that $\mathrm{Sel}_E(F_\infty)_p$ is Λ-cotorsion. It is not hard to show from this that for all but finitely many values of s, $S_{A_s}(F_\infty)^{\Gamma_n}$ will be finite for all $n \geq 0$. Since there is a map $S_{A_s}(F_n) \to S_{A_s}(F_\infty)^{\Gamma_n}$ with finite kernel, it follows that $S_{A_s}(F_n)$ is finite for all $n \geq 0$. There is also a variant of Cassels' theorem for A_s: the cokernel of the global-to-local map for the G_{F_n}-module A_s is isomorphic to $H^0(F_n, A_{-s})$. But this last group is finite and has order bounded by $|E(F_\infty)_p|$. The surjectivity of the global-to-local map for A_s over F_∞ follows just as before. Lemma 4.6 follows since $A_s \cong E[p^\infty]$ as G_{F_∞}-modules. (Note: the variant of Cassels' theorem is a consequence of proposition 4.13. It may be necessary to exclude one more value of s.) ∎

The following lemma, together with lemmas 4.2–4.4, implies theorem 4.1.

Lemma 4.7. *Under the assumptions of theorem 4.1, we have*

$$|\ker(g)| = |\ker(r)|\,|(\mathrm{Sel}_E(F_\infty)_p)_\Gamma|/|E(F)_p|.$$

Proof. By lemma 4.6, the following sequence is exact:

$$0 \to \mathrm{Sel}_E(F_\infty)_p \to H^1(F_\Sigma/F_\infty, E[p^\infty]) \to \mathcal{P}_E^\Sigma(F_\infty) \to 0.$$

Now Γ acts on these groups. We can take the corresponding cohomology sequence obtaining

$$H^1(F_\Sigma/F_\infty, E[p^\infty])^\Gamma \to \mathcal{P}_E^\Sigma(F_\infty)^\Gamma \to (\mathrm{Sel}_E(F_\infty)_p)_\Gamma \to H^1(F_\Sigma/F_\infty, E[p^\infty])_\Gamma.$$

In the appendix, we will give a proof that the last term is zero. Thus we get the following commutative diagram with exact rows and columns.

$$\begin{array}{ccccccc}
H^1(F_\Sigma/F, E[p^\infty]) & \xrightarrow{a} & \mathcal{P}_E^\Sigma(F) & \longrightarrow & \mathcal{P}_E^\Sigma(F)/\mathcal{G}_E^\Sigma(F) & \to & 0 \\
\downarrow & & \downarrow & & \downarrow & & \\
H^1(F_\Sigma/F_\infty, E[p^\infty])^\Gamma & \xrightarrow{b} & \mathcal{P}_E^\Sigma(F_\infty)^\Gamma & \longrightarrow & (\mathrm{Sel}_E(F_\infty)_p)_\Gamma & \to & 0 \\
\downarrow & & \downarrow & & \downarrow & & \\
0 & & 0 & & 0 & &
\end{array}$$

The exactness of the first row is clear. The remark above gives the exactness of the second row. The surjectivity of the first vertical arrow is because Γ has p-cohomological dimension 1. The surjectivity of the second vertical arrow can be verified similarly. One must consider each $v \in \Sigma$ separately, showing that $\mathcal{P}_E^{(v)}(F) \to \mathcal{P}_E^{(v)}(F_\infty)^\Gamma$ is surjective. One must take into account the fact that v can split completely in F_n/F for some n. But then it is easy to see that $\mathcal{P}_E^{(v)}(F) \xrightarrow{\sim} \mathcal{P}_E^{(v)}(F_n)^{\mathrm{Gal}(F_n/F)}$. One then uses the fact that $\mathrm{Gal}((F_\infty)_\eta/(F_n)_{v_n})$ has p-cohomological dimension 1, looking at the maps r_{v_n} for $v \nmid p$ or d_{v_n} for $v|p$. For archimedean v, one easily verifies that $\mathcal{P}_E^{(v)}(F) \xrightarrow{\sim} \mathcal{P}_E^{(v)}(F_\infty)^\Gamma$. The surjectivity of the third vertical arrow follows. It is also clear that $\mathrm{Im}(a)$ is mapped surjectively to $\mathrm{Im}(b)$. We then obtain the following commutative diagram

$$\begin{array}{ccccccccc}
0 & \to & \mathcal{G}_E^\Sigma(F) & \longrightarrow & \mathcal{P}_E^\Sigma(F) & \longrightarrow & \mathcal{P}_E^\Sigma(F)/\mathcal{G}_E^\Sigma(F) & \to & 0 \\
& & \downarrow g & & \downarrow r & & \downarrow t & & \\
0 & \to & \mathrm{Im}(b) & \longrightarrow & \mathcal{P}_E^\Sigma(F_\infty)^\Gamma & \longrightarrow & (\mathrm{Sel}_E(F_\infty)_p)_\Gamma & \to & 0 \\
& & \downarrow & & \downarrow & & \downarrow & & \\
& & 0 & & 0 & & 0 & &
\end{array}$$

From the snake lemma, we then obtain $0 \to \ker(g) \to \ker(r) \to \ker(t) \to 0$. Thus, $|\ker(g)| = |\ker(r)|/|\ker(t)|$. Combining this with Cassels' theorem and the obvious value of $|\ker(t)|$ proves lemma 4.7. ■

The last commutative diagram, together with Cassels' theorem, gives the following consequence which will be quite useful in the discussion of various examples in section 5. A more general result will be proved in the appendix.

Proposition 4.8. *Assume that E is an elliptic curve defined over F with good, ordinary reduction at all primes of F lying over p. Assume that $\mathrm{Sel}_E(F)_p$ is finite and that $E(F)_p = 0$. Then $\mathrm{Sel}_E(F_\infty)_p$ has no proper Λ-submodules of finite index. In particular, if $\mathrm{Sel}_E(F_\infty)_p$ is nonzero, then it must be infinite.*

Proof. We have the map $t: E(F)_p \to \mathrm{Sel}_E(F_\infty)_\Gamma$, which is surjective. Since $E(F)_p = 0$, it follows that $(\mathrm{Sel}_E(F_\infty)_p)_\Gamma = 0$ too. Suppose that $\mathrm{Sel}_E(F_\infty)_p$ has a finite, nonzero Λ-module quotient M. Then M is just a nonzero, finite, abelian p-group on which Γ acts. Obviously, $M_\Gamma \neq 0$. But M_Γ is a homomorphic image of $(\mathrm{Sel}_E(F_\infty)_p)_\Gamma$, which is impossible. ■

Theorem 4.1 gives a conjectural relationship of $f_E(0)$ to the value of the Hasse-Weil L-function $L(E/F,s)$ at $s=1$. This is based on the Birch and Swinnerton-Dyer conjecture for E over F, for the case where $E(F)$ is assumed to be finite. We assume of course that $\text{Ш}_E(F)_p$ is finite and hence so is $\text{Sel}_E(F)_p = \text{Ш}_E(F)_p$. We also assume that $L(E/F,s)$ has an analytic continuation to $s=1$. The conjecture then asserts that $L(E/F,1) \neq 0$ and that for a suitably defined period $\Omega(E/F)$, the value $L(E/F,1)/\Omega(E/F)$ is rational and

$$L(E/F,1)/\Omega(E/F) \sim (\prod_{v \text{bad}} c_v^{(p)})\, |\text{Sel}_E(F)_p| \,/\, |E(F)_p|^2 .$$

As before, $\sim$ means that the two sides have the same p-adic valuation. If $\mathcal{O}$ denotes the ring of integers in F, then one must choose a minimal Weierstrass equation for E over $\mathcal{O}_{(p)}$, the localization of $\mathcal{O}$ at p, to define $\Omega(E/F)$ (as a product of periods over the archimedean primes of F). For $v|p$, the Euler factor for v in $L(E/F,s)$ is

$$(1-\alpha_v N(v)^{-s})(1-\beta_v N(v)^{-s}),$$

where α_v, β_v are as defined just after theorem 4.1. Recall that $\alpha_v \in \mathbb{Z}_p^\times$. (We are assuming that E has good, ordinary reduction at all $v|p$.) Then we have

$$|\widetilde{E}(f_v)_p| \sim (1-\alpha_v) \sim (1-\alpha_v^{-1}) = (1-\beta_v N(v)^{-1}).$$

The last quantity is one factor in the Euler factor for v, evaluated at $s=1$. Thus, theorem 4.1 conjecturally states that

$$f_E(0) \sim (\prod_{v|p}(1-\beta_v N(v)^{-1})^2) L(E/F,1)/\Omega(E/F).$$

For $F=\mathbb{Q}$, one should compare this with conjecture 1.13.

As we mentioned in the introduction, there is a result of P. Schneider (generalizing a result of B. Perrin-Riou for elliptic curves with complex multiplication) which concerns the behavior of $f_E(T)$ at $T=0$. We assume that E is an elliptic curve/F with good, ordinary reduction at all primes of F lying over p, that p is odd and that $F \cap \mathbb{Q}_\infty = \mathbb{Q}$ (to slightly simplify the statement). Let $r = \text{rank}(E(F))$. We will state the result for the case where $r=1$ and $\text{Ш}_E(F)_p$ is finite. (Then $\text{Sel}_E(F)_p$ has $\mathbb{Z}_p$-corank 1.) Since then $T|f_E(T)$, one can write $f_E(T) = Tg_E(T)$, where $g_E(T) \in \Lambda$. The result is that

$$g_E(0) \sim \frac{h_p(P)}{p} (\prod_{v \text{ bad}} c_v^{(p)}) (\prod_{v|p} |\widetilde{E}_v(f_v)_p|^2) |\text{Ш}_E(F)_p| / |E(F)_p^2|.$$

Here $P \in E(F)$ is a generator of $E(F)/E(F)_{\text{tors}}$ and $h_p(P)$ is its analytic p-adic height. (See [Sch2] and the references there for the definition of $h_p(P)$.) The other factors are as in theorem 4.1. Conjecturally, one should have $h_p(P) \neq 0$. This would mean that $f_E(T)$ has a simple zero at $T = 0$. But if $h_p(P) = 0$, the result means that $g_E(0) = 0$, i.e., $T^2 | f_E(T)$. If $F = \mathbb{Q}$ and E is modular, then B. Perrin-Riou [Pe3] has proven an analogue of a theorem of Gross and Zagier for the p-adic L-function $L_p(E/\mathbb{Q}, s)$. Assume that $L(E/\mathbb{Q}, s)$ has a simple zero at $s = 1$. Then a result of Kolyvagin shows that $\text{rank}(E(\mathbb{Q})) = 1$ and $Ш_E(\mathbb{Q})$ is finite. Assume that P generates $E(\mathbb{Q})/E(\mathbb{Q})_{\text{tors}}$. Assume that $h_p(P) \neq 0$. Perrin-Riou's result asserts that $L_p(E/\mathbb{Q}, s)$ also has a simple zero at $s = 1$ and that

$$L'_p(E/\mathbb{Q}, 1)/h_p(P) = (1 - \beta_p p^{-1})^2 L'(E/\mathbb{Q}, 1)/\Omega_E h_\infty(P)$$

where $h_\infty(P)$ is the canonical height of P. If one assumes the validity of the Birch and Swinnerton-Dyer conjecture, then this result and Schneider's result are compatible with conjecture 1.13.

The proof of theorem 4.1 can easily be adapted to the case where E has multiplicative reduction at some primes of F lying over p. One then obtains a special case of a theorem of J. Jones [Jo]. Jones determines the p-adic valuation of $(f_E(T)/T^r)|_{T=0}$, where $r = \text{rank}(E(F))$, generalizing the results of P. Schneider. He studies certain natural Λ-modules which can be larger, in some sense, than $\text{Sel}_E(F_\infty)_p$. Their characteristic ideal will contain $T^e f_E(T)$, where e is the number of primes of F where E has split, multiplicative reduction. This is an example of the phenomenon of "trivial zeros". Another example of this phenomenon is the Λ-module S_∞ in the case where p splits in an imaginary quadratic field F. As we explained in the introduction, S_∞^Γ is infinite. That is, a generator of its characteristic ideal will vanish at $T = 0$. For a general discussion of this phenomenon, we refer the reader to [Gr4].

To state the analogue of theorem 4.1, we assume that $\text{Sel}_E(F)_p$ is finite, that E has either good, ordinary or multiplicative reduction at all primes of F over p, and that $\log_p(N_{F_v/\mathbb{Q}_p}(q_E^{(v)})) \neq 0$ for all v lying over p where E has split, multiplicative reduction. (As in section 3, $q_E^{(v)}$ denotes the Tate period for E over F_v.) Under these assumptions, $\ker(r_v)$ will be finite for all $v|p$. It follows from proposition 3.7 that $\text{Sel}_E(F_\infty)_p^\Gamma$ will be finite and hence $\text{Sel}_E(F_\infty)_p$ will be Λ-cotorsion. In theorem 4.1, the only necessary change is to replace the factor $|\widetilde{E}_v(f_v)_p|^2$ for those $v|p$ where E has multiplicative reduction by the factor $|\ker(r_v)|/c_v^{(p)}$. (Note that the factor $c_v^{(p)}$ for such v will occur in $\prod_{v \text{ bad}} c_v^{(p)}$.) The analogue of theorem 4.1 can be expressed as

$$f_E(0) \sim \Big(\prod_{v|p} l_v\Big)\Big(\prod_{v \text{ bad}} c_v^{(p)}\Big)\, |\text{Sel}_E(F)_p| \,/|E(F)_p|^2.$$

If E has good, ordinary reduction at v, then $l_v = |\widetilde{E}_v(f_v)_p|^2$. Assume that E has nonsplit, multiplicative reduction at v. If p is odd, then both $|\ker(r_v)|$ and $c_v^{(p)}$ are equal to 1. If $p = 2$, then $|\ker(r_v)| = 2c_v^{(p)}$. (Recalling the discussion concerning $\ker(r_v)$ after the proof of proposition 3.6, the 2 corresponds to $|\ker(b_v)|$, and the $c_v^{(p)}$ corresponds to $|\ker(a_v)| = [\mathrm{Im}(\lambda_v) : \mathrm{Im}(\kappa_v)]$. In the case of good, ordinary reduction at v, both $\ker(a_v)$ and $\ker(b_v)$ have order $|\widetilde{E}_v(f_v)_p|$.) Thus, if E has nonsplit, multiplicative reduction at v, one can take $l_v = 2$ (for any prime p). We remark that the Euler factor for v in $L(E/F, s)$ is $(1 + N(v)^{-s})^{-1}$. One should take $\alpha_v = -1$, $\beta_v = 0$. Perhaps this factor $l_v = 2$ should be thought of as $(1 - \alpha_v^{-1})$. (This is suggested by the fact that, for a modular elliptic curve E defined over $F = \mathbb{Q}$, the p-adic L-function constructed in [M-T-T] has a factor $(1 - \alpha_p^{-1})$ in its interpolation property when E has multiplicative reduction at p. This is in place of $(1 - \alpha_p^{-1})^2 = (1 - \beta_p p^{-1})^2$ when E has good, ordinary reduction at p.)

Finally, assume that E has split, multiplicative reduction at v. (Then $(1 - \alpha_v^{-1})$ would be zero.) We have $c_v^{(p)} = \mathrm{ord}_v(q_E^{(v)})$. If we let $\mathbb{Q}_p^{\mathrm{unr}}$ denote the unramified $\mathbb{Z}_p$-extension of $\mathbb{Q}_p$ and $\mathbb{Q}_p^{\mathrm{cyc}}$ denote the cyclotomic $\mathbb{Z}_p$-extension of $\mathbb{Q}_p$, then we should take

$$l_v = \frac{\log_p(N_{F_v/\mathbb{Q}_p}(q_E^{(v)}))}{\mathrm{ord}_p(N_{F_v/\mathbb{Q}_p}(q_E^{(v)})} \cdot \frac{[F_v \cap \mathbb{Q}_p^{\mathrm{unr}} : \mathbb{Q}_p]}{2p[F_v \cap \mathbb{Q}_p^{\mathrm{cyc}} : \mathbb{Q}_p]}.$$

(Again, we refer to the discussion of $\ker(r_v)$ following proposition 3.6. This time, $\ker(a_v) = 0$ and $\ker(r_v) \cong \ker(b_v)$.) We will give another way to define l_v, at least up to a p-adic unit, which comes directly from the earlier discussion of $\ker(r_v)$. Let F_v^{cyc} and F_v^{unr} denote the cyclotomic and the unramified $\mathbb{Z}_p$-extensions of F_v. Fix isomorphisms

$$\theta_{F_v}^{\mathrm{cyc}} : \mathrm{Gal}(F_v^{\mathrm{cyc}}/F_v) \xrightarrow{\sim} \mathbb{Z}_p, \qquad \theta_{F_v}^{\mathrm{unr}} : \mathrm{Gal}(F_v^{\mathrm{unr}}/F_v) \xrightarrow{\sim} \mathbb{Z}_p.$$

Then $l_v \sim \theta_{F_v}^{\mathrm{cyc}}(\mathrm{rec}_{F_v}(q_E^{(v)})|_{F_v^{\mathrm{cyc}}}) / \theta_{F_v}^{\mathrm{unr}}(\mathrm{rec}_{F_v}(q^{(v)})|_{F_v^{\mathrm{unr}}})$. The value of l_v given above comes from choosing specific isomorphisms.

Appendix to Section 4. We will give a proof of the following important result, which will allow us to justify the assertion used in the proof of lemma 4.7 that, under the hypotheses of theorem 4.1, $H^1(F_\Sigma/F_\infty, E[p^\infty])_\Gamma = 0$. Later, we will prove a rather general form of Cassels' theorem as well as a generalization of proposition 4.8.

Proposition 4.9. *Assume that* $\mathrm{Sel}_E(F_\infty)_p$ *is* Λ*-cotorsion. Then the* Λ*-module* $H^1(F_\Sigma/F_\infty, E[p^\infty])$ *has no proper* Λ*-submodules of finite index.*

In the course of the proof, we will show that $H^1(F_\Sigma/F_\infty, E[p^\infty])$ has Λ-corank $[F : \mathbb{Q}]$ and also that $H^2(F_\Sigma/F_\infty, E[p^\infty])$ is Λ-cotorsion. For odd p, these results are contained in [Gr2]. (See section 7 there.) For $p = 2$, one can modify the arguments given in that article. However, we will present a rather different approach here which has the advantage of avoiding the use of a spectral sequence. In either approach, the crucial point is that the group

$$R^2(F_\Sigma/F_\infty, E[p^\infty]) = \ker\big(H^2(F_\Sigma/F_\infty, E[p^\infty]) \to \prod_{\eta|\infty} H^2((F_\infty)_\eta, E[p^\infty])\big)$$

is zero, under the assumption that $\mathrm{Sel}_E(F_\infty)_p$ is Λ-cotorsion. (Note: It probably seems more natural to take the product over all η lying over primes in Σ. However, if η is nonarchimedean, then $G_{(F_\infty)_\eta}$ has p-cohomological dimension 1 and hence $H^2((F_\infty)_\eta, E[p^\infty]) = 0$.)

First of all, we will determine the Λ-corank of $\mathcal{P}_E^\Sigma(F_\infty)$. Now $\mathcal{P}_E^{(v)}(F_\infty)$ is Λ-cotorsion if $v \nmid p$. This is clear if v is archimedean because $\mathcal{P}_E^{(v)}(F_\infty)$ then has exponent 2. (It is zero if p is odd.) If v is nonarchimedean, then $\mathcal{P}_E^{(v)}(F) = H^1(F_v, E[p^\infty])$ is finite. The map $\mathcal{P}_E^{(v)}(F_v) \to \mathcal{P}_E^{(v)}(F_\infty)^\Gamma$ is surjective. Hence $\mathcal{P}_E^{(v)}(F_\infty)^\Gamma$ is finite, which suffices to prove that $\mathcal{P}_E^{(v)}(F_\infty)$ is Λ-cotorsion, using Fact (2) about Λ-modules mentioned in section 3. Alternatively, one can refer to proposition 2 of [Gr2], which gives a more precise result concerning the structure of $\mathcal{P}_E^{(v)}(F_\infty)$. Assume $v|p$. Let $\Gamma_v \subseteq \Gamma$ be the decomposition group for any prime η of F_∞ lying over v. Then by proposition 1 of [Gr2], $H^1((F_\infty)_\eta, E[p^\infty])$ has corank equal to $2[F_v{:}\mathbb{Q}_p]$ over the ring $\mathbb{Z}_p[[\Gamma_v]]$. Also, $H^1((F_\infty)_\eta, C_v)$ has corank $[F_v{:}\mathbb{Q}_p]$. Both of these facts could be easily proved using lemma 2.3, applied to the layers in the $\mathbb{Z}_p$-extension $(F_\infty)_\eta/F_v$. Consequently, $\mathcal{H}_E((F_\infty)_\eta)$ has $\mathbb{Z}_p[[\Gamma_v]]$-corank equal to $[F_v{:}\mathbb{Q}_p]$. It follows that $\mathcal{P}_E^{(v)}(F_\infty)$ has Λ-corank equal to $[F_v:\mathbb{Q}_p]$. Combining these results, we find that

$$\mathrm{corank}_\Lambda(\mathcal{P}_E^\Sigma(F_\infty)) = [F:\mathbb{Q}],$$

using the fact that $\sum_{v|p}[F_v:\mathbb{Q}_p] = [F:\mathbb{Q}]$.

Secondly, we consider the coranks of the Λ-modules $H^1(F_\Sigma/F_\infty, E[p^\infty])$ and $H^2(F_\Sigma/F_\infty, E[p^\infty])$. These are related by the equation

$$\mathrm{corank}_\Lambda(H^1(F_\Sigma/F_\infty, E[p^\infty]) = \mathrm{corank}_\Lambda(H^2(F_\Sigma/F_\infty, E[p^\infty]) + \delta,$$

where $\delta = \sum_{v|\infty}[F_v:\mathbb{R}] = [F:\mathbb{Q}]$. As a consequence, we have the inequalities

$$\mathrm{corank}_\Lambda(H^1(F_\Sigma/F_\infty, E[p^\infty])) \geq [F:\mathbb{Q}].$$

(For more discussion of this relationship, see [Gr2], section 4. It is essentially the fact that $-\delta$ is the Euler characteristic for the $\mathrm{Gal}(F_\Sigma/F_\infty)$-module $E[p^\infty]$ together with the fact that $H^0(F_\Sigma/F_\infty, E[p^\infty])$ is clearly Λ-cotorsion. This Euler characteristic of Λ-coranks is in turn derived from the fact that

$$\sum_{i=0}^{2}(-1)^i \mathrm{corank}_{\mathbb{Z}_p}(H^i(F_\Sigma/F_n, E[p^\infty])) = -\delta p^n$$

for all $n \geq 0$. That is, $-\delta p^n$ is the Euler characteristic for the $\mathrm{Gal}(F_\Sigma/F_n)$-module $E[p^\infty]$.) Recalling the exact sequence

$$0 \to \mathrm{Sel}_E(F_\infty)_p \to H^1(F_\Sigma/F_\infty, E[p^\infty]) \to \mathcal{G}_E^\Sigma(F_\infty) \to 0,$$

we see that $\mathrm{Sel}_E(F_\infty)_p$ is Λ-cotorsion if and only if $H^1(F_\Sigma/F_\infty, E[p^\infty])$ and $\mathcal{G}_E^\Sigma(F_\infty)$ have the same Λ-corank, both equal to $[F:\mathbb{Q}]$. (The last equality is because $[F:\mathbb{Q}]$ is a lower bound for the Λ-corank of $H^1(F_\Sigma/F_\infty, E[p^\infty])$ and an upper bound for the Λ-corank of $\mathcal{G}_E^\Sigma(F_\infty)$ (which is a Λ-submodule of $\mathcal{P}_E^\Sigma(F_\infty)$). Thus, if we assume that $\mathrm{Sel}_E(F_\infty)_p$ is Λ-cotorsion, then it follows that $H^1(F_\Sigma/F_\infty, E[p^\infty])$ has Λ-corank $[F:\mathbb{Q}]$ and that $H^2(F_\Sigma/F_\infty, E[p^\infty])$ has Λ-corank 0 (and hence is Λ-cotorsion). By lemma 4.6, we already would know that $\mathcal{G}_E^\Sigma(F_\infty)$ has Λ-corank $[F:\mathbb{Q}]$.

We will use a version of Shapiro's Lemma. Let $\mathcal{A} = \mathrm{Hom}(\Lambda, E[p^\infty])$. We consider $\mathcal{A}$ as a Λ-module as follows: if $\phi \in \mathcal{A}$ and $\theta \in \Lambda$, then $\theta\phi$ is defined by $(\theta\phi)(\lambda) = \phi(\theta\lambda)$ for all $\lambda \in \Lambda$. The Pontryagin dual of $\mathcal{A}$ is Λ^2 and so $\mathcal{A}$ has Λ-corank 2. We define a Λ-linear action of $\mathrm{Gal}(F_\Sigma/F)$ on $\mathcal{A}$ as follows: if $\phi \in \mathcal{A}$ and $g \in \mathrm{Gal}(F_\Sigma/F)$, then $g(\phi)$ is defined by $g(\phi)(\lambda) = g(\phi(\widetilde{\kappa}(g)^{-1}\lambda))$ for all $\lambda \in \Lambda$. Here $\widetilde{\kappa}$ is defined as the composite

$$\mathrm{Gal}(F_\Sigma/F) \to \Gamma \to \Lambda^\times$$

where the second map is just the natural inclusion of Γ in its completed group ring Λ. The above definition is just the usual way to define the action of a group on $\mathrm{Hom}(*,*)$, where we let $\mathrm{Gal}(F_\Sigma/F)$ act on Λ by $\widetilde{\kappa}$ and on $E[p^\infty]$ as usual. The Λ-linearity is easily verified, using the fact that Λ is a commutative ring. For any $\theta \in \Lambda$, we will let $\mathcal{A}[\theta]$ denote the kernel of the map $\mathcal{A} \xrightarrow{\theta} \mathcal{A}$, which is just multiplication by θ. Then clearly

$$\mathcal{A}[\theta] \cong \mathrm{Hom}(\Lambda/\Lambda\theta, E[p^\infty]).$$

Let $\kappa : \Gamma \to 1 + 2p\mathbb{Z}_p$ be a fixed isomorphism. If $s \in \mathbb{Z}$ (or in $\mathbb{Z}_p$), then the homomorphism $\kappa^s : \Gamma \to 1 + 2p\mathbb{Z}_p$ induces a homomorphism $\sigma_s : \Lambda \to \mathbb{Z}_p$ of $\mathbb{Z}_p$-algebras. If we write $\Lambda = \mathbb{Z}_p[[T]]$, where $T = \gamma - 1$ as before, then σ_s is defined by $\sigma_s(T) = \kappa^s(\gamma) - 1 \in p\mathbb{Z}_p$. We have $\ker(\sigma_s) = (\theta_s)$, where we have let $\theta_s = (T - (\kappa^s(\gamma) - 1))$. Then $\Lambda/\Lambda\theta_s \cong \mathbb{Z}_p(\kappa^s)$, a $\mathbb{Z}_p$-module of rank 1 on which $\mathrm{Gal}(F_\Sigma/F)$ acts by κ^s. Then

$$\mathcal{A}[\theta_s] \cong \mathrm{Hom}(\mathbb{Z}_p(\kappa^s), E[p^\infty]) \cong E[p^\infty] \otimes (\kappa^{-s}) = A_{-s}$$

as $\mathrm{Gal}(F_\Sigma/F)$-modules.

The version of Shapiro's Lemma that we will use is the following.

Proposition 4.10. *For all $i \geq 0$, $H^i(F_\Sigma/F_\infty, E[p^\infty]) \cong H^i(F_\Sigma/F, \mathcal{A})$ as Λ-modules.*

Remark. The first cohomology group is a Λ-module by virtue of the natural action of Γ on $H^i(F_\Sigma/F_\infty, E[p^\infty])$; the second cohomology group is a Λ-module by virtue of the Λ-module structure on $\mathcal{A}$.

Proof. We let A denote $E[p^\infty]$. The map $\phi \to \phi(1)$, for each $\phi \in \mathcal{A}$, defines a $\mathrm{Gal}(F_\Sigma/F_\infty)$-equivariant homomorphism $\mathcal{A} \to A$. The isomorphism in the proposition is defined by

$$H^i(F_\Sigma/F, \mathcal{A}) \xrightarrow{\text{rest.}} H^i(F_\Sigma/F_\infty, \mathcal{A}) \to H^i(F_\Sigma/F_\infty, A).$$

One can verify that this composite map is a Λ-homomorphism as follows. $\mathrm{Gal}(F_\Sigma/F_\infty)$ acts trivially on Λ. We therefore have a canonical isomorphism

$$H^i(F_\Sigma/F_\infty, \mathcal{A}) \cong \mathrm{Hom}(\Lambda, H^i(F_\Sigma/F_\infty, A)) \tag{7}$$

The image of the restriction map in (7) is contained in $H^i(F_\Sigma/F_\infty, \mathcal{A})^\Gamma$, which corresponds under (7) to $\mathrm{Hom}_\Gamma(\Lambda, H^i(F_\Sigma/F_\infty, A))$. The action of Γ on Λ is given by $\tilde{\kappa}$. But this is simply the usual structure of Λ as a Λ-module, restricted to $\Gamma \subseteq \Lambda$. Thus, by continuity, we have

$$H^i(F_\Sigma/F_\infty, \mathcal{A})^\Gamma \cong \mathrm{Hom}_\Lambda(\Lambda, H^i(F_\Sigma/F_\infty, A))$$

under (7). Now $\mathrm{Hom}_\Lambda(\Lambda, H^i(F_\Sigma/F_\infty, A)) \cong H^i(F_\Sigma/F_\infty, A)$ as Λ-modules, under the map defined by evaluating a homomorphism at $\lambda = 1$.

To verify that the map $H^i(F_\Sigma/F, \mathcal{A}) \to H^i(F_\Sigma/F_\infty, A)$ is bijective, note that both groups and the map are direct limits:

$$H^i(F_\Sigma/F, \mathcal{A}) = \varinjlim_n H^i(F_\Sigma/F, \mathcal{A}[\theta^{(n)}]),$$
$$H^i(F_\Sigma/F_\infty, A) = \varinjlim_n H^i(F_\Sigma/F_n, A).$$

Here $\theta^{(n)} = (1+T)^{p^n} - 1$ and so $\mathcal{A}[\theta^{(n)}] = \mathrm{Hom}(\mathbb{Z}_p[\mathrm{Gal}(F_n/F)], A)$. On each term the composite map

$$H^i(F_\Sigma/F, \mathcal{A}[\theta^{(n)}]) \to H^i(F_\Sigma/F_n, \mathcal{A}[\theta^{(n)}]) \to H^i(F_\Sigma/F_n, A)$$

defined analogously to (7) is known to be bijective by the usual version of Shapiro's Lemma. The map (7) is the direct limit of these maps (which are compatible) and so is bijective too. ■

For the proof of proposition 4.9, we may assume that $H^1(F_\Sigma/F, \mathcal{A})$ has Λ-corank $[F:\mathbb{Q}]$ and that $H^2(F_\Sigma/F, \mathcal{A})$ is Λ-cotorsion. Let $s \in \mathbb{Z}$. The exact sequence

$$0 \longrightarrow \mathcal{A}[\theta_s] \longrightarrow \mathcal{A} \xrightarrow{\theta_s} \mathcal{A} \longrightarrow 0$$

induces an exact sequence

$$H^1(F_\Sigma/F, \mathcal{A})/\theta_s H^1(F_\Sigma/F, \mathcal{A}) \xrightarrow{a} H^2(F_\Sigma/F, \mathcal{A}[\theta_s]) \xrightarrow{b} H^2(F_\Sigma/F, \mathcal{A})[\theta_s]$$

where of course a is injective and b is surjective. Let X denote the Pontryagin dual of $H^1(F_\Sigma/F, \mathcal{A})$. Since X is a finitely generated Λ-module, it is clear that $\ker(X \xrightarrow{\theta_s} X)$ will be finite for all but finitely many values of s. (Just choose s so that $\theta_s \nmid f(T)$, where $f(T)$ is a generator of the characteristic ideal of $X_{\Lambda\text{-tors}}$. The θ_s's are irreducible and relatively prime.) Now $\mathrm{Im}(a) = \ker(b)$ is the Pontryagin dual of $\ker(X \xrightarrow{\theta_s} X)$. We will show that $\ker(b)$ is always a divisible group. Hence, for suitable s, $\ker(X \xrightarrow{\theta_s} X) = 0$. Now if Z is a nonzero, finite Λ-module, then $\ker(Z \xrightarrow{\theta_s} Z)$ is also clearly nonzero, since $\theta_s \notin \Lambda^\times$. Therefore, X cannot contain a nonzero, finite Λ-submodule, which is equivalent to the assertion in proposition 4.9.

Assume that p is odd. Then $\mathrm{Gal}(F_\Sigma/F)$ has p-cohomological dimension 2. Since $\mathcal{A}[\theta_s] = A_{-s}$ is divisible, it follows from lemma 4.5 that $H^2(F_\Sigma/F, \mathcal{A}[\theta_s])$ is also divisible. Hence the same is true for $H^2(F_\Sigma/F, \mathcal{A})[\theta_s]$. But since $H^2(F_\Sigma/F, \mathcal{A})$ is Λ-cotorsion, $H^2(F_\Sigma/F, \mathcal{A})[\theta_s]$ will be finite for some value of s. Hence it must be zero. But this implies that $H^2(F_\Sigma/F, \mathcal{A}) = 0$, using Fact 1 about Λ-modules. Thus $\ker(b) = H^2(F_\Sigma/F, \mathcal{A}[\theta_s])$ for all s and this is indeed divisible, proving proposition 4.9 if p is odd.

The difficulty with the prime $p = 2$ is that $\mathrm{Gal}(F_\Sigma/F)$ doesn't have finite p-cohomological dimension (unless F is totally complex, in which case the argument in the preceding paragraph works). But we use the following fact: *the map*

$$\beta_n : H^n(\mathrm{Gal}(F_\Sigma/F), M) \to \prod_{v|\infty} H^n(F_v, M)$$

is an isomorphism for all $n \geq 3$. Here M can be any p-primary $\mathrm{Gal}(F_\Sigma/F)$-module. (This is proved in [Mi], theorem 4.10(c) for the case where M is finite. The general case follows from this.) The groups $H^n(F_v, M)$ have exponent ≤ 2 for all $n \geq 1$. The following lemma is the key to dealing with the prime 2.

Lemma 4.11. *Assume that M is divisible. Then the kernel of the map*

$$\beta_2 : H^2(F_\Sigma/F, M) \to \prod_{v|\infty} H^2(F_v, M)$$

is a divisible group.

Proof. Of course, if p is odd, then $H^2(F_v, M) = 0$ for $v|\infty$. We already know that $H^2(F_\Sigma/F, M)$ is divisible in this case. Let $p = 2$. For any $m \geq 1$, consider the following commutative diagram with exact rows

$$\begin{array}{ccccc}
H^2(F_\Sigma/F, M) & \xrightarrow{2^m} & H^2(F_\Sigma/F, M) & \xrightarrow{\alpha} & H^3(F_\Sigma/F, M[2^m]) \\
\downarrow{\scriptstyle \beta_2} & & \downarrow{\scriptstyle \beta_2} & & \downarrow{\scriptstyle \beta} \\
\prod_{v|\infty} H^2(F_v, M) & \xrightarrow{2^m} & \prod_{v|\infty} H^2(F_v, M) & \xrightarrow{\gamma} & \prod_{v|\infty} H^3(F_v, M[2^m])
\end{array}$$

induced from the exact sequence $0 \to M[2^m] \to M \xrightarrow{2^m} M \to 0$. Since the group $H^2(F_v, M)$ is of exponent ≤ 2, the map γ is injective. Since β is injective too, it follows that $\ker(\alpha) = \ker(\beta_2)$. Thus $\ker(\beta_2) = 2^m H^2(F_\Sigma/F, M)$ for any $m \geq 1$. Using this for $m = 1, 2$, we see that $\ker(\beta_2) = 2\ker(\beta_2)$, which implies that $\ker(\beta_2)$ is indeed divisible. ■

Now we can prove that $b : H^2(F_\Sigma/F, \mathcal{A}[\theta_s]) \to H^2(F_\Sigma/F, \mathcal{A})$ has a divisible kernel even when $p = 2$. We use the following commutative diagram:

$$\begin{array}{ccccccc}
0 \longrightarrow & R^2(F_\Sigma/F, \mathcal{A}[\theta_s]) & \longrightarrow & H^2(F_\Sigma/F, \mathcal{A}[\theta_s]) & \longrightarrow & \prod_{v|\infty} H^2(F_v, \mathcal{A}[\theta_s]) \\
& \downarrow d & & \downarrow b & & \downarrow e \\
0 \longrightarrow & R^2(F_\Sigma/F, \mathcal{A})[\theta_s] & \longrightarrow & H^2(F_\Sigma/F, \mathcal{A})[\theta_s] & \longrightarrow & \prod_{v|\infty} H^2(F_v, \mathcal{A})[\theta_s]
\end{array}$$

The rows are exact by definition. (We define $R^2(F_\Sigma/F, M)$ as the kernel of the map $H^2(F_\Sigma/F, M) \to \prod_{v|\infty} H^2(F_v, M)$.) The map b is surjective. Now $\mathcal{A}[\theta_s] \cong A_{-s}$ is divisible and hence, by lemma 4.11, $R^2(F_\Sigma/F, \mathcal{A}[\theta_s])$ is divisible. Under the assumption that $H^2(F_\Sigma/F, \mathcal{A})$ is Λ-cotorsion, we will show that $\ker(b)$ coincides with the divisible group $R^2(F_\Sigma/F, \mathcal{A}[\theta_s])$, completing the proof of proposition 4.9 for all p. Suppose that $v|\infty$. Since v splits completely in F_∞/F, we have $H^1(F_v, \mathcal{A}) = \mathrm{Hom}(\Lambda, H^1(F_v, E[p^\infty]))$. Of course, this group is zero unless $p = 2$ and $H^1(F_v, E[2^\infty]) \cong \mathbb{Z}/2\mathbb{Z}$, in which case $H^1(F_v, \mathcal{A}) = \mathrm{Hom}(\Lambda, \mathbb{Z}/2\mathbb{Z}) \cong (\Lambda/2\Lambda)\hat{}$. This last group is divisible by θ_s for any s, which implies that the map e must be injective. The snake lemma then implies that the map d is surjective. Thus $R^2(F_\Sigma/F, \mathcal{A})[\theta_s]$ is divisible for all $s \in \mathbb{Z}$. But this group is finite for all but finitely many s, since $H^2(F_\Sigma/F, \mathcal{A})$ is Λ-cotorsion. Hence, for some s, $R^2(F_\Sigma/F, \mathcal{A})[\theta_s] = 0$. This implies that the Λ-module $R^2(F_\Sigma/F, \mathcal{A})$ is zero. Therefore, since e is injective, $\ker(b) = \ker(d) = R^2(F_\Sigma/F, \mathcal{A}[\theta_s])$ for all s, as claimed. ■

The following proposition summarizes several consequences of the above arguments, which we translate back to the traditional form.

Proposition 4.12. *$H^1(F_\Sigma/F_\infty, E[p^\infty])$ has Λ-corank $[F:\mathbb{Q}]$ if and only if $H^2(F_\Sigma/F_\infty, E[p^\infty])$ is Λ-cotorsion. If this is so, then $H^1(F_\Sigma/F_\infty, E[p^\infty])$ has no proper Λ-submodule of finite index. Also, $H^2(F_\Sigma/F_\infty, E[p^\infty])$ will be zero if p is odd and $(\Lambda/2\Lambda)$-cofree if $p = 2$.*

In this form, proposition 4.12 should apply to all primes p, since one conjectures that $H^2(F_\Sigma/F_\infty, E[p^\infty])$ is always Λ-cotorsion. (See conjecture 3 in [Gr2].) If E has potentially good or multiplicative reduction at all primes over p, then, as mentioned in section 1, one expects that $\mathrm{Sel}_E(F_\infty)_p$ is Λ-cotorsion, which suffices to prove that $H^2(F_\Sigma/F_\infty, E[p^\infty])$ is indeed Λ-cotorsion. For any prime p, the conjecture that $\mathrm{Sel}_E(F_n)_p$ has bounded $\mathbb{Z}_p$-corank as n varies can also be shown to suffice.

We must now explain why $H^1(F_\Sigma/F_\infty, E[p^\infty])_\Gamma$ is zero, under the hypotheses of theorem 4.1. We can assume that $\mathrm{Sel}_E(F_\infty)_p$ is Λ-cotorsion and that $H^1(F_\Sigma/F_\infty, E[p^\infty])$ has Λ-corank equal to $[F:\mathbb{Q}]$. By proposition 4.9, it is enough to prove that $H^1(F_\Sigma/F_\infty, E[p^\infty])_\Gamma$ is finite. Let

$$Q = H^1(F_\Sigma/F_\infty, E[p^\infty])/H^1(F_\Sigma/F_\infty, E[p^\infty])_{\Lambda\text{-div}}.$$

Thus, Q is cofinitely generated and cotorsion as a Λ-module. Its Pontryagin dual is the torsion Λ-submodule of the Pontryagin dual of $H^1(F_\Sigma/F_\infty, E[p^\infty])$. We have

$$H^1(F_\Sigma/F_\infty, E[p^\infty])_\Gamma = Q_\Gamma.$$

But Q_Γ and Q^Γ have the same $\mathbb{Z}_p$-corank. Also, $\hat{\Lambda}^\Gamma \cong \mathbb{Q}_p/\mathbb{Z}_p$ has $\mathbb{Z}_p$-corank 1. Since the map $H^1(F_\Sigma/F, E[p^\infty]) \to H^1(F_\Sigma/F_\infty, E[p^\infty])^\Gamma$ is surjective and has finite kernel, we see that

$$\mathrm{corank}_{\mathbb{Z}_p}(H^1(F_\Sigma/F, E[p^\infty])) = [F:\mathbb{Q}] + \mathrm{corank}_{\mathbb{Z}_p}(Q_\Gamma).$$

Now

$$\mathrm{Sel}_E(F)_p = \ker(H^1(F_\Sigma/F, E[p^\infty]) \to \mathcal{P}_E^\Sigma(F)).$$

The $\mathbb{Z}_p$-corank of $\mathcal{P}_E^\Sigma(F)$ is equal to $[F:\mathbb{Q}]$. Since we are assuming that $\mathrm{Sel}_E(F)_p$ is finite, it follows that $H^1(F_\Sigma/F, E[p^\infty])$ has $\mathbb{Z}_p$-corank $[F:\mathbb{Q}]$ and hence that, indeed, Q_Γ is finite, which completes the argument. We should point out that sometimes $H^1(F_\Sigma/F_\infty, E[p^\infty])_\Gamma$ is nonzero. This clearly happens for example when $\mathrm{rank}_{\mathbb{Z}}(E(F)) > [F:\mathbb{Q}]$. For then $H^1(F_\Sigma/F, E[p^\infty])$ must have $\mathbb{Z}_p$-corank at least $[F:\mathbb{Q}]+1$, which implies that Q_Γ is nonzero.

We will now prove a rather general version of Cassels' theorem. Let Σ be a finite set of primes of a number field F, containing at least all primes of F lying above p and ∞. We suppose that M is a $\mathrm{Gal}(F_\Sigma/F)$-module isomorphic to $(\mathbb{Q}_p/\mathbb{Z}_p)^d$ as a group (for any $d \geq 1$). For each $v \in \Sigma$, we assume that L_v is a divisible subgroup of $H^1(F_v, M)$. Then we define a "Selmer group"

$$S_M(F) = \ker\Big(H^1(F_\Sigma/F, M) \to \prod_{v\in\Sigma} H^1(F_v, M)/L_v\Big).$$

This is a discrete, p-primary group which is cofinitely generated over $\mathbb{Z}_p$. Let

$$T^* = \mathrm{Hom}(M, \mu_{p^\infty})$$

which is a free $\mathbb{Z}_p$-module of rank d. For each $v \in \Sigma$, we define a subgroup U_v^* of $H^1(F_v, T^*)$ as the orthogonal complement of L_v under the perfect pairing (from Tate's local duality theorems)

$$H^1(F_v, M) \times H^1(F_v, T^*) \to \mathbb{Q}_p/\mathbb{Z}_p. \tag{8}$$

Since L_v is divisible, it follows that $H^1(F_v,T^*)/U_v^*$ is $\mathbb{Z}_p$-torsion free. Thus U_v^* contains $H^1(F_v,T^*)_{\mathrm{tors}}$. We define the Selmer group

$$S_{T^*}(F) = \ker\big(H^1(F_\Sigma/F,T^*) \to \prod_{v\in\Sigma} H^1(F_v,T^*)/U_v^*\big)$$

which will be a finitely generated $\mathbb{Z}_p$-module. Let $V^* = T^* \otimes \mathbb{Q}_p$. Let $M^* = V^*/T^* = T^* \otimes (\mathbb{Q}_p/\mathbb{Z}_p)$. For each $v \in \Sigma$, we can define a divisible subgroup L_v^* of $H^1(F_v, M^*)$ as follows: Under the map $H^1(F_v,T^*) \to H^1(F_v,V^*)$, the image of U_v^* generates a $\mathbb{Q}_p$-subspace of $H^1(F_v,V^*)$. We define L_v^* as the image of this subspace under the map $H^1(F_v,V^*) \to H^1(F_v,M^*)$. Thus, we can define a Selmer group

$$S_{M^*}(F) = \ker\big(H^1(F_\Sigma/F,M^*) \to \prod_{v\in\Sigma} H^1(F_v,M^*)/L_v^*\big).$$

One can verify that the $\mathbb{Z}_p$-corank of $S_{M^*}(F)$ is equal to the $\mathbb{Z}_p$-rank of $S_{T^*}(F)$.

We will use the following notation. Let

$$P = \prod_{v\in\Sigma} H^1(F_v,M), \qquad P^* = \prod_{v\in\Sigma} H^1(F_v,T^*)$$

$$L = \prod_{v\in\Sigma} L_v, \qquad U^* = \prod_{v\in\Sigma} U_v^*.$$

Then (8) induces a perfect pairing $P \times P^* \to \mathbb{Q}_p/\mathbb{Z}_p$, under which L and U^* are orthogonal complements. Furthermore, we let

$$G = \mathrm{Im}\left(H^1(F_\Sigma/F,M) \to P\right), \qquad G^* = \mathrm{Im}\left(H^1(F_\Sigma/F,T^*) \to P^*\right).$$

The duality theorems of Poitou and Tate imply that G and G^* are also orthogonal complements under the above perfect pairing. Consider the map

$$\gamma : H^1(F_\Sigma/F,M) \to P/L,$$

whose kernel is, by definition, $S_M(F)$. The cokernel of γ is clearly P/GL. But the orthogonal complement of GL under the pairing $P \times P^* \to \mathbb{Q}_p/\mathbb{Z}_p$ must be $G^* \cap U^*$. Thus $\mathrm{coker}(\gamma) \cong (G^* \cap U^*)\hat{}$. Again by definition, $S_{T^*}(F)$ is the inverse image of U^* under the map $H^1(F_\Sigma/F,T^*) \to P^*$. Thus clearly $G^* \cap U^*$ is a homomorphic image of $S_{T^*}(F)$. As we mentioned above, $\mathrm{rank}_{\mathbb{Z}_p}(S_{T^*}(F))$ is equal to $\mathrm{corank}_{\mathbb{Z}_p}(S_{M^*}(F))$. On the other hand, since $H^1(F_v,T^*)_{\mathrm{tors}}$ is contained in U_v^* for all $v \in \Sigma$, it follows that

$$S_{T^*}(F)_{\mathrm{tors}} = H^1(F_\Sigma/F,T^*)_{\mathrm{tors}},$$

which in turn is isomorphic to $H^0(F_\Sigma/F,M^*)/H^0(F_\Sigma/F,M^*)_{\mathrm{div}}$. (This last assertion follows from the cohomology sequence induced from the exact sequence $0 \to T^* \to V^* \to M^* \to 0$.) We denote $H^0(F_\Sigma/F,M^*) = (M^*)^{G_F}$ by $M^*(F)$ as usual. Then, as a $\mathbb{Z}_p$-module, we have

$$S_{T^*}(F) \cong (M^*(F)/M^*(F)_{\mathrm{div}}) \times \mathbb{Z}_p^{\mathrm{corank}_{\mathbb{Z}_p}(S_{M^*}(F))}.$$

The preceding discussion proves that the Pontryagin dual of the cokernel of the map γ is a homomorphic image of $S_{T^*}(F)$. In particular, one important special case is: *if* $S_{M^*}(F)$ *is finite and* $M^*(F) = 0$, *then* $\mathrm{coker}(\gamma) = 0$.

We now make the following slightly restrictive hypothesis: $M^*(F_v) = H^0(F_v, M^*)$ *is finite for at least one* $v \in \Sigma$. This implies that $M^*(F)$ is also finite. Consider the following commutative diagram.

$$\begin{array}{ccc} H^0(F_\Sigma/F, M^*) & \xrightarrow{\sim} & H^1(F_\Sigma/F, T^*)_{\mathrm{tors}} \\ \downarrow & & \downarrow \\ H^0(F_v, M^*) & \xrightarrow{\sim} & H^1(F_v, T^*)_{\mathrm{tors}} \end{array} \tag{9}$$

Since the first vertical arrow is obviously injective, so is the second. Hence the map $H^1(F_\Sigma/F, T^*)_{\mathrm{tors}} \to P^*$ is injective. It follows that if $S_{T^*}(F)$ is finite, then

$$\mathrm{coker}(\gamma) \cong (G^* \cap U^*)\hat{} \cong S_{T^*}(F)\hat{} = H^1(F_\Sigma/F, T^*)_{\mathrm{tors}}.$$

This last group is isomorphic to $M^*(F)$. We obtain the following general version of Cassels' theorem.

Proposition 4.13. *Assume that* $m^* = \mathrm{corank}_{\mathbb{Z}_p}(\mathrm{Sel}_{M^*}(F))$. *Assume also that* $H^0(F_v, M^*)$ *is finite for at least one* $v \in \Sigma$. *Then the cokernel of the map*

$$\gamma : H^1(F_\Sigma/F, M) \to \prod_{v \in \Sigma} H^1(F_v, M)/L_v$$

has $\mathbb{Z}_p$*-corank* $\leq m^*$. *Also,*

$$\dim_{\mathbb{Z}/p\mathbb{Z}}(\mathrm{coker}(\gamma)[p]) \leq m^* + \dim_{\mathbb{Z}/p\mathbb{Z}}(H^0(F, M^*[p])).$$

If $m^* = 0$, *then* $\mathrm{coker}(\gamma) \cong H^0(F, M^*)\hat{}$.

It is sometimes useful to know how $\mathrm{Im}(\gamma)$ sits inside of P/L. We can make the following remark. Let v_0 be any prime in Σ for which $H^0(F_{v_0}, M^*)$ is finite. Assume that $S_{M^*}(F)$ is finite. Then

$$\mathrm{Im}(\gamma)(H^1(F_{v_0}, M)/L_{v_0}) = P/L.$$

Here $H^1(F_{v_0}, M)/L_v$ is a direct factor in P/L. To justify this, one must just show that the map

$$\gamma' : H^1(F_\Sigma/F, M) \to \prod_{\substack{v \in \Sigma \\ v \neq v_0}} H^1(F_v, M)/L_v$$

is surjective under the above assumptions about $S_{M^*}(F)$ and v_0. In the above arguments, one can study $\mathrm{coker}(\gamma')$ by changing L_{v_0} to $L'_{v_0} = H^1(F_{v_0}, M)$

and leaving L_v for $v \neq v_0$ unchanged. Now L'_{v_0} may not be divisible, but we still have $\mathrm{coker}(\gamma') \cong (G^* \cap U'^*)^{\widehat{}}$, where now $U^*_{v_0}$ has been replaced by $U'^*_{v_0} = 0$. Since $U'^* \subseteq U^*$, the corresponding Selmer group $S'_{T^*}(F)$ is still finite. Thus an element σ in $S'_{T^*}(F)$ is in $H^1(F_\Sigma/F, T^*)_{\mathrm{tors}}$ and has the property that $\sigma|_{G_{F_{v_0}}}$ is trivial. But the diagram (9) shows that

$$H^1(F_\Sigma/F, T^*)_{\mathrm{tors}} \to H^1(F_{v_0}, T^*)_{\mathrm{tors}}$$

is injective. Hence σ is trivial. Thus, $S'_{T^*}(F)$ is trivial and hence so is $\mathrm{coker}(\gamma')$.

Cassels' theorem is the following special case of proposition 4.13: $M = E[p^\infty]$, Σ = any finite set of primes of F containing the primes lying over p or ∞ and the primes where E has bad reduction, and $L_v = \mathrm{Im}(\kappa_v)$ for all $v \in \Sigma$. Then $T^* = T_p(E)$ by the Weil pairing. Thus $M^* = E[p^\infty]$, $L^*_v = \mathrm{Im}(\kappa_v)$, and $S_{M^*}(F) = S_M(F) = \mathrm{Sel}_E(F)_p$. It is clear that $H^0(F_v, M^*)$ is finite for any nonarchimedean $v \in \Sigma$. Thus, proposition 4.13 implies that

$$\mathrm{coker}\big(H^1(F_\Sigma/F, E[p^\infty]) \to \prod_{v \in \Sigma} H^1(F_v, E[p^\infty])/\mathrm{Im}(\kappa_v)\big) \cong E(F)_p^{\widehat{}}$$

if $\mathrm{Sel}_E(F)_p$ is finite. (Of course, as a group, $E(F)_p^{\widehat{}} \cong E(F)_p$.) In the proof of lemma 4.6 we need the following case: E is an elliptic curve which we assume has (potentially) good, ordinary or multiplicative reduction at all $v|p$, $M = E[p^\infty] \otimes \kappa^s$ where $s \in \mathbb{Z}$, $L_v = \mathrm{Im}\,\big(H^1(F_v, C_v \otimes \kappa^s) \to H^1(F_v, M)\big)_{\mathrm{div}}$ if $v|p$, $L_v = 0$ if $v \nmid p$. Then $T^* = T_p(E) \otimes \kappa^{-s}$, $M^* = E[p^\infty] \otimes \kappa^{-s}$, and L^*_v is defined just as L_v. Assuming that $\mathrm{Sel}_E(F_\infty)_p$ is Λ-cotorsion, we can choose $s \in \mathbb{Z}$ so that $S_{M^*}(F)$ is finite. The hypothesis that $H^0(F_v, M^*)$ is finite for some $v \in \Sigma$ is also easily satisfied (possibly avoiding one value of s). Then the cokernel of the map γ will be isomorphic to the finite group $H^0(F, M^*)^{\widehat{}}$.

We can now prove the following generalization of proposition 4.8.

Proposition 4.14. *Assume that E is an elliptic curve defined over F and that $\mathrm{Sel}_E(F_\infty)_p$ is Λ-cotorsion. Assume that $E(F)_p = 0$. Then $\mathrm{Sel}_E(F_\infty)_p$ has no proper Λ-submodules of finite index.*

Proof. As in the proof of lemma 4.6, we will use the twisted Galois modules $A_s = E[p^\infty] \otimes (\kappa^s)$, where $s \in \mathbb{Z}$. Since $E(F)_p = 0$, it follows that $E(F_\infty)_p = 0$ too. (One uses the fact that Γ is pro-p.) Since $A_s \cong E[p^\infty]$ as G_{F_∞}-modules, it is clear that $H^0(F, A_s) = 0$ for all s. Now E must have potentially ordinary or multiplicative reduction at all $v|p$, since we are assuming that $\mathrm{Sel}_E(F_\infty)_p$ is Λ-cotorsion. So we can define a Selmer group $S_{A_s}(K)$ for any algebraic extension K of F. If we take K to be a subfield of F_Σ, then $S_{A_s}(K)$ is the kernel of a map $H^1(F_\Sigma/K, A_s) \to \mathcal{P}^\Sigma(A_s, K)$, where this last group is defined in a way analogous to $\mathcal{P}^\Sigma_E(K)$. As we pointed out in the proof of lemma 4.6, we have

$$S_{A_s}(F_\infty) \cong \mathrm{Sel}_E(F_\infty)_p \otimes (\kappa^s)$$

as Λ-modules. We also have $\mathcal{P}^{\Sigma}(A_s, F_\infty) \cong \mathcal{P}_E^{\Sigma}(F_\infty) \otimes (\kappa^s)$ as Λ-modules. The hypothesis that $\mathrm{Sel}_E(F_\infty)_p$ is Λ-cotorsion implies that $S_{A_s}(F_\infty)^{\Gamma}$, and hence $S_{A_s}(F)$, will be finite for all but finitely many values of s. (We will add another requirement on s below.) We let $M = A_s$, where $s \in \mathbb{Z}$ has been chosen so that $S_{A_{-s}}(F)$ is finite. Note that $M^* = A_{-s}$. Since $S_{M^*}(F)$ is finite and $M^*(F) = 0$, we can conclude that the map

$$\gamma : H^1(F_\Sigma/F, M) \to \mathcal{P}^{\Sigma}(M, F)$$

is surjective. Since Γ has cohomological dimension 1, the restriction maps $H^1(F_\Sigma/F, M) \to H^1(F_\Sigma/F_\infty, M)^{\Gamma}$ and $\mathcal{P}^{\Sigma}(M, F) \to \mathcal{P}^{\Sigma}(M, F_\infty)^{\Gamma}$ are both surjective. Hence it follows that the map

$$H^1(F_\Sigma/F_\infty, M)^{\Gamma} \to \mathcal{P}^{\Sigma}(M, F_\infty)^{\Gamma}$$

must be surjective. We have the exact sequence defining $S_M(F_\infty)$:

$$0 \to S_M(F_\infty) \to H^1(F_\Sigma/F_\infty, M) \to \mathcal{P}^{\Sigma}(M, F_\infty) \to 0.$$

This is just the exact sequence defining $\mathrm{Sel}_E(F_\infty)_p$, twisted by κ^s. The corresponding cohomology sequence induces an injective map

$$S_M(F_\infty)_{\Gamma} \to H^1(F_\Sigma/F_\infty, M)_{\Gamma}.$$

If we let $Q = H^1(F_\Sigma/F_\infty, E[p^\infty])/H^1(F_\Sigma/F_\infty, E[p^\infty])_{\Lambda\text{-div}}$, as before, then

$$H^1(F_\Sigma/F_\infty, M)_{\Gamma} \cong (Q \otimes (\kappa^s))_{\Gamma}$$

and, since Q is Λ-cotorsion, we can choose s so that $(Q \otimes (\kappa^s))_{\Gamma}$ is finite. (This will be true for all but finitely many values of s.) But since $H^1(F_\Sigma/F_\infty, E[p^\infty])$ has no proper Λ-submodules of finite index, neither does $H^1(F_\Sigma/F_\infty, M)$. It follows that, for suitably chosen s, $H^1(F_\Sigma/F_\infty, M)_{\Gamma} = 0$. Hence $S_M(F_\infty)_{\Gamma} = 0$. This implies that $S_M(F_\infty)$ has no proper Λ-submodules of finite index, from which proposition 4.14 follows. ∎

We will give two other sufficient conditions for the nonexistence of proper Λ-submodules of finite index in $\mathrm{Sel}_E(F_\infty)_p$. We want to mention that a rather different proof of proposition 4.14 and part of the following proposition has been found by Hachimori and Matsuno [HaMa]. This proof is based on the Cassels-Tate pairing for $Ш_E(F_n)_p$. This topic will be pursued much more generally in [Gr6].

Proposition 4.15. *Assume that E is an elliptic curve defined over F and that $\mathrm{Sel}_E(F_\infty)_p$ is Λ-cotorsion. Assume that at least one of the following two hypotheses holds:*

(i) *There is a prime v_0 of F, $v_0 \nmid p$, where E has additive reduction.*

(ii) *There exists a prime v_0 of F, $v_0|p$, such that the ramification index e_{v_0} of $F_{v_0}/\mathbb{Q}_p$ satisfies $e_{v_0} \leq p-2$ and such that E has good, ordinary or multiplicative reduction at v_0.*

Then $\mathrm{Sel}_E(F_\infty)_p$ has no proper Λ-submodules of finite index.

Remark. If condition (i) holds, then $H^0(I_{\mathbb{Q}_{v_0}}, E[p^\infty])$ is finite. This group will be zero if $p \geq 5$. Then $E(F)_p = 0$ and we are in the situation of proposition 4.14.

Proof. We will modify the proof of proposition 4.14. In addition to the requirements on $M = A_s$ occurring in that proof, we also require that $H^0(F_{v_0}, M^*)$ be finite, which is true for all but finitely many values of $s \in \mathbb{Z}$. Here v_0 is the prime of F satisfying (i) or (ii). (If E has additive reduction at v_0, $v_0 \nmid p$, then this holds for all s.) Assume first that (i) holds. In this case, let $S'_M(F)$, $S'_M(F_\infty)$ denote the Selmer groups where one omits the local condition at v_0 (or the primes above v_0). If η is a prime of F_∞ lying over v_0, then $H^0((F_\infty)_\eta, M^*)$ is finite. This implies that $H^1((F_\infty)_\eta, M) = 0$. Thus, $S'_M(F_\infty) = S_M(F_\infty)$. The remark following proposition 4.13 shows that the map

$$\gamma' : H^1(F_\Sigma/F, M) \to \mathcal{P}^{\Sigma'}(M, F)$$

is surjective, where $\Sigma' = \Sigma - \{v_0\}$ and $\mathcal{P}^{\Sigma'}(M, F)$ is the product over all primes of Σ'. The proof then shows that $S'_M(F_\infty)$ has no proper Λ-submodules of finite index. This obviously gives the same statement for $\mathrm{Sel}_E(F_\infty)_p$.

Now assume (ii). We again define $S'_M(F_\infty)$ by omitting the local condition at all primes η of F_∞ lying over v_0. Just as above, we see that $S'_M(F_\infty)$ has no proper Λ-submodules of finite index. Thus, the same is true for $\mathrm{Sel}'_E(F_\infty)_p$. By lemma 4.6, we see that

$$\mathrm{Sel}'_E(F_\infty)_p/\mathrm{Sel}_E(F_\infty)_p \cong \prod_{\eta|v_0} \mathcal{H}_E((F_\infty)_\eta).$$

But $\mathcal{H}_E((F_\infty)_\eta) \cong H^1((F_\infty)_\eta, E[p^\infty])/\mathrm{Im}(\kappa_\eta) \cong H^1((F_\infty)_\eta, D_{v_0})$ by proposition 2.4 and the analogous statement proved in section 2 for the case where E has multiplicative reduction at v. Here $D_{v_0} = E[p^\infty]/C_{v_0}$ is an unramified $G_{F_{v_0}}$-module isomorphic to $\mathbb{Q}_p/\mathbb{Z}_p$. We can use a remark made in section 2 (preceding proposition 2.4) to conclude that $H^1((F_\infty)_\eta, D_{v_0})$ is $\mathbb{Z}_p[[\mathrm{Gal}((F_\infty)_\eta/F_{v_0})]]$-cofree. Proposition 4.15 in case (ii) is then a consequence of the following fact about finitely generated Λ-modules: *Suppose that X' is a finitely generated Λ-module which has no nonzero, finite Λ-submodules. Assume that Y is a free Λ-submodule of X'. Then $X = X'/Y$ has no nonzero, finite Λ-submodules.* The proof is quite easy. By induction, one can assume that $Y \cong \Lambda$. Suppose that X does have a nonzero, finite

Λ-submodule. Then $Y \subseteq Y_0$, where $[Y_0 : Y] < \infty$, $Y \neq Y_0$, and Y_0 is a Λ-submodule of X'. Then Y_0 is pseudo-isomorphic to Λ and has no nonzero, finite Λ-submodules. Hence Y_0 would be isomorphic to a submodule of Λ of finite index. It would follow that Λ contains a proper ideal of finite index which is isomorphic to Λ, i.e., a principal ideal. But if $f \in \Lambda$, then (f) can't have finite index unless $f \in \Lambda^\times$, in which case $(f) = \Lambda$. Hence in fact X has no nonzero, finite Λ-submodules. ■

5. Conclusion

In this final section we will discuss the structure of $\mathrm{Sel}_E(F_\infty)_p$ in various special cases, making full use of the results of sections 3 and 4. In particular, we will see that each of the invariants μ_E, $\lambda_E^{M\text{-}W}$, and $\lambda_E^{Ш}$ can be positive. We will assume (usually) that the base field F is $\mathbb{Q}$ and that $E/\mathbb{Q}$ has good, ordinary reduction at p. Our examples will be based on the predicted order of the Shafarevich-Tate groups given in Cremona's tables. In principle, these orders can be verified by using results of Kolyvagin.

We start with the following corollary to proposition 3.8.

Proposition 5.1. *Assume that E is an elliptic curve/$\mathbb{Q}$ and that both $E(\mathbb{Q})$ and $Ш_E(\mathbb{Q})$ are finite. Let p vary over the primes where E has good, ordinary reduction. Then $\mathrm{Sel}_E(\mathbb{Q}_\infty)_p = 0$ except for p in a set of primes of zero density. This set of primes is finite if E is $\mathbb{Q}$-isogenous to an elliptic curve E' such that $|E'(\mathbb{Q})| > 1$.*

Remark. Recall that if p is a prime where E has supersingular (or potentially supersingular) reduction, then $\mathrm{Sel}_E(\mathbb{Q}_\infty)_p$ has positive Λ-corank. Under the hypothesis that $E(\mathbb{Q})$ and $Ш_E(\mathbb{Q})$ are finite, this Λ-corank can be shown to equal 1, agreeing with the conjecture stated after theorem 1.7. If E doesn't have complex multiplication, the set of supersingular primes for E also has zero density.

Proof. We are assuming that $\mathrm{Sel}_E(\mathbb{Q})$ is finite. Thus, excluding finitely many primes, we can assume that $\mathrm{Sel}_E(\mathbb{Q})_p = 0$. If we also exclude the finite set of primes dividing $\prod_l c_l$, where l varies over the primes where E has bad reduction and c_l is the corresponding Tamagawa factor, then hypotheses (ii) and (iii) in proposition 3.8 are satisfied. As for hypothesis (i), it is equivalent to $a_p \equiv 1 \pmod{p}$, where $a_p = 1 + p - |\widetilde{E}(\mathbb{F}_p)|$. Now we have Hasse's result that $|a_p| < 2\sqrt{p}$ and hence $a_p \equiv 1 \pmod{p} \Rightarrow a_p = 1$ if $p > 5$. By using the Chebotarev Density Theorem, one can show that $\{p \mid a_p = 1\}$ has zero density. (That is, the cardinality of $\{p \mid a_p = 1, p < x\}$ is $o(x/\log(x))$ as $x \to \infty$.) The argument is a standard one, using the l-adic representation attached to E for any fixed prime l. The trace of a Frobenius element for p $(\neq l)$ is a_p. One considers separately the cases where E does or does not have complex multiplication. For the non-CM case, see [Se1], IV–13, exercise 1. These remarks show that the hypotheses in proposition 3.8 hold if p is outside a set of primes of zero density. For such p, $\mathrm{Sel}_E(\mathbb{Q}_\infty)_p = 0$. The final part of proposition 5.1 follows from the next lemma.

Lemma 5.2. *Suppose that E is an elliptic curve defined over $\mathbb{Q}$ and that p is a prime where E has good reduction. If $E(\mathbb{Q})$ has a point of order 2 and $p > 5$, then $a_p \not\equiv 1 \pmod{p}$. If E is $\mathbb{Q}$-isogenous to an elliptic curve E' such that $E'(\mathbb{Q})_{\mathrm{tors}}$ has a subgroup of order $q > 2$ and if $p \nmid q$, then $a_p \not\equiv 1 \pmod{p}$.*

Proof. $\mathbb{Q}$-isogenous elliptic curves have the same set of primes of bad reduction. If E has good reduction at p, then the prime-to-p part of $E'(\mathbb{Q})_{\mathrm{tors}}$ maps injectively into $\widetilde{E}'(\mathbb{F}_p)$, which has the same order as $\widetilde{E}(\mathbb{F}_p)$. For the first part, $a_p \equiv 1 \pmod{p}$ implies that $2p$ divides $|\widetilde{E}(\mathbb{F}_p)|$. Hence $2p < 1 + p + 2\sqrt{p}$, which is impossible for $p > 5$. For the second part, if $a_p \equiv 1 \pmod{p}$ and $p \nmid q$, then qp divides $|\widetilde{E}(\mathbb{F}_p)|$. Hence $qp < 1 + p + 2\sqrt{p}$, which again is impossible since $q > 2$. ∎

Here are several specific examples.

$\mathbf{E = X_0(11)}$. The equation $y^2 + y = x^3 - x^2 - 10x - 20$ defines this curve, which is 11(A1) in [Cre]. E has split, multiplicative reduction at $p = 11$ and good reduction at all other primes. We have $\mathrm{ord}_{11}(j_E) = -5$, $E(\mathbb{Q}) \cong \mathbb{Z}/5\mathbb{Z}$, and we will assume that $\mathrm{Sel}_E(\mathbb{Q}) = 0$ as predicted. If $p \neq 11$, then $a_p \equiv 1 \pmod{p}$ happens only for $p = 5$. Therefore, if E has good, ordinary reduction at $p \neq 5$, then $\mathrm{Sel}_E(\mathbb{Q}_\infty)_p = 0$ according to proposition 3.8. We will discuss the case $p = 5$ later, showing that $\mathrm{Sel}_E(\mathbb{Q}_\infty)_p \cong \mathrm{Hom}(\Lambda/p\Lambda, \mathbb{Z}/p\mathbb{Z})$ and hence that $\mu_E = 1$, $\lambda_E = 0$. We just mention now that, by theorem 4.1, $f_E(0) \sim 5$. We will also discuss quite completely the other two elliptic curves/$\mathbb{Q}$ of conductor 11 for the case $p = 5$. If $p = 11$, then $\mathrm{Sel}_E(\mathbb{Q}_\infty)_p = 0$. This is verified in [Gr3], example 3.

$\mathbf{E = X_0(32)}$. This curve is defined by $y^2 = x^3 - 4x$ and is 32(A1) in [Cre]. It has complex multiplication by $\mathbb{Z}[i]$. E has potentially supersingular reduction at 2. For an odd prime p, E has good, ordinary reduction at p if and only if $p \equiv 1 \pmod{4}$. We have $E(\mathbb{Q}) \cong \mathbb{Z}/4\mathbb{Z}$, $Ш_E(\mathbb{Q}) = 0$ (as verified in Rubin's article in this volume), and $c_2 = 4$. By lemma 5.2, there are no anomalous primes for E. Therefore, $\mathrm{Sel}_E(\mathbb{Q}_\infty)_p = 0$ for all primes p where E has good, ordinary reduction.

$\mathbf{E_1 : y^2 = x^3 + x^2 - 7x + 5}$ and $\mathbf{E_2 : y^2 = x^3 + x^2 - 647x - 6555}$. Both of these curves have conductor 768. They are 768(D1) and 768(D3) in [Cre]. They are related by a 5-isogeny defined over $\mathbb{Q}$. We will assume that $\mathrm{Sel}_{E_1}(\mathbb{Q})$ is trivial as predicted by the Birch and Swinnerton-Dyer conjecture. This implies that $\mathrm{Sel}_{E_2}(\mathbb{Q})_p = 0$ for all primes $p \neq 5$. We will verify later that this is true for $p = 5$ too. Both curves have additive reduction at $p = 2$, and split, multiplicative reduction at $p = 3$. For E_1, the Tamagawa factors are

$c_2 = 2$, $c_3 = 1$. For E_2, they are $c_2 = 2$, $c_3 = 5$. We have $E_1(\mathbb{Q}) \cong \mathbb{Z}/2\mathbb{Z} \cong E_2(\mathbb{Q})$. By lemma 5.2, no prime $p > 5$ is anomalous for E_1 or E_2. If E_1 (and hence E_2) have ordinary reduction at a prime $p > 5$, then proposition 3.8 implies that $\mathrm{Sel}_{E_1}(\mathbb{Q}_\infty)_p = 0 = \mathrm{Sel}_{E_2}(\mathbb{Q}_\infty)_p$. Both of these curves have good, ordinary reduction at $p = 5$. (In fact, $\widetilde{E}_1 = \widetilde{E}_2 : y^2 = x^3 + x^2 + \widetilde{3}x$ and one finds 4 points. That is, $a_5 = 2$ and so $p = 5$ is not anomalous for E_1 or E_2.) The hypotheses of proposition 3.8 are satisfied for E_1 and $p = 5$. Hence $\mathrm{Sel}_{E_1}(\mathbb{Q}_\infty)_5 = 0$. But, by using either the results of section 3 or theorem 4.1, one sees that $\mathrm{Sel}_{E_2}(\mathbb{Q}_\infty)_5 \neq 0$. (One can either point out that $\mathrm{coker}(\mathrm{Sel}_{E_2}(\mathbb{Q})_5 \to \mathrm{Sel}_{E_2}(\mathbb{Q}_\infty)_5^\Gamma)$ is nonzero or that $f_{E_2}(0) \sim 5$. We remark that proposition 4.8 tells us that $\mathrm{Sel}_{E_2}(\mathbb{Q}_\infty)$ cannot just be finite if it is nonzero.) Now if $\phi : E_1 \to E_2$ is a 5-isogeny defined over $\mathbb{Q}$, the induced map $\Phi : \mathrm{Sel}_{E_1}(\mathbb{Q}_\infty)_5 \to \mathrm{Sel}_{E_2}(\mathbb{Q}_\infty)_5$ will have kernel and cokernel of exponent 5. Hence $\lambda_{E_2} = \lambda_{E_1} = 0$ (for $p = 5$). Since $f_{E_2}(0) \sim 5$, it is clear that $\mu_{E_2} = 1$. Below we will verify directly that $\mathrm{Sel}_{E_2}(\mathbb{Q}_\infty)_5 \cong \mathrm{Hom}(\Lambda/5\Lambda, \mathbb{Z}/5\mathbb{Z})$. Note that this example illustrates conjecture 1.11.

$\mathbf{E : y^2 + y = x^3 + x^2 - 12x - 21}$. This is 67(A1) in [Cre]. It has split, multiplicative reduction at $p = 67$, good reduction at all other primes. We have $E(\mathbb{Q}) = 0$ and $c_{67} = 1$. It should be true that $\mathrm{Sel}_E(\mathbb{Q}) = 0$, which we will assume. According to proposition 3.8, $\mathrm{Sel}_E(\mathbb{Q}_\infty)_p = 0$ for any prime $p \neq 67$ where $a_p \not\equiv 0, 1 \pmod{p}$. If $a_p \equiv 0 \pmod{p}$, then E has supersingular reduction at p, and hence $\mathrm{Sel}_E(\mathbb{Q}_\infty)_p$ is not even Λ-cotorsion. (In fact, the Λ-corank will be 1.) If $a_p \equiv 1 \pmod{p}$, then $\mathrm{Sel}_E(\mathbb{Q}_\infty)_p$ must be nonzero and hence infinite. (Proposition 4.8 applies.) By proposition 4.1, we in fact have $f_E(0) \sim |\widetilde{E}(\mathbb{F}_p)|^2 \sim p^2$ for any such prime p. (Here we use Hasse's estimate on $|\widetilde{E}(\mathbb{F}_p)|$, noting that $1 + p + 2\sqrt{p} < p^2$ for $p \geq 3$. The prime $p = 2$ is supersingular for this elliptic curve.) Now it seems reasonable to expect that E has infinitely many anomalous primes. The first such p is $p = 3$ (and the only such $p < 100$). Conjecture 1.11 implies that $\mu_E = 0$. Assuming this, we will later see that $\lambda_E^{M\text{-}W} = 0$ and $\lambda_E^{Ш} = 2$.

$\mathbf{E : y^2 + y = x^3 - x^2 - 460x - 11577}$. This curve has conductor 915. It is 915(A1) in [Cre]. It has split, multiplicative reduction at 5 and 61, nonsplit at 3. We have $c_3 = c_{61} = 1$ and $c_5 = 7$. $\mathrm{Sel}_E(\mathbb{Q}) = 0$, conjecturally. $E(\mathbb{Q}) = 0$. Proposition 3.8 implies that $\mathrm{Sel}_E(\mathbb{Q}_\infty)_p = 0$ for any prime p where E has good, ordinary reduction, unless either $p = 7$ or $a_p \equiv 1 \pmod{p}$. In these two cases, $\mathrm{Sel}_E(\mathbb{Q}_\infty)_p$ must be infinite by proposition 4.8. More precisely, theorem 4.1 implies the following: Let $p = 7$. Then $f_E(0) \sim 7$. (One must note that $a_7 = 3 \not\equiv 1 \pmod{7}$.) This implies that $f_E(T)$ is an irreducible element of Λ. On the other hand, suppose $a_p \equiv 1 \pmod{p}$ but $p \neq 5$ or 61. Then $f_E(0) \sim p^2$ The only such anomalous prime $p \leq 100$ is $p = 43$. Assuming the validity of conjecture 1.11 for E, we will see later that $\lambda_E^{M\text{-}W} = 0$ and $\lambda_E^{Ш} = 2$ for $p = 43$.

$\mathbf{E : y^2 + xy = x^3 - 3x + 1}$. This is 34(A1) in [Cre]. $\mathrm{Sel}_E(\mathbb{Q})$ should be trivial. E has multiplicative reduction at 2 and 17, $c_2 = 6$, $c_{17} = 1$. Also, $E(\mathbb{Q}) \cong \mathbb{Z}/6\mathbb{Z}$. The prime $p = 3$ is anomalous: $a_3 = -2$ and so $|\widetilde{E}(\mathbb{F}_3)| = 6$. If p is any other prime where E has good, ordinary reduction, then $a_p \not\equiv 1 \pmod{p}$ and we clearly have $\mathrm{Sel}_E(\mathbb{Q}_\infty)_p = 0$. For $p = 3$, proposition 4.1 gives $f_E(0) \sim 3$. Thus, $f_E(T)$ is irreducible. Let F be the first layer of the cyclotomic $\mathbb{Z}_3$-extension of $\mathbb{Q}$. Then $F = \mathbb{Q}(\beta)$, where $\beta = \zeta + \zeta^{-1}$, ζ denoting a primitive 9-th root of unity. Notice that β is a root of $x^3 - 3x + 1$. Thus $(\beta, -\beta)$ is a point in $E(F)$, which is not in $E(\mathbb{Q})$. Now the residue field for $\mathbb{Q}_\infty$ at the unique prime η above 3 is $\mathbb{F}_3$. The prime-to-3 torsion of $E(\mathbb{Q}_\infty)$ is mapped by reduction modulo η injectively into $\widetilde{E}(\mathbb{F}_3)$, and thus is $\mathbb{Z}/2\mathbb{Z}$. It is defined over $\mathbb{Q}$. The discussion preceding proposition 3.9 shows that $E(\mathbb{Q}_\infty)_3 = E(\mathbb{Q})_3$. Thus, $E(\mathbb{Q}_\infty)_{\mathrm{tors}} = E(\mathbb{Q})_{\mathrm{tors}}$. It follows that $(\beta, -\beta)$ has infinite order. Now $\mathrm{Gal}(F/\mathbb{Q})$ acts faithfully on $E(F) \otimes \mathbb{Q}_3$. It is clear that this $\mathbb{Q}_3$-representation must be isomorphic to ρ^t where ρ is the unique 2-dimensional irreducible $\mathbb{Q}_3$-representation of $\mathrm{Gal}(F/\mathbb{Q})$ and $t \geq 1$. If γ generates $\Gamma = \mathrm{Gal}(\mathbb{Q}_\infty/\mathbb{Q})$ topologically, then $\rho(\gamma|_F)$ is given by a matrix with trace -1, determinant 1. Regarding ρ as a representation of Γ and letting $T = \gamma - 1$, $\rho(\gamma - 1)$ has characteristic polynomial

$$\theta_1 = (1+T)^2 + (1+T) + 1 = T^2 + 3T + 3.$$

Since $E(F) \otimes (\mathbb{Q}_3/\mathbb{Z}_3)$ is a Λ-submodule of $\mathrm{Sel}_E(\mathbb{Q}_\infty)_3$, it follows that θ_1^t divides $f_E(T)$. Comparing the valuation of $\theta_1(0)$ and $f_E(0)$, we clearly have $t = 1$ and $f_E(T) = \theta_1$, up to a factor in $\Lambda^\times$. Therefore $\lambda_E^{M\text{-}W} = 2$, $\lambda_E^{\text{Ш}} = 0$, and $\mu_E = 0$.

When is $\mathrm{Sel}_E(\mathbb{Q}_\infty)_p$ infinite? A fairly complete answer is given by the following partial converse to proposition 3.8.

Proposition 5.3. *Assume that E has good, ordinary reduction at p and that $E(\mathbb{Q})$ has no element of order p. Assume also that at least one of the following statements is true:*

(i) $\mathrm{Sel}_E(\mathbb{Q})_p \neq 0$.
(ii) $a_p \equiv 1 \pmod{p}$.
(iii) *There exists at least one prime l where E has multiplicative reduction such that $a_l \equiv 1 \pmod{p}$ and $\mathrm{ord}_l(j_E) \equiv 0 \pmod{p}$.*
(iv) *There exists at least one prime l where E has additive reduction such that $E(\mathbb{Q}_l)$ has a point of order p.*

Then $\mathrm{Sel}_E(\mathbb{Q}_\infty)_p$ is infinite.

Remark. If E has multiplicative reduction at l, then $a_l = \pm 1$. Thus, in (iii), $a_l \equiv 1 \pmod{p}$ is always true if $p = 2$ and is equivalent to $a_l = 1$ if p is odd. Also, (iii) and (iv) simply state that there exists an l such that $p | c_l$. If E has additive reduction at l, then the only possible prime factors of c_l are 2, 3, or l. Since E has good reduction at p, (iv) can only occur for $p = 2$ or 3.

Proof. If $\mathrm{Sel}_E(\mathbb{Q})_p$ is infinite, the conclusion follows from theorem 1.2, or more simply from lemma 3.1. If $\mathrm{Sel}_E(\mathbb{Q})_p$ is finite, then we can apply proposition 4.1 to say that $f_E(0)$ is not in $\mathbb{Z}_p^\times$. Hence $f_E(T)$ is not invertible and so $X_E(\mathbb{Q}_\infty)$ must indeed be infinite. (The characteristic ideal of a finite Λ-module is Λ.) Alternatively, one can point out that since $E(\mathbb{Q})_p = 0$, it follows that $\ker(h_0) = 0$ and $\mathcal{G}_E^\Sigma(\mathbb{Q}) = \mathcal{P}_E^\Sigma(\mathbb{Q})$, where Σ consists of p, ∞, and all primes of bad reduction. Hence, if (i) holds, then $\mathrm{Sel}_E(\mathbb{Q}_\infty)_p \neq 0$. If (ii), (iii), or (iv) holds, then $\ker(g_0) \neq 0$. Therefore, since $\ker(h_0)$ and $\mathrm{coker}(h_0)$ are both zero, we have $\mathrm{coker}(s_0) \neq 0$. This implies again that $\mathrm{Sel}_E(\mathbb{Q}_\infty)_p \neq 0$. Finally, proposition 4.8 then shows that $\mathrm{Sel}_E(\mathbb{Q}_\infty)_p$ must be infinite. ■

As our examples show, quite a variety of possibilities for the data going into theorem 4.1 can arise. This is made even more clear from the following observation, where is a variant on lemma 8.19 of [Maz1].

Proposition 5.4. *Let P and L be disjoint, finite sets of primes. Let Q be any finite set of primes. For each $p \in P$, let a_p^* be any integer satisfying $|a_p^*| < 2\sqrt{p}$. For each $l \in L$, let $a_l^* = +1$ or -1. If $a_l^* = +1$, let c_l^* be any positive integer. If $a_l^* = -1$, let $c_l^* = 1$ or 2. Then there exist infinitely many non-isomorphic elliptic curves E defined over $\mathbb{Q}$ such that*

(i) *For each $p \in P$, E has good reduction at p and $a_p = a_p^*$.*
(ii) *For each $l \in L$, E has multiplicative reduction at l, $a_l = a_l^*$, and $c_l = c_l^*$.*
(iii) *For each $q \in Q$, $E[q]$ is irreducible as a $\mathbb{F}_q$-representation space of $G_{\mathbb{Q}}$.*

Proof. This is an application of the Chinese Remainder Theorem. For each $p \in P$, a theorem of Deuring states that an elliptic curve $\widetilde{E}_p$ defined over $\mathbb{F}_p$ exists such that $|\widetilde{E}_p(\mathbb{F}_p)| = 1 + p - a_p^*$. One can then choose arbitrarily a lifting E_p^* of $\widetilde{E}_p$ defined by a Weierstrass equation (as described in Tate's article [Ta]). We write this equation as $f_p^*(x, y) = 0$ where $f_p^*(x, y) \in \mathbb{Z}_p[x, y]$. Let $l \in L$. If $a_l^* = +1$, we let E_l^* denote the Tate curve over $\mathbb{Q}_l$ with $j_{E_l^*} = l^{-c_l^*}$. Then E_l^* has split, multiplicative reduction at l and $\mathrm{ord}_l(j_E) = -c_l^*$. If $a_l^* = -1$, then we instead take E_l^* as the unramified quadratic twist of this Tate curve, so that E_l^* has non-split, multiplicative reduction. The index $[E_l^*(\mathbb{Q}_l) : E_{l,0}^*(\mathbb{Q}_l)]$ is then 1 or 2, depending on the parity of c_l^*. In either case, we let $f_l^*(x, y) = 0$ be a Weierstrass equation for E_l^*, where $f_l^*(x, y) \in \mathbb{Z}_l[x, y]$. Let $q \in Q$. Then we can choose a prime $r = r_q \neq q$ and an elliptic curve $\widetilde{E}_r$ defined over $\mathbb{F}_r$ such that $\widetilde{E}_r[q]$ is irreducible for the action of $G_{\mathbb{F}_r}$. If $q = 2$, this is easy. We take r to be an odd prime and define $\widetilde{E}_r$ by $y^2 = g(x)$, where $g(x) \in \mathbb{F}_r[x]$ is an irreducible cubic polynomial. Then $\widetilde{E}_r(\mathbb{F}_r)$ has no element of order 2, which suffices. If q is odd, we choose r to be an odd prime such that $-r$ is a quadratic nonresidue mod q. We can choose $\widetilde{E}/\mathbb{F}_r$ to be supersingular. Then the action of $\mathrm{Frob}_r \in G_{\mathbb{F}_r}$ on $\widetilde{E}_r[q]$ has characteristic polynomial $t^2 + r$. Since this has no roots in $\mathbb{F}_q$, $\widetilde{E}_r[q]$ indeed has no proper invariant subspaces under the action of Frob_r.

We can choose a lifting E_r^* of $\widetilde{E}_r$ defined over $\mathbb{Q}_r$ by a Weierstrass equation $f_r^*(x,y) = 0$, where $f_r^*(x,y) \in \mathbb{Z}_r[x,y]$. For each $q \in Q$, infinitely many suitable r_q's exist. Hence we can also require that the r_q's are distinct and outside of $P \cap L$. We let $R = \{r_q\}_{q \in Q}$, choosing one r_q for each $q \in Q$. We then choose an equation $f(x,y) = 0$ in Weierstrass form, where $f(x,y) \in \mathbb{Z}[x,y]$ and satisfies $f(x,y) \equiv f_m^*(x,y) \pmod{m^{t_m}}$ for all $m \in P \cup L \cup R$, where t_m is chosen sufficiently large. The equation $f(x,y) = 0$ determines an elliptic curve defined over $\mathbb{Q}$. If $p \in P$, we just take $t_p = 1$. Then E has good reduction at p, $\widetilde{E} = \widetilde{E}_p$, and hence $a_p = a_p^*$, as desired. If $r \in R$, then $r = r_q$ for some $q \in Q$. We take $t_r = 1$ again. E has good reduction at r, $\widetilde{E} = \widetilde{E}_r$, and hence the action of a Frobenius automorphism in $\mathrm{Gal}(\mathbb{Q}(E[q])/\mathbb{Q})$ (for any prime above r) on $E[q]$ has no invariant subspaces. Hence obviously $E[q]$ is irreducible as an $\mathbb{F}_q$-representation space of $G_{\mathbb{Q}}$. Finally, suppose $l \in L$. If we take t_l sufficiently large, then clearly j_E will be close enough to $j_{E_l^*}$ to guarantee that $\mathrm{ord}_l(j_E) = \mathrm{ord}_l(j_{E_l^*}) = -c_l^*$. In terms of the coefficients of a Weierstrass equation over $\mathbb{Z}_l$, there is a simple criterion for an elliptic curve to have split or nonsplit reduction at l. (It involves the coset in $\mathbb{Q}_l^\times/(\mathbb{Q}_l^\times)^2$ containing the quantity $-c_4/c_6$ in the notation of Tate.) Hence it is clear that E will have multiplicative reduction at l and that $a_l = a_l^*$ if t_l is taken large enough. Thus E will have the required properties. The fact that infinitely many non-isomorphic E's exist is clear, since we can vary L and thus the set of primes where E has bad reduction. ■

Remark. We can assume that $P \cup L$ contains 3 and 5. Any elliptic curve E defined over $\mathbb{Q}$ and satisfying (i) and (ii) will be semistable at 3 and 5 and therefore will be modular. This follows from a theorem of Diamond [D]. Furthermore, let E_d denote the quadratic twist of E by some square-free integer d. If we assume that all primes in $P \cup L$ split in $\mathbb{Q}(\sqrt{d})$, then E_d also satisfies (i), (ii), and (iii). One can choose such d so that $L(E_d/\mathbb{Q}, 1) \neq 0$. (See [B-F-H] for a discussion of this result which was first proved by Waldspurger.) A theorem of Kolyvagin then would imply that $E_d(\mathbb{Q})$ and $\text{Ш}_{E_d}(\mathbb{Q})$ are finite. Thus, there in fact exist infinitely many non-isomorphic modular elliptic curves E satisfying (i), (ii), and (iii) and such that $\mathrm{Sel}_E(\mathbb{Q})$ is finite.

Corollary 5.5. *Let P be any finite set of primes. Then there exist infinitely many elliptic curves $E/\mathbb{Q}$ such that E has good, ordinary reduction at p, $a_p = 1$, and $E[p]$ is an irreducible $\mathbb{F}_p$-representation space for $G_{\mathbb{Q}}$, for all $p \in P$.*

Proof. This follows immediately from proposition 5.4. One takes $P = Q$, $a_p^* = 1$ for all $p \in P$, and $L = \varnothing$.

Corollary 5.6. *Let p be any prime. Assume that conjecture 1.11 is true when $F = \mathbb{Q}$. Then λ_E is unbounded as E varies over elliptic curves defined over $\mathbb{Q}$ with good, ordinary reduction at p.*

Proof. Take $P = \{p\} = Q$. Let a_p^* be such that $p \nmid a_p^*$. Take L to be a large finite set of primes. For each $l \in L$, let $a_l^* = +1$, $c_l^* = p$. Let $E/\mathbb{Q}$ be any elliptic curve satisfying the statements in proposition 5.4. As remarked above, we can assume E is modular. Now E has good, ordinary reduction at p. According to Theorem 1.5, $\mathrm{Sel}_E(\mathbb{Q}_\infty)_p$ is Λ-cotorsion. (Alternatively, we could assume that $\mathrm{Sel}_E(\mathbb{Q})_p$ is finite and then use theorem 1.4. The rest of this proof becomes somewhat easier if we make this assumption on E.) Also $\mu_E = 0$ by conjecture 1.11. We will show that $\lambda_E \geq |L|$, which certainly implies the corollary. Let $t = |L|$. Let $n = \mathrm{corank}_{\mathbb{Z}_p}(\mathrm{Sel}_E(\mathbb{Q})_p)$. Of course, $\lambda_E \geq n$ by theorem 1.2. So we can assume now that $n \leq t$. Let Σ be the set of primes p, ∞, and all primes where E has bad reduction. Then, by proposition 4.13, there are at most p^n elements of order p in $\mathcal{P}_E^\Sigma(\mathbb{Q})/\mathcal{G}_E^\Sigma(\mathbb{Q})$. Also, for each $l \in L$, we have $|\ker(r_l)| = c_l = p$. Thus, the kernel of the restriction map $\mathcal{P}_E^\Sigma(\mathbb{Q}) \to \mathcal{P}_E^\Sigma(\mathbb{Q}_\infty)$ contains a subgroup isomorphic to $(\mathbb{Z}/p\mathbb{Z})^t$. It follows that $\ker(g_0)$ contains a subgroup isomorphic to $(\mathbb{Z}/p\mathbb{Z})^{t-n}$. Now $\ker(h_0) = \mathrm{coker}(h_0) = 0$. Thus it follows that $\mathrm{coker}(s_0)$ contains a subgroup isomorphic to $(\mathbb{Z}/p\mathbb{Z})^{t-n}$. By proposition 4.14, and the assumption that $\mu_E = 0$, we have

$$\mathrm{Sel}_E(\mathbb{Q}_\infty)_p \cong (\mathbb{Q}_p/\mathbb{Z}_p)^{\lambda_E}.$$

But $\mathrm{Sel}_E(\mathbb{Q}_\infty)_p$ contains a subgroup $\mathrm{Im}(s_0)_{\mathrm{div}}$ isomorphic to $(\mathbb{Q}_p/\mathbb{Z}_p)^n$ and the corresponding quotient has a subgroup isomorphic to $(\mathbb{Z}/p\mathbb{Z})^{t-n}$. It follows that $\lambda_E \geq t$, as we claimed. ■

Remark. If we don't assume conjecture 1.11, then one still gets the weaker result that $\lambda_E + \mu_E$ is unbounded as E varies over modular elliptic curves with good, ordinary reduction at a fixed prime p. For the above argument shows that $\dim_{\Lambda/\mathbf{m}\Lambda}(X_E(\mathbb{Q}_\infty)/\mathbf{m}X_E(\mathbb{Q}_\infty))$ is unbounded, where $\mathbf{m}$ denotes the maximal ideal of Λ. We then use the following result about Λ-modules: *Suppose X is a finitely generated, torsion Λ-module and that X has no nonzero, finite Λ-submodules. Let λ and μ denote the corresponding invariants. Then*

$$\lambda + \mu \geq \dim_{\Lambda/\mathbf{m}\Lambda}(X/\mathbf{m}X).$$

The proof is not difficult. One first notes that the right-hand side, which is just the minimal number of generators of X as a Λ-module, is "sub-additive" in an exact sequence $0 \to X_1 \to X_2 \to X_3 \to 0$ of Λ-modules. Both λ and μ are additive. One then reduces to the special cases where either (a) X has exponent p and has no finite, nonzero Λ-submodule or (b) X has no $\mathbb{Z}_p$-torsion. In the first case, X is a $(\Lambda/p\Lambda)$-module. One then uses the fact that $\Lambda/p\Lambda$ is a PID. In the second case, λ is the minimal number of generators of X as a $\mathbb{Z}_p$-module. The inequality is clear.

We will now discuss the μ-invariant μ_E of $\mathrm{Sel}_E(\mathbb{Q}_\infty)_p$. We always assume that E is defined over $\mathbb{Q}$ and has either good, ordinary or multiplicative reduction at p. According to conjecture 1.11, we should have $\mu_E = 0$ if $E[p]$ is irreducible as a $G_\mathbb{Q}$-module. Unfortunately, it seems very difficult to verify this even for specific examples. In this discussion we will assume that $E[p]$ is reducible as a $G_{\mathbb{Q}}$-module, i.e., that E admits a cyclic $\mathbb{Q}$-isogeny of degree p. In [Maz2], Mazur proves that this can happen only for a certain small set of primes p. With the above restriction on the reduction type of E at p, then p is limited to the set $\{2,3,5,7,13,37\}$. For $p = 2,3,5,7$, or 13, there are infinitely many possible E's, even up to quadratic twists. For $p = 37$, E must be the elliptic curve defined by $y^2 + xy + y = x^3 + x^2 - 8x + 6$ (which has conductor 35^2) or another elliptic curve related to this by a $\mathbb{Q}$-isogeny of degree 37, up to a quadratic twist.

Assume at first that $E[p]$ contains a $G_{\mathbb{Q}}$-invariant subgroup Φ isomorphic to μ_p. We will let Σ be the finite set consisting of p, ∞, and all primes where E has bad reduction. Then we have a natural map

$$\epsilon : H^1(\mathbb{Q}_\Sigma/\mathbb{Q}_\infty, \Phi) \to H^1(\mathbb{Q}_\Sigma/\mathbb{Q}_\infty, E[p^\infty]).$$

It is easy to verify that $\ker(\epsilon)$ is finite. We also have the Kummer homomorphism

$$\beta : \mathcal{U}_\infty/\mathcal{U}_\infty^p \to H^1(\mathbb{Q}_\Sigma/\mathbb{Q}_\infty, \mu_p),$$

where $\mathcal{U}_\infty$ denotes the unit group of $\mathbb{Q}_\infty$. The map β is injective. Dirichlet's unit theorem implies that the $(\Lambda/p\Lambda)$-module $\mathcal{U}_\infty/\mathcal{U}_\infty^p$ has corank 1. Consider a prime $l \neq p$. Let η be a prime of $\mathbb{Q}_\infty$ lying over l. Then $(\mathbb{Q}_\infty)_\eta$ is the unramified $\mathbb{Z}_p$-extension of $\mathbb{Q}_l$ (which is the only $\mathbb{Z}_p$-extension of $\mathbb{Q}_l$). All units of $(\mathbb{Q}_\infty)_\eta$ are p-th powers. Thus, if $u \in \mathcal{U}_\infty$, then u is a p-th power in $(\mathbb{Q}_\infty)_\eta$. Therefore, if $\varphi \in \mathrm{Im}(\beta)$, then $\varphi|_{G_{(\mathbb{Q}_\infty)_\eta}}$ is trivial. If we fix an isomorphism $\Phi \cong \mu_p$, then it follows that the elements of $\mathrm{Im}(\epsilon \circ \beta)$ satisfy the local conditions defining $\mathrm{Sel}_E(\mathbb{Q}_\infty)_p$ at all primes η of $\mathbb{Q}_\infty$ not lying over p or ∞. Now assume that p is odd. We can then ignore the archimedean primes of $\mathbb{Q}_\infty$. Since the inertia subgroup $I_{\mathbb{Q}_p}$ acts nontrivially on μ_p (because p is odd) and acts trivially on $E[p^\infty]/C_p$, it follows that $\Phi \subseteq C_p$. If π denotes the unique prime of $\mathbb{Q}_\infty$ lying over p, recall that $\mathrm{Im}(\kappa_\pi) = \mathrm{Im}(\lambda_\pi)$, where λ_π is the map

$$H^1((\mathbb{Q}_\infty)_\pi, C_p) \to H^1((\mathbb{Q}_\infty)_\pi, E[p^\infty]).$$

Therefore, it is obvious that if $\varphi \in \mathrm{Im}(\epsilon)$, then $\varphi|_{G_{(\mathbb{Q}_\infty)_\pi}} \in \mathrm{Im}(\kappa_\pi)$. Combining the above observations, it follows that $\mathrm{Im}(\epsilon \circ \beta) \subseteq \mathrm{Sel}_E(\mathbb{Q}_\infty)_p$ if p is odd. Thus, $\mathrm{Sel}_E(\mathbb{Q}_\infty)_p$ contains a Λ-submodule of exponent p with $(\Lambda/p\Lambda)$-corank equal to 1, which implies that either $\mu_E \geq 1$ or $\mathrm{Sel}_E(\mathbb{Q}_\infty)_p$ is not Λ-cotorsion.

We will prove a more general result. Suppose that $E[p^\infty]$ has a $G_{\mathbb{Q}}$-invariant subgroup Φ which is cyclic of order p^m, with $m \geq 1$. If E has semistable reduction at p, then it actually follows that E has either good, ordinary reduction or multiplicative reduction at p. Φ has a $G_{\mathbb{Q}}$-composition series with composition factors isomorphic to $\Phi[p]$. We assume again that p is an odd prime. Then the action of $I_{\mathbb{Q}_p}$ on $\Phi[p]$ is either trivial or given by the

Teichmüller character ω. In the first case, Φ is isomorphic as a $G_{\mathbb{Q}_p}$-module to a subgroup of $E[p^\infty]/C_p$. The action of $I_{\mathbb{Q}_p}$ on Φ is trivial and so we say that Φ is unramified at p. In the second case, we have $\Phi \subseteq C_p$ and we say that Φ is ramified at p. The action of $\mathrm{Gal}(\mathbb{C}/\mathbb{R})$ on $\Phi[p]$ determines its action on Φ. We say that Φ is even or odd, depending on whether the action of $\mathrm{Gal}(\mathbb{C}/\mathbb{R})$ is trivial or nontrivial. With this terminology, we can state the following result.

Proposition 5.7. *Assume that p is odd and that E is an elliptic curve/$\mathbb{Q}$ with good, ordinary or multiplicative reduction at p. Assume that $\mathrm{Sel}_E(\mathbb{Q}_\infty)_p$ is Λ-cotorsion. Assume also that $E[p^\infty]$ contains a cyclic $G_{\mathbb{Q}}$-invariant subgroup Φ of order p^m which is ramified at p and odd. Then $\mu_E \geq m$.*

Proof. We will show that $\mathrm{Sel}_E(\mathbb{Q}_\infty)_p$ contains a Λ-submodule pseudo-isomorphic to $\widehat{\Lambda}[p^m]$. Consider the map

$$\epsilon : H^1(\mathbb{Q}_\Sigma/\mathbb{Q}_\infty, \Phi) \to H^1(\mathbb{Q}_\Sigma/\mathbb{Q}_\infty, E[p^\infty]).$$

The kernel is finite. Let $l \in \Sigma$, $l \neq p$ or ∞. There are just finitely many primes η of $\mathbb{Q}_\infty$ lying over l. For each η, $H^1((\mathbb{Q}_\infty)_\eta, \Phi)$ is finite. (An easy way to verify this is to note that any Sylow pro-p subgroup V of $G_{(\mathbb{Q}_\infty)_\eta}$ is isomorphic to $\mathbb{Z}_p$ and that the restriction map $H^1((\mathbb{Q}_\infty)_\eta, \Phi) \to H^1(V, \Phi)$ is injective.) Therefore

$$\ker\big(H^1(\mathbb{Q}_\Sigma/\mathbb{Q}_\infty, \Phi) \to \prod_{\substack{l \in \Sigma \\ l \neq p,\infty}} \prod_{\eta | l} H^1((\mathbb{Q}_\infty)_\eta, \Phi)\big)$$

has finite index in $H^1(\mathbb{Q}_\Sigma/\mathbb{Q}_\infty, \Phi)$. On the other hand, $\Phi \subseteq C_p$. Hence, elements in $\mathrm{Im}(\epsilon)$ automatically satisfy the local condition at π occurring in the definition of $\mathrm{Sel}_E(\mathbb{Q}_\infty)_p$. These remarks imply that $\mathrm{Im}(\epsilon) \cap \mathrm{Sel}_E(\mathbb{Q}_\infty)_p$ has finite index in $\mathrm{Im}(\epsilon)$ and therefore $\mathrm{Sel}_E(\mathbb{Q}_\infty)_p$ contains a Λ-submodule pseudo-isomorphic to $H^1(\mathbb{Q}_\Sigma/\mathbb{Q}_\infty, \Phi)$.

One can study the structure of $H^1(\mathbb{Q}_\Sigma/\mathbb{Q}_\infty, \Phi)$ either by restriction to a subgroup of finite index in $\mathrm{Gal}(\mathbb{Q}_\Sigma/\mathbb{Q}_\infty)$ which acts trivially on Φ or by using Euler characteristics. We will sketch the second approach. The restriction map

$$H^1(\mathbb{Q}_\Sigma/\mathbb{Q}_n, \Phi) \to H^1(\mathbb{Q}_\Sigma/\mathbb{Q}_\infty, \Phi)^{\Gamma_n}$$

is surjective and its kernel is finite and has bounded order as $n \to \infty$. The Euler characteristic of the $\mathrm{Gal}(\mathbb{Q}_\Sigma/\mathbb{Q}_n)$-module Φ is $\prod_{v|\infty} |\Phi/\Phi^{D_v}|^{-1}$, where v runs over the infinite primes of the totally real field $\mathbb{Q}_n$ and $D_v = G_{(\mathbb{Q}_n)_v}$. By assumption, $\Phi^{D_v} = 0$ and hence this Euler characteristic is p^{-mp^n} for all $n \geq 0$. Therefore, $H^1(\mathbb{Q}_\Sigma/\mathbb{Q}_n, \Phi)$ has order divisible by p^{mp^n}. It follows that the Λ-module $H^1(\mathbb{Q}_\Sigma/\mathbb{Q}_\infty, \Phi)$, which is of exponent p^m and hence certainly Λ-cotorsion, must have μ-invariant $\geq m$. This suffices to prove that

$\mu_E \geq m$. Under the assumptions that $E[p]$ is reducible as a $G_{\mathbb{Q}}$-module and that $\mathrm{Sel}_E(\mathbb{Q}_\infty)_p$ is Λ-cotorsion, it follows from the next proposition that $\mathrm{Sel}_E(\mathbb{Q}_\infty)_p$ contains a Λ-submodule pseudo-isomorphic to $\widehat{\Lambda}[p^{\mu_E}]$ and that the corresponding quotient has finite $\mathbb{Z}_p$-corank. Also, $\mathrm{Im}(\epsilon)$ must almost coincide with $H^1(\mathbb{Q}_\Sigma/\mathbb{Q}_\infty, E[p^\infty])[p^m]$. (That is, the intersection of the two groups must have finite index in both.) This last Λ-module is pseudo-isomorphic to $\widehat{\Lambda}[p^m]$ according to the proposition below.

If E is any elliptic curve/$\mathbb{Q}$ and p is any prime, the weak Leopoldt conjecture would imply that $H^1(\mathbb{Q}_\Sigma/\mathbb{Q}_\infty, E[p^\infty])$ has Λ-corank equal to 1. That is, $H^1(\mathbb{Q}_\Sigma/\mathbb{Q}_\infty, E[p^\infty])_{\Lambda\text{-div}}$ should be pseudo-isomorphic to $\widehat{\Lambda}$. (This has been proven by Kato if E is modular.) Here we will prove a somewhat more precise statement under the assumption that $E[p]$ is reducible as a $G_{\mathbb{Q}}$-module. It will be a rather simple consequence of the Ferrero-Washington theorem mentioned in the introduction. As usual, Σ is a finite set of primes of $\mathbb{Q}$ containing p, ∞, and all primes where E has bad reduction.

Proposition 5.8. *Assume that E is an elliptic curve defined over $\mathbb{Q}$ and that E admits a $\mathbb{Q}$-isogeny of degree p for some prime p. Then $H^1(\mathbb{Q}_\Sigma/\mathbb{Q}_\infty, E[p^\infty])$ has Λ-corank 1. Furthermore, $H^1(\mathbb{Q}_\Sigma/\mathbb{Q}_\infty, E[p^\infty])/H^1(\mathbb{Q}_\Sigma/\mathbb{Q}_\infty, E[p^\infty])_{\Lambda\text{-div}}$ has μ-invariant equal to 0 if p is odd. If $p=2$, this quotient has μ-invariant equal to 0 or 1, depending on whether $E(\mathbb{R})$ has 1 or 2 connected components.*

Proof. First assume that p is odd. Then we have an exact sequence

$$0 \to \Phi \to E[p] \to \Psi \to 0$$

where $\mathrm{Gal}(\mathbb{Q}_\Sigma/\mathbb{Q})$ acts on the cyclic groups Φ and Ψ of order p by characters $\varphi, \psi : \mathrm{Gal}(\mathbb{Q}_\Sigma/\mathbb{Q}) \to (\mathbb{Z}/p\mathbb{Z})^\times$. We know that $H^1(\mathbb{Q}_\Sigma/\mathbb{Q}_\infty, E[p^\infty])$ has Λ-corank ≥ 1. Also, the exact sequence

$$0 \to E[p] \to E[p^\infty] \xrightarrow{p} E[p^\infty] \to 0$$

induces a surjective map $H^1(\mathbb{Q}_\Sigma/\mathbb{Q}_\infty, E[p]) \to H^1(\mathbb{Q}_\Sigma/\mathbb{Q}_\infty, E[p^\infty])[p]$ with finite kernel. Thus, it clearly is sufficient to prove that $H^1(\mathbb{Q}_\Sigma/\mathbb{Q}_\infty, E[p])$ has $(\Lambda/p\Lambda)$-corank 1. Now the determinant of the action of $G_{\mathbb{Q}}$ on $E[p]$ is the Teichmüller character ω. Hence, $\varphi\psi = \omega$. Since ω is an odd character, one of the characters φ or ψ is odd, the other even. We have the following exact sequence:

$$H^1(\mathbb{Q}_\Sigma/\mathbb{Q}_\infty, \Phi) \to H^1(\mathbb{Q}_\Sigma/\mathbb{Q}_\infty, E[p]) \to H^1(\mathbb{Q}_\Sigma/\mathbb{Q}_\infty, \Psi)$$

and hence proposition 5.8 (for odd primes p) is a consequence of the following lemma.

Lemma 5.9. *Let p be any prime. Let Θ be a $\mathrm{Gal}(\mathbb{Q}_\Sigma/\mathbb{Q})$-module which is cyclic of order p. Then $H^1(\mathbb{Q}_\Sigma/\mathbb{Q}_\infty, \Theta)$ has $(\Lambda/p\Lambda)$-corank 1 if Θ is odd or if $p=2$. Otherwise, $H^1(\mathbb{Q}_\Sigma/\mathbb{Q}_\infty, \Theta)$ is finite. If $p=2$, then the map*

$$\alpha : H^1(\mathbb{Q}_\Sigma/\mathbb{Q}_\infty, \Theta) \to \mathcal{P}_\Theta^{(\infty)}(\mathbb{Q}_\infty)$$

is surjective and has finite kernel. Here $\mathcal{P}_{\Theta}^{(\infty)}(\mathbb{Q}_{\infty}) = \varinjlim_{n} \prod_{v_n|\infty} H^1((\mathbb{Q}_n)_{v_n}, \Theta)$.

Remark. We will use a similar notation to that introduced in the remark following lemma 4.6. For example, $\mathcal{P}_C^{(\ell)}(\mathbb{Q}_\infty)$, which occurs in the following proof, is defined as $\varinjlim_{n} \prod_{v_n|\ell} H^1((\mathbb{Q}_n)_{v_n}, C)$. If ℓ is a nonarchimedean prime, then $\mathcal{P}_C^{(\ell)}(\mathbb{Q}_\infty) = \prod_{\eta|\ell} H^1((\mathbb{Q}_\infty)_\eta, C)$, since there are only finitely many primes η of $\mathbb{Q}_\infty$ lying over ℓ.

Proof. Let θ be the character (with values in $(\mathbb{Z}/p\mathbb{Z})^\times$) which gives the action of $\mathrm{Gal}(\mathbb{Q}_\Sigma/\mathbb{Q})$ on Θ. Let $C = (\mathbb{Q}_p/\mathbb{Z}_p)(\theta)$, where we now regard θ as a character of $\mathrm{Gal}(\mathbb{Q}_\Sigma/\mathbb{Q})$ with values in $\mathbb{Z}_p^\times$. Then $\Theta = C[p]$. We have an isomorphism

$$H^1(\mathbb{Q}_\Sigma/\mathbb{Q}_\infty, \Theta) \xrightarrow{\sim} H^1(\mathbb{Q}_\Sigma/\mathbb{Q}_\infty, C)[p].$$

(The surjectivity is clear. The injectivity follows from the fact that $H^0(\mathbb{Q}_\Sigma/\mathbb{Q}_\infty, C)$ is either C or 0, depending on whether θ is trivial or nontrivial.) We will relate the structure of $H^1(\mathbb{Q}_\Sigma/\mathbb{Q}_\infty, C)$ to various classical Iwasawa modules. Let $\Sigma' = \Sigma - \{p\}$. Consider

$$S'_C(\mathbb{Q}_\infty) = \ker\Big(H^1(\mathbb{Q}_\Sigma/\mathbb{Q}_\infty, C) \to \prod_{\ell \in \Sigma'} \mathcal{P}_C^{(\ell)}(\mathbb{Q}_\infty)\Big).$$

If $\ell \in \Sigma'$ is nonarchimedean, then $H^1((\mathbb{Q}_\infty)_\eta, C)$ is either trivial or isomorphic to $\mathbb{Q}_p/\mathbb{Z}_p$, for any prime η of $\mathbb{Q}_\infty$ lying over ℓ. $\mathcal{P}_C^{(\ell)}(\mathbb{Q}_\infty)$ is then a cotorsion Λ-module with μ-invariant 0. If $\ell = \infty$, then $(\mathbb{Q}_\infty)_\eta = \mathbb{R}$ for any $\eta|\ell$. $H^1(\mathbb{R}, C)$ is, of course, trivial if p is odd. But if $p = 2$, then θ is trivial and $H^1(\mathbb{R}, C) \cong \mathbb{Z}/2\mathbb{Z}$. Thus, in this case, $\mathcal{P}_C^{(\infty)}(\mathbb{Q}_\infty)$ is isomorphic to $\mathrm{Hom}(\Lambda/2\Lambda, \mathbb{Z}/2\mathbb{Z}) = \widehat{\Lambda}[2]$, which is Λ-cotorsion and has μ-invariant 1. It follows that $H^1(\mathbb{Q}_\Sigma/\mathbb{Q}_\infty, C)/S'_C(\mathbb{Q}_\infty)$ is Λ-cotorsion and has μ-invariant 0 if p is odd. If $p = 2$, then the μ-invariant is ≤ 1.

Assume that p is odd. Let F be the cyclic extension of $\mathbb{Q}$ corresponding to θ. (Thus, $F \subseteq \mathbb{Q}_\Sigma$ and θ is a faithful character of $\mathrm{Gal}(F/\mathbb{Q})$.) Then $F_\infty = F\mathbb{Q}_\infty$ is the cyclotomic $\mathbb{Z}_p$-extension of F. We let $\Delta = \mathrm{Gal}(F_\infty/\mathbb{Q}_\infty) \cong \mathrm{Gal}(F/\mathbb{Q})$. Let

$$X = \mathrm{Gal}(L_\infty/F_\infty), \qquad Y = \mathrm{Gal}(M_\infty/F_\infty)$$

where M_∞ is the maximal abelian pro-p extension of F_∞ unramified at all primes of F_∞ not lying over p and L_∞ is the maximal subfield of M_∞ unramified at the primes of F_∞ over p too. Now $\mathrm{Gal}(F_\infty/\mathbb{Q}) \cong \Delta \times \Gamma$ acts on both X and Y by inner automorphisms. Thus, they are both Λ-modules on which Δ acts Λ-linearly. That is, X and Y are $\Lambda[\Delta]$-modules.

The restriction map $H^1(\mathbb{Q}_\Sigma/\mathbb{Q}_\infty, C) \to H^1(\mathbb{Q}_\Sigma/F_\infty, C)^\Delta$ is an isomorphism. Also, $\mathrm{Gal}(\mathbb{Q}_\Sigma/F_\infty)$ acts trivially on C. Hence the elements of $H^1(\mathbb{Q}_\Sigma/F_\infty, C)$ are homomorphisms. Taking into account the local conditions, the restriction map induces an isomorphism

$$S'_C(\mathbb{Q}_\infty) \xrightarrow{\sim} \mathrm{Hom}_\Delta(Y, C) = \mathrm{Hom}(Y^\theta, C)$$

as Λ-modules, where $Y^\theta = e_\theta Y$, the θ-component of the Δ-module Y. (Here e_θ denotes the idempotent for θ in $\mathbb{Z}_p[\Delta]$.) Iwasawa proved that Y^θ is Λ-torsion if θ is even and has Λ-rank 1 if θ is odd. One version of the Ferrero-Washington theorem states that the μ-invariant of Y^θ vanishes if θ is even. Thus, in this case, $H^1(\mathbb{Q}_\Sigma/\mathbb{Q}_\infty, C)$ must be Λ-cotorsion and have μ-invariant 0. It then follows that $H^1(\mathbb{Q}_\Sigma/\mathbb{Q}_\infty, \Theta)$ must be finite. On the other hand, if θ is odd, then $S'_C(\mathbb{Q}_\infty)$ will have Λ-corank 1. Hence, the same is true of $H^1(\mathbb{Q}_\Sigma/\mathbb{Q}_\infty, C)$ and so $H^1(\mathbb{Q}_\Sigma/\mathbb{Q}_\infty, C)[p]$ will have $(\Lambda/p\Lambda)$-corank ≥ 1. We will prove that equality holds and, therefore, $H^1(\mathbb{Q}_\Sigma/\mathbb{Q}_\infty, \Theta)$ indeed has $(\Lambda/p\Lambda)$-corank 1. It is sufficient to prove that $S'_C(\mathbb{Q}_\infty)[p]$ has $(\Lambda/p\Lambda)$-corank 1. We will deduce this from another version of the Ferrero-Washington theorem—the assertion that the torsion Λ-module X has μ-invariant 0. Let π be the unique prime of $\mathbb{Q}_\infty$ lying over p. Consider

$$S_C(\mathbb{Q}_\infty) = \ker(S'_C(\mathbb{Q}_\infty) \to H^1((\mathbb{Q}_\infty)_\pi, C)).$$

In the course of proving lemma 2.3, we actually determined the structure of $H^1((\mathbb{Q}_\infty)_\pi, C)$. (See also section 3 of [Gr2].) It has Λ-corank 1 and the quotient $H^1((\mathbb{Q}_\infty)_\pi, C)/H^1((\mathbb{Q}_\infty)_\pi, C)_{\Lambda\text{-div}}$ is either trivial or isomorphic to $\mathbb{Q}_p/\mathbb{Z}_p$ as a group. To show that $S'_C(\mathbb{Q}_\infty)[p]$ has $(\Lambda/p\Lambda)$-corank 1, it suffices to prove that $S_C(\mathbb{Q}_\infty)[p]$ is finite. Now the restriction map identifies $S_C(\mathbb{Q}_\infty)$ with the subgroup of $\mathrm{Hom}_\Delta(Y, C)$ which is trivial on all the decomposition subgroups of Y corresponding to primes of F_∞ lying over p. Thus, $S_C(\mathbb{Q}_\infty)$ is isomorphic to a Λ-submodule of $\mathrm{Hom}_\Delta(X, C) = \mathrm{Hom}(X^\theta, C)$. Since the μ-invariant of X vanishes, it is clear that $S_C(\mathbb{Q}_\infty)[p]$ is indeed finite. This completes the proof of lemma 5.9 when p is odd.

Now assume that $p = 2$. Thus, θ is trivial. We let $F_\infty = \mathbb{Q}_\infty$. Let M_∞ be as defined above. Then it is easy to see that $M_\infty = \mathbb{Q}_\infty$. For let M_0 be the maximal abelian extension of $\mathbb{Q}$ contained in M_∞. Thus, $\mathrm{Gal}(M_0/\mathbb{Q}_\infty) \cong Y/TY$. We must have $M_0 \subseteq \mathbb{Q}(\mu_{2^\infty})$. But M_0 is totally real and so clearly $M_0 = \mathbb{Q}_\infty$. Hence $Y/TY = 0$. This implies that $Y = 0$ and hence that $M_\infty = \mathbb{Q}_\infty$. Therefore, $S'_C(\mathbb{Q}_\infty) = 0$. It follows that $H^1(\mathbb{Q}_\Sigma/\mathbb{Q}_\infty, C)$ is Λ-cotorsion and has μ-invariant ≤ 1. In fact, the μ-invariant is 1 and arises in the following way. Let $\mathcal{U}_\infty$ denote the unit group of $\mathbb{Q}_\infty$. Let $K_\infty = \mathbb{Q}_\infty(\{\sqrt{u} \,|\, u \in \mathcal{U}_\infty\})$. Then $K_\infty \subseteq M^*_\infty$, the maximal abelian pro-2 extension of $\mathbb{Q}_\infty$ unramified outside of the primes over p and ∞. Also, one can see that $\mathrm{Gal}(K_\infty/\mathbb{Q}_\infty) \cong \Lambda/2\Lambda$. Thus, clearly $H^1(\mathbb{Q}_\Sigma/\mathbb{Q}_\infty, C)[2] = H^1(\mathbb{Q}_\Sigma/\mathbb{Q}_\infty, \Theta)$ contains the Λ-submodule $\mathrm{Hom}(\mathrm{Gal}(K_\infty/\mathbb{Q}_\infty), \Theta)$ which has μ-invariant 1. To complete the proof of lemma 5.9, we point out that K_∞ can't contain any totally real subfield larger than $\mathbb{Q}_\infty$, since $M_\infty = \mathbb{Q}_\infty$. That is,

$\ker(\alpha) \cap \mathrm{Hom}(\mathrm{Gal}(K_\infty/\mathbb{Q}_\infty), \Theta)$ is trivial. It follows that $\ker(\alpha)$ is finite. We also see that α must be surjective because $\mathcal{P}_{\Theta}^{(\infty)}(\mathbb{Q}_\infty)$ is isomorphic to $\widehat{\Lambda}[2]$. ■

We must complete the proof of proposition 5.8 for $p = 2$. Consider the following commutative diagram with exact rows:

$$\begin{array}{ccccc} H^1(\mathbb{Q}_\Sigma/\mathbb{Q}_\infty, \Phi) & \xrightarrow{a} & H^1(\mathbb{Q}_\Sigma/\mathbb{Q}_\infty, E[2]) & \xrightarrow{b} & H^1(\mathbb{Q}_\Sigma/\mathbb{Q}_\infty, \Psi) \\ \downarrow{\scriptstyle \alpha_1} & & \downarrow{\scriptstyle \alpha_2} & & \downarrow{\scriptstyle \alpha_3} \\ \mathcal{P}_{\Phi}^{(\infty)}(\mathbb{Q}_\infty) & \xrightarrow{c} & \mathcal{P}_{E[2]}^{(\infty)} & \xrightarrow{d} & \mathcal{P}_{\Psi}^{(\infty)}(\mathbb{Q}_\infty) \end{array}$$

By lemma 5.9, both α_1 and α_3 are surjective and have finite kernel. Also, $\ker(a)$ is finite. We see that $H^1(\mathbb{Q}_\Sigma/\mathbb{Q}_\infty, E[2])$ has $(\Lambda/2\Lambda)$-corank equal to 1 or 2. First assume that $E(\mathbb{R})$ is connected, i.e., that the discriminant of a Weierstrass equation for E is negative. Then $H^1(\mathbb{R}, E[2]) = 0$, and so $\mathcal{P}_{E[2]}^{(\infty)}(\mathbb{Q}_\infty) = 0$. It follows that $d \circ \alpha_2$ is the zero map and hence $\mathrm{Im}(b)$ is finite. Thus, $H^1(\mathbb{Q}_\Sigma/\mathbb{Q}_\infty, E[2])$ is pseudo-isomorphic to $H^1(\mathbb{Q}_\Sigma/\mathbb{Q}_\infty, \Phi)$ and so has $(\Lambda/2\Lambda)$-corank 1. In this case, $H^1(\mathbb{Q}_\Sigma/\mathbb{Q}_\infty, E[2^\infty])$ must have Λ-corank 1 and its maximal Λ-cotorsion quotient must have μ-invariant 0. This proves proposition 5.8 in the case that $E(\mathbb{R})$ is connected.

Now assume that $E(\mathbb{R})$ has two components, i.e., that a Weierstrass equation for E has positive discriminant. Then $E[2] \subseteq E(\mathbb{R})$ and $H^1(\mathbb{R}, E[2]) \cong (\mathbb{Z}/2\mathbb{Z})^2$. The $(\Lambda/2\Lambda)$-module $\mathcal{P}_{E[2]}^{(\infty)}(\mathbb{Q}_\infty)$ is isomorphic to $\widehat{\Lambda}[2]^2$. In this case, we will see that $H^1(\mathbb{Q}_\Sigma/\mathbb{Q}_\infty, E[2])$ has $(\Lambda/2\Lambda)$-corank 2. This is clear if $E[2] \cong \Phi \times \Psi$ as a $G_{\mathbb{Q}}$-module. If $E[2]$ is a nonsplit extension of Ψ by Φ, then $F = \mathbb{Q}(E[2])$ is a real quadratic field contained in $\mathbb{Q}_\Sigma$. Let $F_\infty = F\mathbb{Q}_\infty$. Considering the field $K_\infty = F_\infty(\{\sqrt{u} \mid u \in \mathcal{U}_{F_\infty}\})$, where $\mathcal{U}_{F_\infty}$ is the group of units of F_∞, one finds that $H^1(\mathbb{Q}_\Sigma/F_\infty, \Phi)$ and $H^1(\mathbb{Q}_\Sigma/F_\infty, \Psi)$ have $(\Lambda/2\Lambda)$-corank 2. Now $E[2] \cong \Phi \times \Psi$ as a G_F-module and so $H^1(\mathbb{Q}_\Sigma/F_\infty, E[2])$ has $(\Lambda/2\Lambda)$-corank 4. The inflation-restriction sequence then will show that $H^1(\mathbb{Q}_\Sigma/\mathbb{Q}_\infty, E[2])$ is pseudo-isomorphic to $H^1(\mathbb{Q}_\Sigma/F_\infty, E[2])^\Delta$, where $\Delta = \mathrm{Gal}(F_\infty/\mathbb{Q}_\infty)$. One then sees that $H^1(\mathbb{Q}_\Sigma/\mathbb{Q}_\infty, E[2])$ must have $(\Lambda/2\Lambda)$-corank 2. The fact that c is injective and that both α_1 and α_3 have finite kernel implies that α_2 has finite kernel too. The map α_2 must therefore be surjective. Now consider the commutative diagram

$$\begin{array}{ccc} H^1(\mathbb{Q}_\Sigma/\mathbb{Q}_\infty, E[2]) & \xrightarrow{\alpha_2} & \mathcal{P}_{E[2]}^{(\infty)}(\mathbb{Q}_\infty) \\ \downarrow & & \downarrow{\scriptstyle e} \\ H^1(\mathbb{Q}_\Sigma/\mathbb{Q}_\infty, E[2^\infty]) & \xrightarrow{\alpha_E} & \mathcal{P}_{E[2^\infty]}^{(\infty)}(\mathbb{Q}_\infty) \end{array}$$

Note that $\mathcal{P}^{(\infty)}_{E[2^\infty]}(\mathbb{Q}_\infty)$ is what we denoted by $\mathcal{P}^{(\infty)}_E(\mathbb{Q}_\infty)$ in section 4. The map $H^1(\mathbb{R}, E[2]) \to H^1(\mathbb{R}, E[2^\infty])$ is surjective. (But it's not injective since $H^1(\mathbb{R}, E[2^\infty]) \cong \mathbb{Z}/2\mathbb{Z}$ when $E(\mathbb{R})$ has two components.) Hence the map e is surjective. Thus $e \circ \alpha_2$ is surjective and this implies that α_E is surjective. In fact, more precisely, the above diagram shows that the restriction of α_E to $H^1(\mathbb{Q}_\Sigma/\mathbb{Q}_\infty, E[2^\infty])[2]$ is surjective.

We can now easily finish the proof of proposition 5.8. Clearly

$$H^1(\mathbb{Q}_\Sigma/\mathbb{Q}_\infty, E[2^\infty])_{\Lambda\text{-div}} \subseteq \ker(\alpha_E).$$

Since $\ker(\alpha_E)[2]$ has $(\Lambda/2\Lambda)$-corank 1, it is clear that $H^1(\mathbb{Q}_\Sigma/\mathbb{Q}_\infty, E[2^\infty])$ has Λ-corank 1 and that $\ker(\alpha_E)/H^1(\mathbb{Q}_\Sigma/\mathbb{Q}_\infty, E[2^\infty])_{\Lambda\text{-div}}$ has μ-invariant 0. Hence the maximal Λ-cotorsion quotient of $H^1(\mathbb{Q}_\Sigma/\mathbb{Q}_\infty, E[2^\infty])$ has μ-invariant 1. ∎

Remark. Assume that E is an elliptic curve/$\mathbb{Q}$ which has a $\mathbb{Q}$-isogeny of degree p. Assuming that $\mathrm{Sel}_E(\mathbb{Q}_\infty)_p$ is Λ-cotorsion, the above results show that $\mathrm{Sel}_E(\mathbb{Q}_\infty)_p$ contains a Λ-submodule pseudo-isomorphic to $\widehat{\Lambda}[p^{\mu_E}]$. Thus the μ-invariant of $\mathrm{Sel}_E(\mathbb{Q}_\infty)_p$ arises "non-semisimply" if $\mu_E > 1$. For odd p, we already noted this before. For $p = 2$, it follows from the above discussion of $\ker(\alpha_E)$ and the fact that $\mathrm{Sel}_E(\mathbb{Q}_\infty)_p \subseteq \ker(\alpha_E)$. If E has no $\mathbb{Q}$-isogeny of degree p, then μ_E is conjecturally 0, although there has been no progress on proving this.

Before describing various examples where μ_E is positive, we will prove another consequence of lemma 5.9 (and its proof).

Proposition 5.10. *Assume that p is odd and that E is an elliptic curve/$\mathbb{Q}$ with good, ordinary or multiplicative reduction at p. Assume also that $E[p^\infty]$ contains a $G_\mathbb{Q}$-invariant subgroup Φ of order p which is either ramified at p and even or unramified at p and odd. Then* $\mathrm{Sel}_E(\mathbb{Q}_\infty)_p$ *is Λ-cotorsion and* $\mu_E = 0$.

Proof. We will show that $\mathrm{Sel}_E(\mathbb{Q}_\infty)[p]$ is finite. This obviously implies the conclusion. We have the exact sequence

$$H^1(\mathbb{Q}_\Sigma/\mathbb{Q}_\infty, \Phi) \xrightarrow{a} H^1(\mathbb{Q}_\Sigma/\mathbb{Q}_\infty, E[p]) \xrightarrow{b} H^1(\mathbb{Q}_\Sigma/\mathbb{Q}_\infty, \Psi)$$

as before. Under the above hypotheses, both $\Phi^{G_{\mathbb{Q}_\infty}}$ and $\Psi^{G_{\mathbb{Q}_\infty}}$ are trivial. Hence $H^0(\mathbb{Q}_\Sigma/\mathbb{Q}_\infty, E[p]) = 0$. This implies that

$$H^1(\mathbb{Q}_\Sigma/\mathbb{Q}_\infty, E[p]) \xrightarrow{\sim} H^1(\mathbb{Q}_\Sigma/\mathbb{Q}_\infty, E[p^\infty])[p]$$

under the natural map. Thus we can regard $\mathrm{Sel}_E(\mathbb{Q}_\infty)[p]$ as a subgroup of $H^1(\mathbb{Q}_\Sigma/\mathbb{Q}_\infty, E[p])$. Assume that $\mathrm{Sel}_E(\mathbb{Q}_\infty)[p]$ is infinite. Hence either $B = b(\mathrm{Sel}_E(\mathbb{Q}_\infty)[p])$ or $A = \mathrm{Im}(a) \cap \mathrm{Sel}_E(\mathbb{Q}_\infty)[p]$ is infinite. Assume first that B is infinite. Then, by lemma 5.9, Ψ must be odd. Hence Ψ is unramified,

Φ is ramified at p. Let $\tilde{\pi}$ be any prime of $\mathbb{Q}_\Sigma$ lying over the prime π of $\mathbb{Q}_\infty$ over p. Then $\Phi = C_{\tilde{\pi}}[p]$, where $C_{\tilde{\pi}}$ is the subgroup of $E[p^\infty]$ occurring in propositions 2.2, 2.4. (For example, if E has good reduction at p, then $C_{\tilde{\pi}}$ is the kernel of reduction modulo $\tilde{\pi} : E[p^\infty] \to \widetilde{E}[p^\infty]$.) The inertia subgroup $I_{\tilde{\pi}}$ of $\mathrm{Gal}(\mathbb{Q}_\Sigma/\mathbb{Q}_\infty)$ for $\tilde{\pi}$ acts trivially on $D_{\tilde{\pi}} = E[p^\infty]/C_{\tilde{\pi}}$. Thus, Ψ can be identified with $D_{\tilde{\pi}}[p]$. Let σ be a 1-cocycle with values in $E[p]$ representing a class in $\mathrm{Sel}_E(\mathbb{Q}_\infty)[p]$. Let $\tilde{\sigma}$ be the induced 1-cocycle with values in Ψ. Since $H^1(I_{\tilde{\pi}}, D_{\tilde{\pi}}) = \mathrm{Hom}(I_{\tilde{\pi}}, D_{\tilde{\pi}})$, it is clear that $\tilde{\sigma}|_{I_{\tilde{\pi}}} = 0$. Thus, $\tilde{\sigma} \in H^1(\mathbb{Q}_\Sigma/\mathbb{Q}_\infty, \Psi)$ is unramified at $\tilde{\pi}$. Now for each of the finite number of primes η of $\mathbb{Q}_\infty$ lying over some $\ell \in \Sigma$, $\ell \neq p$, $H^1((\mathbb{Q}_\infty)_\eta, \Psi)$ is finite. Thus, it is clear that $B \cap H^1_{\mathrm{unr}}(\mathbb{Q}_\Sigma/\mathbb{Q}_\infty, \Psi)$ is of finite index in B and is therefore infinite, where $H^1_{\mathrm{unr}}(\mathbb{Q}_\Sigma/\mathbb{Q}_\infty, \Psi)$ denotes the group of everywhere unramified cocycle classes. However, if we let F denote the extension of $\mathbb{Q}$ corresponding to ψ, then we see that

$$H^1_{\mathrm{unr}}(\mathbb{Q}_\Sigma/\mathbb{Q}_\infty, \Psi) = \mathrm{Hom}(X^\psi, \Psi),$$

where we are using the same notation as in the proof of proposition 5.9. The Ferrero-Washington theorem implies that $H^1_{\mathrm{unr}}(\mathbb{Q}_\Sigma/\mathbb{Q}_\infty, \Psi)$ is finite. Hence in fact B must be finite. Similarly, if A is infinite, then Φ must be odd and hence unramified. Thus, $\Phi \cap C_{\tilde{\pi}} = 0$. If σ is as above, then $\sigma|_{I_{\tilde{\pi}}}$ must have values in $C_{\tilde{\pi}}$. But if σ represents a class in A, then we can assume that its values are in Φ. Thus $\sigma|_{I_{\tilde{\pi}}} = 0$. Now the map $H^1(I_{\tilde{\pi}}, \Phi) \to H^1(I_{\tilde{\pi}}, E[p])$ is injective. Thus, we see just as above, that $H^1_{\mathrm{unr}}(\mathbb{Q}_\Sigma/\mathbb{Q}_\infty, \Phi)$ is infinite, again contradicting the Ferrero-Washington theorem. ■

Later we will prove analogues of propositions 5.7 and 5.10 for $p = 2$. One can pursue the situation of proposition 5.10 much further, obtaining for example a simple formula for λ_E in terms of the λ-invariant of X^θ, where θ is the odd character in the pair φ, ψ. (Remark: Obviously, $\varphi\psi = \omega$. It is known that X^θ and $Y^{\omega\theta^{-1}}$ have the same λ-invariants, when θ is odd. Both Λ-modules occur in the arguments.) As mentioned in the introduction, one can prove conjecture 1.13 when $E/\mathbb{Q}$ has good, ordinary reduction at p and satisfies the other hypotheses in proposition 5.10. The key ingredients are Kato's theorem and a comparison of λ-invariants based on a congruence between p-adic L-functions. We will pursue these ideas fully in [GrVa].

Another interesting idea, which we will pursue more completely elsewhere, is to study the relationship between $\mathrm{Sel}_E(\mathbb{Q}_\infty)_p$ and $\mathrm{Sel}_{E'}(\mathbb{Q}_\infty)_p$ when E and E' are elliptic curves/$\mathbb{Q}$ such that $E[p] \cong E'[p]$ as $G_{\mathbb{Q}}$-modules. If E and E' have good, ordinary or multiplicative reduction at p and if p is odd, then it is not difficult to prove the following result: *if* $\mathrm{Sel}_E(\mathbb{Q}_\infty)_p[p]$ *is finite, then so is* $\mathrm{Sel}_{E'}(\mathbb{Q}_\infty)_p[p]$. It follows that if $\mathrm{Sel}_E(\mathbb{Q}_\infty)_p$ is Λ-cotorsion and if $\mu_E = 0$, then $\mathrm{Sel}_{E'}(\mathbb{Q}_\infty)_p$ is also Λ-cotorsion and $\mu_{E'} = 0$. Furthermore, it is then possible to relate the λ-invariants λ_E and $\lambda_{E'}$ to each other. (They usually will not be equal. The relationship involves the sets of primes of bad reduction and the Euler factors at those primes.)

A theorem of Washington [Wa1] as well as a generalization due to E. Friedman [F], which are somewhat analogous to the Ferrero-Washington theorem, can also be used to obtain nontrivial results. This idea was first exploited in [R-W] to prove that $E(K)$ is finitely generated for certain elliptic curves E and certain infinite abelian extensions K of $\mathbb{Q}$. The proof of proposition 5.10 can be easily modified to prove some results of this kind. Here is one.

Proposition 5.11. *Assume that E and p satisfy the hypotheses of proposition 5.10. Let K denote the cyclotomic $\mathbb{Z}_q$-extension of $\mathbb{Q}$, where q is any prime different than p. Then $\mathrm{Sel}_E(K)[p]$ is finite. Hence*

$$\mathrm{Sel}_E(K)_p \cong (\mathbb{Q}_p/\mathbb{Z}_p)^t \times (\text{a finite group})$$

for some $t \geq 0$.

Washington's theorem would state that the power of p dividing the class number of the finite layers in the $\mathbb{Z}_q$-extension FK/F is bounded. To adapt the proof of proposition 5.10, one can replace $\mathrm{Im}(\kappa_\eta)$ by $\mathrm{Im}(\lambda_\eta)$ for each prime η of K lying over p, obtaining a possibly larger subgroup of $H^1(K, E[p^\infty])$. The arguments also work if $\mathrm{Gal}(K/\mathbb{Q}) \cong \prod_{i=1}^{n} \mathbb{Z}_{q_i}$, where $q_1, q_2, \ldots, q_n$ are distinct primes, possibly including p. Then one uses the main result of [F]. One consequence is that $E(K)$ is finitely generated. If E is any modular elliptic curve/$\mathbb{Q}$, this same statement is a consequence of the work of Kato and Rohrlich.

We will now discuss various examples where $\mu_E > 0$. We will take the base field to be $\mathbb{Q}$ and assume always that E is an elliptic curve/$\mathbb{Q}$ with good, ordinary or multiplicative reduction at p. We assume first that p is odd. Since $V_p(E)$ is irreducible as a representation space for $G_\mathbb{Q}$, there is a maximal subgroup Φ of $E[p^\infty]$ such that Φ is cyclic, $G_\mathbb{Q}$-invariant, ramified at p, and odd. Define m_E by $|\Phi| = p^{m_E}$. Thus, $m_E \geq 0$. Proposition 5.8 states that $\mu_E \geq m_E$. It is not hard to see that conjecture 1.11 is equivalent to the assertion that $\mu_E = m_E$. For $p = 2$, m_E can be $0, 1, 2, 3$, or 4. For $p = 3$ or 5, m_E can be 0, 1, or 2. For $p = 7$, 13, or 37, m_E can be 0 or 1. For other odd primes (where E has the above reduction type), there are no $\mathbb{Q}$-isogenies of degree p and so $m_E = 0$. In [Maz1], there is a complete discussion of conductor 11 and numerous other examples having non-trivial p-isogenies.

Conductor = 11. If E has conductor 11, then $E[p]$ is irreducible except for $p = 5$. Let E_1, E_2, and E_3 denote the curves 11A1, 11A2, and 11A3 in Cremona's tables. Thus $E_1 = X_0(11)$ and one has $E_1[5] \cong \mu_5 \times \mathbb{Z}/5\mathbb{Z}$ as a $G_\mathbb{Q}$-module. For E_2 (which is $E_1/(\mathbb{Z}/5\mathbb{Z})$), one has the nonsplit exact sequence

$$0 \to \mu_5 \to E_2[5] \to \mathbb{Z}/5\mathbb{Z} \to 0.$$

Now $E_2/\mu_5 = E_1$ and so one sees that $E_2[5^\infty]$ contains a subgroup Φ which is cyclic of order 25, $G_{\mathbb{Q}}$-invariant, ramified at 5, and odd. (Φ is an extension of μ_5 by μ_5.) For E_3 (which is E_1/μ_5), one has a nonsplit exact sequence

$$0 \to \mathbb{Z}/5\mathbb{Z} \to E_3[5] \to \mu_5 \to 0.$$

All of these statements follow from the data about isogenies and torsion subgroups given in [Cre]. One then sees easily that $m_{E_1} = 1$, $m_{E_2} = 2$, and $m_{E_3} = 0$. We will show that $\mathrm{Sel}_{E_1}(\mathbb{Q}_\infty)_5 \cong \hat{\Lambda}[5]$, $\mathrm{Sel}_{E_2}(\mathbb{Q}_\infty)_5 \cong \hat{\Lambda}[5^2]$, and $\mathrm{Sel}_{E_3}(\mathbb{Q}_\infty)_5 = 0$. Thus, $\lambda_{E_i} = 0$ and $\mu_{E_i} = m_{E_i}$ for $1 \le i \le 3$.

We will let $\Phi_i = \mu_5$ and $\Psi_i = \mathbb{Z}/5\mathbb{Z}$ as $G_{\mathbb{Q}}$-modules for $1 \le i \le 3$. Then we have the following exact sequences of $G_{\mathbb{Q}}$-modules

$$0 \longrightarrow \Phi_2 \longrightarrow E_2[5] \longrightarrow \Psi_2 \longrightarrow 0$$

$$0 \longrightarrow \Psi_3 \longrightarrow E_3[5] \longrightarrow \Phi_3 \longrightarrow 0.$$

These exact sequences are nonsplit. For E_1, we have $E_1[5] = \Phi_1 \times \Psi_1$. As $G_{\mathbb{Q}_5}$-modules, we have exact sequences

$$0 \longrightarrow C_5 \longrightarrow E_i[5^\infty] \longrightarrow D_5 \longrightarrow 0$$

where D_5 is unramified and $C_5 \cong \mu_{5^\infty}$ for the action of $I_{\mathbb{Q}_5}$, the inertia subgroup of $G_{\mathbb{Q}_5}$. There will be no need to index C_5 and D_5 by i. As $G_{\mathbb{Q}_{11}}$-modules, we have exact sequences

$$0 \longrightarrow C_{11} \longrightarrow E_i[5^\infty] \longrightarrow D_{11} \longrightarrow 0$$

where $C_{11} \cong \mu_{5^\infty}$ and $D_{11} \cong \mathbb{Q}_5/\mathbb{Z}_5$ for the action of $G_{\mathbb{Q}_{11}}$. It will again not be necessary to include an index i on these groups. The homomorphisms $E_i[5^\infty] \to D_5$ and $E_i[5^\infty] \to D_{11}$ induce natural identifications. As $G_{\mathbb{Q}_5}$-modules, Ψ_1, Ψ_2, Ψ_3 are all identified with $D_5[5]$. This is clear from the action of $G_{\mathbb{Q}_5}$ on these groups (which is trivial). But, as $G_{\mathbb{Q}_{11}}$-modules, Φ_1, Ψ_1, Ψ_2, and Φ_3 are all identified with $D_{11}[5]$. One verifies this by using the isogeny data and the fact that the Tate periods for the E_i's in $\mathbb{Q}_{11}^\times$ have valuations 5, 1, 1, respectively. For example, if Φ_1 or Ψ_1 were contained in C_{11}, then the Tate period for E_2 or E_3 would have valuation divisible by 5. We will use the fact that the maps

$$H^1(\mathbb{Q}_{11}, D_{11}[5]) \to H^1(\mathbb{Q}_{11}, D_{11}), \qquad H^1(I_{\mathbb{Q}_5}, D_5[5]) \to H^1(I_{\mathbb{Q}_5}, D_5)$$

are both injective. This is so because $G_{\mathbb{Q}_{11}}$ acts trivially on D_{11} and $I_{\mathbb{Q}_5}$ acts trivially on D_5. Our calculations of the Selmer groups will be in several steps and depend mostly on the results of section 2 and 3. We take $\Sigma = \{\infty, 5, 11\}$.

$\mathrm{Sel}_{E_3}(\mathbb{Q})_5 = 0$. Suppose $[\sigma] \in \mathrm{Sel}_{E_3}(\mathbb{Q})[5]$. It is enough to prove that $[\sigma] = 0$. We can assume that σ has values in $E_3[5]$. (But note that in this case the map $H^1(\mathbb{Q}_\Sigma/\mathbb{Q}, E_3[5]) \to H^1(\mathbb{Q}_\Sigma/\mathbb{Q}, E_3[5^\infty])$ has a nontrivial kernel.) The image of σ in $H^1(\mathbb{Q}_{11}, \Phi_3) = H^1(\mathbb{Q}_{11}, D_{11}[5])$ must become trivial in $H^1(\mathbb{Q}_{11}, D_{11})$. Thus this image must be trivial. Now $\Phi_3 = \mu_5$ and $H^1(\mathbb{Q}_\Sigma/\mathbb{Q}, \mu_5) \cong (\mathbb{Z}/5\mathbb{Z})^2$, the classes for the 1-cocycles associated to $\sqrt[5]{5^i 11^j}$, $0 \le i, j \le 4$. The restriction of such a 1-cocycle to $G_{\mathbb{Q}_{11}}$ is trivial when $5^i 11^j \in (\mathbb{Q}_{11}^\times)^5$, which happens only when $i = j = 0$. Thus, the image of $[\sigma]$ in $H^1(\mathbb{Q}_\Sigma/\mathbb{Q}, \Phi_3)$ must be trivial. Hence we can assume that σ has values in $\Psi_3 = \mathbb{Z}/5\mathbb{Z}$.

Now $H^1(\mathbb{Q}_\Sigma/\mathbb{Q}, \Psi_3) \cong (\mathbb{Z}/5\mathbb{Z})^2$ by class field theory, but its image in $H^1(\mathbb{Q}_\Sigma/\mathbb{Q}, E[5^\infty])$ is of order 5. Since $[\sigma] \in \mathrm{Sel}_{E_3}(\mathbb{Q})_5$, it has a trivial image in $H^1(I_{\mathbb{Q}_5}, D_5)$. Hence, regarding σ as an element of $\mathrm{Hom}(\mathrm{Gal}(\mathbb{Q}_\Sigma/\mathbb{Q}), \Psi_3)$, it must be unramified at 5 and hence factor through $\mathrm{Gal}(K/\mathbb{Q})$, where K is the cyclic extension of $\mathbb{Q}$ of conductor 11. But this implies that $[\sigma] = 0$ in $H^1(\mathbb{Q}_\Sigma/\mathbb{Q}, E[5^\infty])$ because $\mathrm{Hom}(\mathrm{Gal}(K/\mathbb{Q}), \Psi_3)$ is the kernel of the map $H^1(\mathbb{Q}_\Sigma/\mathbb{Q}, \Psi_3) \to H^1(\mathbb{Q}_\Sigma/\mathbb{Q}, E[5^\infty])$. To see this, note that this kernel has order 5 and that the map $H^1(I_{\mathbb{Q}_5}, \Psi_3) \to H^1(I_{\mathbb{Q}_5}, D_5)$ is injective. Hence $\mathrm{Sel}_{E_3}(\mathbb{Q})_5 = 0$.

$\mathrm{Sel}_{E_2}(\mathbb{Q})_5 = 0$. We have $H^1(\mathbb{Q}_\Sigma/\mathbb{Q}, E_2[5]) \cong H^1(\mathbb{Q}_\Sigma/\mathbb{Q}, E_2[5^\infty])[5]$. Let $[\sigma] \in \mathrm{Sel}_{E_2}(\mathbb{Q})[5]$. We can assume that σ has values in $E_2[5]$. The image of σ in $H^1(\mathbb{Q}_\Sigma/\mathbb{Q}, \Psi_2)$ must have a trivial restriction to $G_{\mathbb{Q}_{11}}$. But

$$H^1(\mathbb{Q}_\Sigma/\mathbb{Q}, \Psi_2) = \mathrm{Hom}(\mathrm{Gal}(KL/\mathbb{Q}), \mathbb{Z}/5\mathbb{Z}),$$

where K is as above and L is the first layer of the cyclotomic $\mathbb{Z}_5$-extension of $\mathbb{Q}$. Now 11 is inert in $L/\mathbb{Q}$ and ramified in KL/L. Thus it is clear that σ has trivial image in $H^1(\mathbb{Q}_\Sigma/\mathbb{Q}, \Psi_2)$ and hence has values in $\Phi_2 = \mu_5$.

Now $H^1(\mathbb{Q}_\Sigma/\mathbb{Q}, \mu_5) \cong (\mathbb{Z}/5\mathbb{Z})^2$, but the map

$$\epsilon_0 : H^1(\mathbb{Q}_\Sigma/\mathbb{Q}, \mu_5) \to H^1(\mathbb{Q}_\Sigma/\mathbb{Q}, E_2[5]) = H^1(\mathbb{Q}_\Sigma/\mathbb{Q}, E_2[5^\infty])[5]$$

has $\ker(\epsilon_0) \cong \mathbb{Z}/5\mathbb{Z}$. Now $[\sigma] \in \mathrm{Im}(\epsilon_0)$, which we will show is not contained in $\mathrm{Sel}_{E_2}(\mathbb{Q})_5$. This will imply that $\mathrm{Sel}_{E_2}(\mathbb{Q})_5 = 0$. Consider the commutative diagram

$$\begin{array}{ccccc}
H^1(\mathbb{Q}_\Sigma/\mathbb{Q}, \mu_5) & \xrightarrow{a} & H^1(\mathbb{Q}_5, \mu_5) & \xrightarrow{b} & H^1(\mathbb{Q}_5, C_5) \\
\downarrow{\scriptstyle \epsilon_0} & & & & \downarrow{\scriptstyle \lambda_5} \\
H^1(\mathbb{Q}_\Sigma/\mathbb{Q}, E_2[5^\infty]) & & \xrightarrow{\quad c \quad} & & H^1(\mathbb{Q}_5, E_2[5^\infty])
\end{array}$$

One sees easily that a is an isomorphism. Also, $H^1(\mathbb{Q}_5, \mu_5) \cong (\mathbb{Z}/5\mathbb{Z})^2$ and b induces an isomorphism $H^1(\mathbb{Q}_5, \mu_5) \cong H^1(\mathbb{Q}_5, C_5)[5]$. Referring to (2) following the proof of lemma 2.3, one sees that $H^1(\mathbb{Q}_5, C_5) \cong (\mathbb{Q}_5/\mathbb{Z}_5) \times \mathbb{Z}/5\mathbb{Z}$.

(One needs the fact that $|\widetilde{E}_2(\mathbb{Z}/5\mathbb{Z})|$ is divisible by 5, but not by 5^2.) In section 2, one also finds a proof that the map

$$H^1(\mathbb{Q}_5, C_5)/H^1(\mathbb{Q}_5, C_5)_{\text{div}} \to \text{Im}(\lambda_5)/\text{Im}(\lambda_5)_{\text{div}}$$

is an isomorphism. (See (3) in the proof of proposition 2.5.) If we had $\text{Im}(\epsilon_0) \subseteq \text{Sel}_{E_2}(\mathbb{Q})_5$, then we must have $\text{Im}(c \circ \epsilon_0) \subseteq \text{Im}(\lambda_5)_{\text{div}}$, which is the image of the local Kummer homomorphism κ_5. But this can't be so because clearly $\text{Im}(b \circ a) \not\subset H^1(\mathbb{Q}_5, C_5)_{\text{div}}$. It follows that $\text{Sel}_{E_2}(\mathbb{Q})_5 = 0$.

Although we don't need it, we will determine $\ker(\epsilon_0)$. The discussion in the previous paragraph shows that $\ker(\epsilon_0)$ is the inverse image under $b \circ a$ of $H^1(\mathbb{Q}_5, C_5)_{\text{div}}[5]$. One can use proposition 3.11 to determine this. Let φ be the unramified character of $G_{\mathbb{Q}_5}$ giving the action in $D_5 = \widetilde{E}_2[5^\infty]$. Since 5 is an anomalous prime for E_2, one gets an isomorphism

$$\varphi : \text{Gal}(M_\infty/\mathbb{Q}_5) \to 1 + 5\mathbb{Z}_5$$

where M_∞ denotes the unramified $\mathbb{Z}_5$-extension of $\mathbb{Q}_5$. One has

$$H^1(M_\infty, \mu_{5^\infty}) \cong \widehat{R} \times \mathbb{Q}_5/\mathbb{Z}_5,$$

where now $R = \mathbb{Z}_p[[G]]$, $G = \text{Gal}(M_\infty/\mathbb{Q}_5)$. We have $C_5 \cong \mu_{5^\infty} \otimes \varphi^{-1}$ and $H^1(M_\infty, C_5) = H^1(M_\infty, \mu_{5^\infty}) \otimes \varphi^{-1}$. Now $H^1(\mathbb{Q}_5, C_5) \xrightarrow{\sim} H^1(M_\infty, C_5)^G$, by the inflation-restriction sequence. The image of $H^1(\mathbb{Q}_5, C_5)_{\text{div}}$ under the restriction map is $H^1(M_\infty, C_5)^G_{R\text{-div}}$. But $H^1(M_\infty, C_5)_{R\text{-div}}$ coincides with $H^1(M_\infty, \mu_{5^\infty})_{R\text{-div}}$, with the action of G twisted by φ^{-1}. Let $q \in \mathbb{Q}_5^\times$ and let σ_q be the 1-cocycle with values in μ_5 associated to $\sqrt[5]{q}$. Then $\sigma_q \in H^1(\mathbb{Q}_5, C_5)_{\text{div}}$ if and only if $\sigma_q|_{G_{M_\infty}} \in H^1(M_\infty, \mu_{5^\infty})_{R\text{-div}}$. By proposition 3.11, this means that q is a universal norm for $M_\infty/\mathbb{Q}_5$, i.e., $q \in \mathbb{Z}_5^\times$. Now $H^1(\mathbb{Q}_\Sigma/\mathbb{Q}, \mu_5)$ consists of the classes of 1-cocycles associated to $\sqrt[5]{u}$, where $u = 5^i 11^j$, $0 \le i, j \le 4$. It follows that $\ker(\epsilon_0)$ is generated by the 1-cocycle corresponding to $\sqrt[5]{11}$. There are other ways to interpret this result. The extension class of $\mathbb{Z}/5\mathbb{Z}$ by μ_5 given by $E_2[5]$ corresponds to the 1-cocycle associated to $\sqrt[5]{11}$. The field $\mathbb{Q}(E_2[5])$ is $\mathbb{Q}(\mu_5, \sqrt[5]{11})$. The Galois module $E_2[5]$ is "peu ramifiée" at 5, in the sense of Serre. (This of course must be so because E_2 has good reduction at 5.)

$\text{Sel}_{E_1}(\mathbb{Q})_5 = 0$. We have an exact sequence

$$0 \to H^1(\mathbb{Q}_\Sigma/\mathbb{Q}, \Psi_1) \to H^1(\mathbb{Q}_\Sigma/\mathbb{Q}, E_1[5^\infty]) \to H^1(\mathbb{Q}_\Sigma/\mathbb{Q}, E_2[5^\infty]).$$

Since $\text{Sel}_{E_2}(\mathbb{Q})_5 = 0$, it is clear that $\text{Sel}_{E_1}(\mathbb{Q})_5 \subseteq \text{Im}(H^1(\mathbb{Q}_\Sigma/\mathbb{Q}, \Psi_1))$. But $\Psi_1 = \mathbb{Z}/5\mathbb{Z}$ and $H^1(\mathbb{Q}_\Sigma/\mathbb{Q}, \Psi_1) = \text{Hom}(\text{Gal}(KL/\mathbb{Q}), \mathbb{Z}/5\mathbb{Z})$, where K and L are as defined before. Since the decomposition group for 11 in $\text{Gal}(KL/\mathbb{Q})$ is the entire group and since Ψ_1 is mapped to $D_{11}[5]$, we see as before that $H^1(\mathbb{Q}_\Sigma/\mathbb{Q}, \Psi_1) \to H^1(\mathbb{Q}_{11}, E_1[5^\infty])$ is injective. Hence $\text{Sel}_{E_1}(\mathbb{Q})_5 = 0$.

$f_{E_i}(T) = 5^{m_{E_i}}$. We can now apply theorem 4.1 to see that $f_{E_1}(0) \sim 5$, $f_{E_2}(0) \sim 5^2$, and $f_{E_3}(0) \sim 1$, using the fact that $\widetilde{E}_i(\mathbb{Z}/5\mathbb{Z})$ has order 5. But we know that $5^{m_{E_i}}$ divides $f_{E_i}(T)$. Hence it follows that, after multiplication by a factor in $\Lambda^\times$, we can take $f_{E_1}(T) = 5$, $f_{E_2}(T) = 5^2$, and $f_{E_3}(T) = 1$. We now determine directly the precise structure of the Selmer groups $\mathrm{Sel}_{E_i}(\mathbb{Q}_\infty)_5$ as Λ-modules.

$\mathrm{Sel}_{E_3}(\mathbb{Q}_\infty)_5 = 0$. The fact that $f_{E_3}(T) = 1$ shows that $\mathrm{Sel}_{E_3}(\mathbb{Q}_\infty)_5$ is finite. Proposition 4.15 then implies that $\mathrm{Sel}_{E_3}(\mathbb{Q}_\infty)_5 = 0$. However, it is interesting to give a more direct argument. We will show that the restriction map $s_0^{(3)} : \mathrm{Sel}_{E_3}(\mathbb{Q})_5 \to \mathrm{Sel}_{E_3}(\mathbb{Q}_\infty)_5^\Gamma$ is surjective, which then implies that $\mathrm{Sel}_{E_3}(\mathbb{Q}_\infty)_5^\Gamma$ and hence $\mathrm{Sel}_{E_3}(\mathbb{Q}_\infty)_5$ are both zero. Here and in the following discussions, we will let $s_0^{(i)}$, $h_0^{(i)}$, $g_0^{(i)}$, and $r_v^{(i)}$ for $v \in \{5, 11\}$ denote the maps considered in sections 3 and 4 for the elliptic curve E_i, $1 \le i \le 3$. Thus, $\ker(s_0^{(i)}) = 0$ for $1 \le i \le 3$, by proposition 3.9. But $|\ker(h_0^{(3)})| = 5$. We have the exact sequence

$$0 \to \ker(h_0^{(3)}) \to \ker(g_0^{(3)}) \to \mathrm{coker}(s_0^{(3)}) \to 0.$$

Thus it suffices to show that $|\ker(g_0^{(3)})| = 5$. We let

$$A = \ker(\mathcal{P}_{E_3}^{\Sigma}(\mathbb{Q}) \to \mathcal{P}_{E_3}^{\Sigma}(\mathbb{Q}_\infty)).$$

Now $\mathcal{P}_{E_3}^{\Sigma}(\mathbb{Q}) = \mathcal{H}_{E_3}(\mathbb{Q}_5) \times \mathcal{H}_{E_3}(\mathbb{Q}_{11})$, $\mathcal{P}_{E_3}^{\Sigma}(\mathbb{Q}_\infty) = \mathcal{H}_{E_3}(\mathbb{Q}_5^{\mathrm{cyc}}) \times \mathcal{H}_{E_3}(\mathbb{Q}_{11}^{\mathrm{cyc}})$. The local duality theorems easily imply that

$$\mathcal{H}_{E_3}(\mathbb{Q}_5) = H^1(\mathbb{Q}_5, E_3[5^\infty])/\mathrm{Im}(\kappa_5) \cong (\mathbb{Q}_5/\mathbb{Z}_5) \times \mathbb{Z}/5\mathbb{Z}$$

$$\mathcal{H}_{E_3}(\mathbb{Q}_{11}) = H^1(\mathbb{Q}_{11}, E_3[5^\infty]) \cong \mathbb{Z}/5\mathbb{Z}.$$

The kernels of the maps $r_v^{(i)} : \mathcal{H}_{E_i}(\mathbb{Q}_v) \to \mathcal{H}_{E_i}(\mathbb{Q}_v^{\mathrm{cyc}})$ can be determined by the results in section 3. In particular, one finds that $|\ker(r_5^{(3)})| = 5^2$, while $r_{11}^{(3)}$ is injective. Also, $\mathcal{H}_{E_i}(\mathbb{Q}_5^{\mathrm{cyc}}) \cong H^1(\mathbb{Q}_5^{\mathrm{cyc}}, D_5) \cong \widehat{\Lambda}$ for $1 \le i \le 3$. Thus, $\mathrm{Im}(r_5^{(i)})$ is obviously isomorphic to $\mathbb{Q}_5/\mathbb{Z}_5$ for each i. It follows that A has order 5^2, $A \subseteq \mathcal{H}_{E_3}(\mathbb{Q}_5)$, $A \cap \mathcal{H}_{E_3}(\mathbb{Q}_5)_{\mathrm{div}}$ has order 5, and $A\mathcal{H}_{E_3}(\mathbb{Q}_5)_{\mathrm{div}} = \mathcal{H}_{E_3}(\mathbb{Q}_5)$. Now $\mathcal{G}_{E_3}^{\Sigma}(\mathbb{Q})$ has index 5 in $\mathcal{P}_{E_3}^{\Sigma}(\mathbb{Q})$ and projects onto $\mathcal{H}_{E_3}(\mathbb{Q}_{11})$. It follows easily that

$$|\ker(g_0^{(3)})| = |A \cap \mathcal{G}_{E_3}^{\Sigma}(\mathbb{Q})| = 5.$$

As we said, this implies that $\mathrm{Sel}_{E_3}(\mathbb{Q}_\infty)_5 = 0$.

$\mathrm{Sel}_{E_2}(\mathbb{Q}_\infty)_5 \cong \widehat{\Lambda}[5^2]$. Let Φ be the $G_{\mathbb{Q}}$-invariant subgroup of $E_2[5^\infty]$ which is cyclic of order 5^2. (This Φ is an extension of Φ_1 by Φ_2.) We have $E_2/\Phi \cong E_3$. Since $\mathrm{Sel}_{E_3}(\mathbb{Q}_\infty)_5 = 0$, it follows that

$$\mathrm{Sel}_{E_2}(\mathbb{Q}_\infty)_5 \subseteq \mathrm{Im}(H^1(\mathbb{Q}_\Sigma/\mathbb{Q}_\infty, \Phi) \to H^1(\mathbb{Q}_\Sigma/\mathbb{Q}_\infty, E_2[5^\infty])).$$

The index is finite by proposition 5.7. Thus it is clear that $\mathrm{Sel}_{E_2}(\mathbb{Q}_\infty)_5$ is pseudo-isomorphic to $\widehat{\Lambda}[5^2]$ and has exponent 5^2. Since $E_2(\mathbb{Q}) = 0$, we have $\mathcal{G}^{\Sigma}_{E_2}(\mathbb{Q}) = \mathcal{P}^{\Sigma}_{E_2}(\mathbb{Q})$ and $\ker(h_0^{(2)}) = 0$. Hence

$$\mathrm{coker}(s_0^{(2)}) \cong \ker(g_0^{(2)}) \cong \ker(r_5^{(2)}) \times \ker(r_{11}^{(2)}).$$

Now $\ker(r_{11}^{(2)}) = 0$ because $5 \nmid \mathrm{ord}_{11}(q_{E_2}^{(11)})$, where $q_{E_2}^{(11)}$ denotes the Tate period for E_2 in $\mathbb{Q}_{11}^\times$. Also, $|\ker(r_5^{(2)})| = 5^2$. We pointed out earlier that the $G_{\mathbb{Q}}$-module $E_2[5]$ is the nonsplit extension of $\mathbb{Z}/5\mathbb{Z}$ by μ_5 corresponding to $\sqrt[5]{11}$. Since $11 \notin (\mathbb{Q}_5^\times)^5$, this extension remains nonsplit as a $G_{\mathbb{Q}_5}$-module. Thus, $H^0(\mathbb{Q}_5, E_2[5^\infty]) = 0$. One deduces from this that $H^1(\mathbb{Q}_5, E_2[5^\infty]) \cong \mathbb{Q}_5/\mathbb{Z}_5$ and $\mathcal{H}_{E_2}(\mathbb{Q}_5) \cong \mathbb{Q}_5/\mathbb{Z}_5$. This implies that $\ker(r_5^{(2)}) \cong \mathbb{Z}/5^2\mathbb{Z}$. Hence $\ker(g_0^{(2)})$, $\mathrm{coker}(s_0^{(2)})$ and hence $\mathrm{Sel}_{E_2}(\mathbb{Q}_\infty)_5^\Gamma$ are all cyclic of order 5^2. Therefore, $X_{E_2}(\mathbb{Q}_\infty) = \mathrm{Sel}_{E_2}(\mathbb{Q}_\infty)_5\widehat{\ }$ is a cyclic Λ-module of exponent 5^2. That is, $X_{E_2}(\mathbb{Q}_\infty)$ is a quotient of $\Lambda/5^2\Lambda$ and, since the two are pseudo-isomorphic, it follows easily that $X_{E_2}(\mathbb{Q}_\infty) \cong \Lambda/5^2\Lambda$. This gives the stated result about the structure of $\mathrm{Sel}_{E_2}(\mathbb{Q}_\infty)_5$.

$\mathrm{Sel}_{E_1}(\mathbb{Q}_\infty)_5 \cong \widehat{\Lambda}[5]$. Since $E_1/\Phi_1 \cong E_3$, it follows that

$$\mathrm{Sel}_{E_1}(\mathbb{Q}_\infty)_5 \subseteq \mathrm{Im}(H^1(\mathbb{Q}_\Sigma/\mathbb{Q}, \Phi_1) \to H^1(\mathbb{Q}_\Sigma/\mathbb{Q}_\infty, E_1[5^\infty])).$$

Hence $\mathrm{Sel}_{E_1}(\mathbb{Q}_\infty)_5$ has exponent 5 and is pseudo-isomorphic to $\widehat{\Lambda}[5]$. Also, by proposition 4.15, $\mathrm{Sel}_{E_1}(\mathbb{Q}_\infty)_5$ has no proper Λ-submodules of finite index. Thus, $X_{E_1}(\mathbb{Q}_\infty)$ is a $(\Lambda/5\Lambda)$-module pseudo-isomorphic to $(\Lambda/5\Lambda)$ and with no nonzero, finite Λ-submodules. Since $\Lambda/5\Lambda$ is a PID, it follows that $X_{E_1}(\mathbb{Q}_\infty) \cong \Lambda/5\Lambda$, which gives the stated result concerning the structure of $\mathrm{Sel}_{E_1}(\mathbb{Q}_\infty)_5$.

Twists. Let ξ be a quadratic character for $\mathbb{Q}$. Then ξ corresponds to a quadratic field $\mathbb{Q}(\sqrt{d})$, where $d = d_\xi \in \mathbb{Z}$ and $|d|$ is the conductor of ξ. We consider separately the cases where ξ is even or odd. For even ξ, the following conjecture seems reasonable. It can be deduced from conjecture 1.11, but may be more approachable. We let E^ξ denote the quadratic twist of E by d.

Conjecture 5.12. *Let E be an elliptic curve/$\mathbb{Q}$ with potentially ordinary or multiplicative reduction at p, where p is an odd prime. Let ξ be an even quadratic character. Then $\mathrm{Sel}_{E^\xi}(\mathbb{Q}_\infty)_p$ and $\mathrm{Sel}_E(\mathbb{Q}_\infty)_p$ have the same μ-invariants.*

We remark that the λ-invariants can certainly be different. For example, if E is any one of the three elliptic curves of conductor 11, then $\lambda_E = 0$ for any prime p satisfying the hypothesis in the above conjecture. But if ξ is the quadratic character corresponding to $\mathbb{Q}(\sqrt{2})$ (of conductor 8), then

$\operatorname{rank}(E^{\xi}(\mathbb{Q})) = 1$. (In fact, E^{ξ} is 704(A1, 2, or 3) in [Cre].) Then of course $\lambda_{E^{\xi}} \geq 1$ for all such p.

Assume now that ξ is an odd character and that $\xi(5) \neq 0$. Let E be any one of the elliptic curves of conductor 11. Let $p = 5$. Then $E^{\xi}[5]$ is $G_{\mathbb{Q}}$-reducible with composition factors $\mu_5 \otimes \xi$ and $(\mathbb{Z}/5\mathbb{Z}) \otimes \xi$. The hypotheses in proposition 5.10 are satisfied and so the μ-invariant of $\mathrm{Sel}_{E^{\xi}}(\mathbb{Q}_{\infty})_5$ is zero. The λ-invariant $\lambda_{E^{\xi}}$ is unchanged by isogeny and so doesn't depend on the choice of E. It follows from proposition 4.14 that $\mathrm{Sel}_{E^{\xi}}(\mathbb{Q}_{\infty})_5$ is divisible. Hence $\lambda_{E^{\xi}}$, which is the $\mathbb{Z}_5$-corank of $\mathrm{Sel}_{E^{\xi}}(\mathbb{Q}_{\infty})_5$, is obviously equal to the $(\mathbb{Z}/5\mathbb{Z})$-dimension of $\mathrm{Sel}_{E^{\xi}}(\mathbb{Q}_{\infty})_5[5]$. We will not give the verification (which we will discuss more generally elsewhere), but one finds the following formula:

$$\lambda_{E^{\xi}} = 2\lambda_{\xi} + \epsilon_{\xi},$$

where $d = d_{\xi}$ and λ_{ξ} denotes the classical λ-invariant $\lambda(F_{\infty}/F)$ for the imaginary quadratic field $F = \mathbb{Q}(\sqrt{d})$ and for the prime $p = 5$ and where $\epsilon_{\xi} = 1$ if 11 splits in $\mathbb{Q}(\sqrt{d})/\mathbb{Q}$, $\epsilon_{\xi} = 0$ if 11 is inert or ramified. By proposition 3.10, it follows that $\operatorname{corank}_{\mathbb{Z}_5}(\mathrm{Sel}_{E^{\xi}}(\mathbb{Q})_5) \equiv \epsilon_{\xi} \pmod 2$, which is in agreement with the Birch and Swinnerton-Dyer conjecture since the sign in the functional equation for the Hasse-Weil L-function $L(E^{\xi}/\mathbb{Q}, 5) = L(E/\mathbb{Q}, \xi, 5)$ is $(-1)^{\epsilon_{\xi}}$. As an example, consider the case where ξ corresponds to $\mathbb{Q}(\sqrt{-2})$. Then E^{ξ} is 704(K1, 2, or 3). The class number of $\mathbb{Q}(\sqrt{-2})$ is 1. The prime $p = 5$ is inert in $F = \mathbb{Q}(\sqrt{-2})$. Hence the discussion of Iwasawa's theorem in the introduction shows that the λ-invariant for this quadratic field is 0. But 11 splits in F. Therefore, $\lambda_{E^{\xi}} = 1$. Since $\operatorname{rank}(E^{\xi}(\mathbb{Q})) = 1$, it is clear that $\mathrm{Sel}_{E^{\xi}}(\mathbb{Q}_{\infty})_5 = E^{\xi}(\mathbb{Q}) \otimes (\mathbb{Q}_p/\mathbb{Z}_p)$. As another example, suppose that ξ corresponds to $F = \mathbb{Q}(\sqrt{-1})$. Then E^{ξ} is 176(B1, 2, or 3) in [Cre]. The prime 5 splits in $F/\mathbb{Q}$ and so $\lambda(F_{\infty}/F) \geq 1$. In fact, $\lambda(F_{\infty}/F) = 1$. Since 11 is inert in F, we have $\lambda_{E^{\xi}} = 2$. But $E^{\xi}(\mathbb{Q})$ is trivial. If $E^{\xi}(\mathbb{Q}_1)$ had positive rank, one would have $\operatorname{rank}(E^{\xi}(\mathbb{Q}_1)) \geq 4$ (because the nontrivial irreducible $\mathbb{Q}$-representation of $\mathrm{Gal}(\mathbb{Q}_1/\mathbb{Q})$ has degree 4). Hence it is clear that $\lambda_{E^{\xi}}^{\text{Ш}} = 2$, $\lambda_{E^{\xi}}^{M\text{-}W} = 0$. T. Fukuda has done extensive calculations of $\lambda(F_{\infty}/F)$ when F is an imaginary quadratic field and $p = 3, 5$, or 7. Some of these λ-invariants are quite large. Presumably they are unbounded as F varies. For $p = 5$, he finds that $\lambda_{\xi} = 10$ if ξ corresponds to $F = \mathbb{Q}(\sqrt{-3,624,233})$. Since 11 splits in $F/\mathbb{Q}$, we have $\lambda_{E^{\xi}} = 21$ in this case. However, we don't know the values of $\lambda_{E^{\xi}}^{M\text{-}W}$ and $\lambda_{E^{\xi}}^{\text{Ш}}$.

We will briefly explain in the case of E^{ξ} (where E and ξ are as in the previous paragraph and $p = 5$) how to prove conjecture 1.13. Kato's theorem states that $f_E^{\xi}(T)$ divides $f_{E^{\xi}}^{\text{anal}}(T)$, up to multiplication by a power of p. Thus, $\lambda(f_{E^{\xi}}^{\text{anal}}) \geq \lambda_{E^{\xi}}$. Now it is known that λ_{ξ} is equal to the λ-invariant of the Kubota-Leopoldt 5-adic L-function $L_5(\omega\xi, s)$. The μ-invariant is zero (by [Fe-Wa]). In [Maz3], Mazur proves the following congruence formula

$$L_5(E^{\xi}/\mathbb{Q}, s) \equiv (1 - \xi(11)11^{1-s})L_5(\omega\xi, s-1)L_5(\omega\xi, 1-s) \pmod{5\mathbb{Z}_5}$$

for all $s \in \mathbb{Z}_5$. More precisely, one can interpret this as a congruence in the Iwasawa algebra Λ modulo the ideal 5Λ. The left side corresponds to $f_{E^\xi}^{\mathrm{anal}}(T)$, and each factor on the right side corresponds to an element of Λ. The two sides are congruent modulo 5Λ. Now, if $f(T) \in \Lambda$ is any power series with $\mu(f) = 0$, then one has $f(T) \equiv uT^{\lambda(f)} \pmod{p\Lambda}$, where $u \in \Lambda^\times$. Applying this, we obtain $\lambda(f_{E^\xi}^{\mathrm{anal}}) = 2\lambda_\xi + \epsilon_\xi$ and therefore $\lambda(f_{E^\xi}^{\mathrm{anal}}) = \lambda_{E^\xi}$. Since both $f_{E^\xi}^{\mathrm{anal}}(T)$ and $f_E(T)$ have μ-invariant equal to zero, it follows that indeed $(f_E(T)) = (f_{E^\xi}^{\mathrm{anal}}(T))$.

Theorems 4.8, 4.14, and 4.15 give sufficient conditions for the nonexistence of proper Λ-submodules of finite index in $\mathrm{Sel}_E(F_\infty)_p$. In particular, if $F = \mathbb{Q}$ and if E has good, ordinary or multiplicative reduction at p, where p is any odd prime, then no such Λ-submodule of $\mathrm{Sel}_E(\mathbb{Q}_\infty)_p$ can exist. (This is also true for $p = 2$, although the above results don't cover this case completely.) The following example shows that in general some restrictive hypotheses are needed. We let $F = \mathbb{Q}(\mu_5)$, $F_\infty = \mathbb{Q}(\mu_{5^\infty})$. Let $E = E_2$, the elliptic curve of conductor 11 with $E(\mathbb{Q}) = 0$. We shall show that $\mathrm{Sel}_E(F_\infty)_5$ has a Λ-submodule of index 5. To be more precise, note that $\mathrm{Gal}(F_\infty/\mathbb{Q}) = \Delta \times \Gamma$, where $\Delta = \mathrm{Gal}(F_\infty/\mathbb{Q}_\infty)$ and $\Gamma = \mathrm{Gal}(F_\infty/F)$. Now Δ has order 4 and its characters are ω^i, $0 \le i \le 3$. We can decompose $\mathrm{Sel}_E(F_\infty)_5$ as a Λ-module by the action of Δ:

$$\mathrm{Sel}_E(F_\infty)_5 = \bigoplus_{i=0}^{3} \mathrm{Sel}_E(F_\infty)_5^{\omega^i}.$$

As we will see, it turns out that $\mathrm{Sel}_E(F_\infty)_5^{\omega^3} \cong \mathbb{Z}/5\mathbb{Z}$, which of course is a Λ-module quotient of $\mathrm{Sel}_E(F_\infty)_5$. This component is $(\mathrm{Sel}_E(F_\infty)_5 \otimes \omega^{-3})^\Delta$, which can be identified with a subgroup of $H^1(\mathbb{Q}_\Sigma/\mathbb{Q}_\infty, E[5^\infty]\otimes\omega^{-3})$, where $\Sigma = \{\infty, 5, 11\}$. For brevity, we let $A = E[5^\infty]\otimes\omega^{-3}$. We let $S_A(\mathbb{Q}_\infty)$ denote the subgroup of $H^1(\mathbb{Q}_\Sigma/\mathbb{Q}_\infty, A)$ which is identified with $\mathrm{Sel}_E(F_\infty)_5^{\omega^3}$ by the restriction map. Noting that $\omega^{-3} = \omega$, we have a nonsplit exact sequence of $G_\mathbb{Q}$-modules

$$0 \to \mu_5^{\otimes 2} \to A[5] \to \mu_5 \to 0.$$

This is even nonsplit as a sequence of $G_{\mathbb{Q}_5}$-modules or $G_{\mathbb{Q}_{11}}$-modules. The $G_\mathbb{Q}$-submodule $\mu_5^{\otimes 2}$ of $A[5]$ is just $\Phi_2 \otimes \omega$, which we will denote simply by Φ. We let $\Psi = A[5]/\Phi$. We will show that

$$S_A(\mathbb{Q}_\infty) \cong H^1(\mathbb{Q}_\Sigma/\mathbb{Q}_\infty, \Phi)$$

where the isomorphism is by the map $\epsilon : H^1(\mathbb{Q}_\Sigma/\mathbb{Q}_\infty, \Phi) \to H^1(\mathbb{Q}_\Sigma/\mathbb{Q}_\infty, A)$. This map is clearly injective. Since $\Phi \subseteq C_5 \otimes \omega$, it follows that the local condition defining $S_A(\mathbb{Q}_\infty)$ at the prime of $\mathbb{Q}_\infty$ lying over 5 is satisfied by the elements of $\mathrm{Im}(\epsilon)$. We now verify that the local condition at the prime of $\mathbb{Q}_\infty$ over 11 is also satisfied. This is because $\Phi \subseteq C_{11}\otimes\omega$, which is true because the above exact sequence is nonsplit over $\mathbb{Q}_{11}$. Since 11 splits completely in

$F/\mathbb{Q}$, $\omega|G_{\mathbb{Q}_{11}}$ is trivial. Thus $A = E[5^\infty]$ as $G_{\mathbb{Q}_{11}}$-modules. One then easily sees that the map

$$H^1(\mathbb{Q}_{11}^{\mathrm{cyc}}, C_{11}) \to H^1(\mathbb{Q}_{11}^{\mathrm{cyc}}, E[5^\infty])$$

is the zero map. That is, the map $H^1(\mathbb{Q}_{11}^{\mathrm{cyc}}, A) \to H^1(\mathbb{Q}_{11}^{\mathrm{cyc}}, D_{11})$ is an isomorphism. Elements of $\mathrm{Im}(\epsilon)$ are mapped to 0 and hence are trivial already in $H^1(\mathbb{Q}_{11}^{\mathrm{cyc}}, A)$, therefore satisfying the local condition defining $S_A(\mathbb{Q}_\infty)$ at the prime over 11.

So it is clear that $\mathrm{Im}(\epsilon) \subseteq S_A(\mathbb{Q}_\infty)$. We will prove that equality holds and that $H^1(\mathbb{Q}_\Sigma/\mathbb{Q}_\infty, \Phi) \cong \mathbb{Z}/5\mathbb{Z}$. This last assertion is rather easy to verify. Let $F^+ = \mathbb{Q}(\sqrt{5})$, the maximal real subfield of F. By class field theory, one finds that there is a unique cyclic extension K/F^+ of degree 5 such that $K/\mathbb{Q}$ is dihedral and $K \subseteq \mathbb{Q}_\Sigma$. Thus, $H^1(\mathbb{Q}_\Sigma/\mathbb{Q}, \Phi) = \mathrm{Hom}(\mathrm{Gal}(K/F^+), \Phi)$ has order 5. It follows that $H^1(\mathbb{Q}_\Sigma/\mathbb{Q}_\infty, \Phi)$ is nontrivial. Also, one can see that K/F^+ is ramified at the primes of F^+ lying over 5 and 11. Let $\Sigma' = \{\infty, 5\}$. Then Φ is a $\mathrm{Gal}(\mathbb{Q}_{\Sigma'}/\mathbb{Q})$-module and one can verify that $H^1(\mathbb{Q}_{\Sigma'}/\mathbb{Q}_\infty, \Phi) = 0$. (It is enough to show that $H^1(\mathbb{Q}_{\Sigma'}/\mathbb{Q}_\infty, \Phi)^\Gamma = H^1(\mathbb{Q}_{\Sigma'}/\mathbb{Q}, \Phi)$ vanishes. This is clear since K/F^+ is ramified at 11.) Therefore, the restriction map $H^1(\mathbb{Q}_\Sigma/\mathbb{Q}_\infty, \Phi) \to H^1(\mathbb{Q}_{11}^{\mathrm{cyc}}, \Phi)$ must be injective. But $H^1(\mathbb{Q}_{11}^{\mathrm{cyc}}, \Phi) \cong \mathbb{Z}/5\mathbb{Z}$, from which it follows that $H^1(\mathbb{Q}_\Sigma/\mathbb{Q}_\infty, \Phi)$ indeed has order 5.

It remains to show that $S_A(\mathbb{Q}_\infty) \subseteq \mathrm{Im}(\epsilon)$. Let $B = A/\Phi$. Then $B \cong E_1[5^\infty] \otimes \omega$ and $B[5] \cong \mu_5^{\otimes 2} \times \mu_5$ as $G_{\mathbb{Q}}$-modules. We will prove that $S_B(\mathbb{Q}_\infty) = 0$, from which it follows that $S_A(\mathbb{Q}_\infty) \subseteq \mathrm{Im}(\epsilon)$. Consider $S_B(\mathbb{Q}_\infty)[5]$, any element of which is represented by a 1-cocycle σ with values in $B[5]$. The map $B[5] \to \mu_5$ sends σ to a 1-cocycle $\tilde{\sigma}$ such that $[\tilde{\sigma}|_{G_{\mathbb{Q}_5^{\mathrm{cyc}}}}]$ is trivial as an element of $H^1(\mathbb{Q}_5^{\mathrm{cyc}}, D_5 \otimes \omega)$. Thus, $[\tilde{\sigma}]$ is in the kernel of the composite map

$$H^1(\mathbb{Q}_\Sigma/\mathbb{Q}_\infty, \mu_5) \to H^1(\mathbb{Q}_5^{\mathrm{cyc}}, \mu_5) \to H^1(\mathbb{Q}_5^{\mathrm{cyc}}, D_5 \otimes \omega).$$

The second map is clearly injective. If the kernel of the first map were nontrivial, then it would have a nonzero intersection with $H^1(\mathbb{Q}_\Sigma/\mathbb{Q}_\infty, \mu_5)^\Gamma = H^1(\mathbb{Q}_\Sigma/\mathbb{Q}, \mu_5)$. One then sees that the map $a : H^1(\mathbb{Q}_\Sigma/\mathbb{Q}, \mu_5) \to H^1(\mathbb{Q}_5, \mu_5)$ would have a nonzero kernel. But, as we already used before, the map a is injective. (The elements of $H^1(\mathbb{Q}_\Sigma/\mathbb{Q}, \mu_5)$ are represented by the 1-cocycles associated to $\sqrt[5]{5^i 11^j}$, $0 \le i, j \le 3$. But $5^i 11^j \in (\mathbb{Q}_5^\times)^5 \Leftrightarrow i = j = 0$.) Thus, the first map is injective too. Thus $[\tilde{\sigma}] = 0$. Hence we may assume that σ has values in $\mu_5^{\otimes 2}$. Now, in contrast to A, we have $\mu_5^{\otimes 2} \not\subseteq C_{11} \otimes \omega$. That is, the map $B \to D_{11}$ induces an isomorphism $\mu_5^{\otimes 2} \xrightarrow{\sim} D_{11}[5]$. The composite map

$$H^1(\mathbb{Q}_{11}^{\mathrm{cyc}}, \mu_5^{\otimes 2}) \to H^1(\mathbb{Q}_{11}^{\mathrm{cyc}}, B) \to H^1(\mathbb{Q}_{11}^{\mathrm{cyc}}, D_{11})$$

is clearly injective. Since $[\sigma]$ becomes trivial in $H^1(\mathbb{Q}_{11}^{\mathrm{cyc}}, B)$, it follows that

$$[\sigma] \in \ker\left(H^1(\mathbb{Q}_\Sigma/\mathbb{Q}_\infty, \mu_5^{\otimes 2}) \to H^1(\mathbb{Q}_{11}^{\mathrm{cyc}}, \mu_5^{\otimes 2})\right)$$

But we already showed that this kernel is trivial. (Recall that $\mu_5^{\otimes 2} \cong \Phi$.) Hence $[\sigma] = 0$, proving that $S_B(\mathbb{Q}_\infty) = 0$ as claimed.

Conductor = 768. We return now to the elliptic curves 768(D1, D3) which we denoted previously by E_1 and E_2. We take $p = 5$. As we mentioned earlier, $\mathrm{Sel}_{E_1}(\mathbb{Q})_5 = 0$ and $\mathrm{Sel}_{E_1}(\mathbb{Q}_\infty)_5 = 0$. Also, E_1 and E_2 are related by an isogeny of degree 5. Let Φ denote the $G_{\mathbb{Q}}$-invariant subgroup of $E_2[5^\infty]$ such that $E_2/\Phi \cong E_1$, $|\Phi| = 5$. Let $\Psi = E_2[5]/\Phi$. Then $G_{\mathbb{Q}}$ acts on Φ and Ψ by characters φ, ψ with values in $(\mathbb{Z}/5\mathbb{Z})^\times$ which factor through $\mathrm{Gal}(\mathbb{Q}_\Sigma/\mathbb{Q})$, where now $\sigma = \{\infty, 2, 3, 5\}$. Since E_2 has good, ordinary reduction at 5, one of the characters φ, ψ will be unramified at 5. Denote this character by θ. By looking at the Fourier coefficients for the modular form associated to E_2 (which are given in [Cre]), one finds that θ is the even character of conductor 16 determined by $\theta(5) = 2 + 5\mathbb{Z}$. Then $\theta(3) = 3 + 5\mathbb{Z}$. Now E_1 and E_2 have split, multiplicative reduction at 3. One has a nonsplit exact sequence of $G_{\mathbb{Q}}$-modules

$$0 \to \Psi \to E_1[5] \to \Phi \to 0$$

which remains nonsplit for the action of $G_{\mathbb{Q}_3}$ since the Tate period for E_1 over $\mathbb{Q}_3$ has valuation not divisible by 5. Thus, $\Psi \cong \mu_5$ as $G_{\mathbb{Q}_3}$-modules. Thus, $\psi(3) = 3 + 5\mathbb{Z}$, $\varphi(3) = 1 + 5\mathbb{Z}$. Hence we have $\theta = \psi$. Therefore, Ψ is even and unramified at 5, Φ is odd and ramified at 5. By theorem 5.7, we see that $\mathrm{Sel}_{E_2}(\mathbb{Q}_\infty)_5$ has positive μ-invariant. But since $\mathrm{Sel}_{E_1}(\mathbb{Q}_\infty)_5 = 0$, it is clear that $\mathrm{Sel}_{E_2}(\mathbb{Q}_\infty)_5 \subseteq H^1(\mathbb{Q}_\Sigma/\mathbb{Q}_\infty, \Phi)$. Thus, $\mu_{E_2} = 1$ and $\mathrm{Sel}_{E_2}(\mathbb{Q}_\infty)_5$ has exponent 5. One then sees easily (using proposition 4.8 and the fact that $\Lambda/5\Lambda$ is a PID) that $\mathrm{Sel}_{E_2}(\mathbb{Q}_\infty)_5 \cong \widehat{\Lambda}[5]$, as we stated earlier. Theorem 4.1 then implies that $\mathrm{Sel}_{E_2}(\mathbb{Q})_5 = 0$.

Conductor = 14. Let $p = 3$. The situation is quite analogous to that for elliptic curves of conductor 11 and for $p = 5$. The μ-invariants of $\mathrm{Sel}_E(\mathbb{Q}_\infty)_3$ if E has conductor 14 can be 0, 1, or 2. The λ-invariant is 0.

Conductor = 34. Let $p = 3$. There are four isogenous curves of conductor 34. We considered earlier the curve $E = 34(\mathrm{A}1)$, showing that $f_E(T) = \theta_1$, up to a factor in $\Lambda^\times$, where $\theta_1 = T^2 + 3T + 3$. The curve 34(A2) is related to E by a $\mathbb{Q}$-isogeny of degree 2 and so again has μ-invariant 0 and λ-invariant equal to 2. The two other curves of conductor 34 have $\mathbb{Q}$-isogenies of degree 3 with kernel isomorphic to μ_3 as a $G_{\mathbb{Q}}$-module. It then follows that they have μ-invariant 1. Denoting either of them by E', the characteristic ideal of the Pontryagin dual of $\mathrm{Sel}_{E'}(\mathbb{Q}_\infty)_3$ is generated by $3\theta_1$.

Conductor = 306. Take $p = 3$. We will consider just the elliptic curve E defined by $y^2 + xy = x^3 - x^2 - 927x + 11097$. This is 306(B3) in [Cre]. It is the quadratic twist of 34(A3) by the character ω of conductor 3. The Mordell-Weil group $E(\mathbb{Q})$ is of rank 1, isomorphic to $\mathbb{Z} \times (\mathbb{Z}/6\mathbb{Z})$. E has potentially ordinary reduction at 3, and has good ordinary reduction over $K = \mathbb{Q}(\mu_3)$ at the prime $\mathfrak{p}$ lying over 3. The unique subgroup Φ of $E(\mathbb{Q})$ of order 3 is contained in the kernel of reduction modulo $\mathfrak{p}$ for $E(K)$. Although the hypotheses of proposition 5.10 are not satisfied by E, the proof can still be followed to show that the μ-invariant of $\mathrm{Sel}_E(\mathbb{Q}_\infty)_3$ is 0. Let F denote the first layer of $\mathbb{Q}_\infty$, $F = \mathbb{Q}(\beta)$ where $\beta = \zeta + \zeta^{-1}$, ζ being a primitive 9-th root of unity. The prime 17 splits completely in $F/\mathbb{Q}$. Using this fact, it is easy to verify directly that

$$\mathrm{Hom}(\mathrm{Gal}(F/\mathbb{Q}), \Phi) \subseteq \mathrm{Sel}_E(\mathbb{Q})_3.$$

This clearly implies that $\ker(\mathrm{Sel}_E(\mathbb{Q})_3 \to \mathrm{Sel}_E(\mathbb{Q}_\infty)_3)$ is nontrivial. (Contrast this with proposition 3.9.) However, we can explain this in the following more concrete way, using the results of a calculation carried out by Karl Rubin. The point $P = (9, 54)$ on $E(\mathbb{Q})$ is a generator of $E(\mathbb{Q})/E(\mathbb{Q})_{\mathrm{tors}}$. But $P = 3Q$, where $Q = (-6\beta^2 + 9\beta + 15, 15\beta^2 - 48\beta + 9)$ is in $E(F)$. This implies that the map $E(\mathbb{Q}) \otimes (\mathbb{Q}_3/\mathbb{Z}_3) \to E(F) \otimes (\mathbb{Q}_3/\mathbb{Z}_3)$ has a nontrivial kernel. Let ϕ be the 1-cocycle defined by $\phi(g) = g(Q) - Q$ for $g \in G_{\mathbb{Q}}$. This cocycle has values in $E(F)[3]$, which is easily seen to be just $\Phi = E(\mathbb{Q})[3]$, and factors through $\mathrm{Gal}(F/\mathbb{Q})$. Thus it generates $\mathrm{Hom}(\mathrm{Gal}(F/\mathbb{Q}), \Phi)$ and is certainly contained in $\mathrm{Sel}_E(\mathbb{Q})_3$.

For $n \geq 1$, it turns out that $\ker(\mathrm{Sel}_E(\mathbb{Q}_n)_3 \to \mathrm{Sel}_E(\mathbb{Q}_\infty)_3) = 0$. This can be seen by checking that the local condition at any prime of $\mathbb{Q}_n$ lying above 17 (which will be inert in $\mathbb{Q}_\infty/\mathbb{Q}_n$) fails to be satisfied by a nontrivial element of $\mathrm{Hom}(\mathrm{Gal}(\mathbb{Q}_\infty/\mathbb{Q}_n), \Phi)$. The fact that E has split, multiplicative reduction at 17 helps here. The argument given in [HaMa] then shows that $\mathrm{Sel}_E(\mathbb{Q}_\infty)_3$ has no proper Λ-submodule of finite index. As remarked above, the μ-invariant is 0. A calculation of McCabe for the p-adic L-function associated to E combined with Kato's theorem implies that the λ-invariant of $\mathrm{Sel}_E(\mathbb{Q}_\infty)_3$ is 1. It follows that $\mathrm{Sel}_E(\mathbb{Q}_\infty)_3 = E(\mathbb{Q}_\infty) \otimes (\mathbb{Q}_3/\mathbb{Z}_3) \cong \mathbb{Q}_3/\mathbb{Z}_3$, on which Γ acts trivially.

Conductor = 26. Consider 26(B1, B2). These curves are related by isogenies with kernels isomorphic to μ_7 and $\mathbb{Z}/7\mathbb{Z}$. Let E_1 be 26(B1). Then $\mathrm{Sel}_{E_1}(\mathbb{Q})_7$ should be zero. From [Cre], we have $c_2 = 7$, $c_{13} = 1$, $a_7 = 1$, and $|E_1(\mathbb{Q})| = 7$. Take $p = 7$. Theorem 4.1 then implies that $f_{E_1}(0) \sim 7$. Thus, $f_{E_1}(T)$ is an irreducible element of Λ. The only nonzero, proper, $G_{\mathbb{Q}}$-invariant subgroup of $E_1[7]$ is $E_1(\mathbb{Q}) \cong \mathbb{Z}/7\mathbb{Z}$. Although we haven't verified it, it seems likely that $\mu_{E_1} = 0$. (Conjecture 1.11 would predict this.) If this is so, then $\lambda_{E_1} > 0$. Let E_2 be 26(B2). Then $c_2 = c_{13} = 1$, $a_7 = 1$, and $E_2(\mathbb{Q}) = 0$. One can verify that $\mathrm{Sel}_{E_2}(\mathbb{Q})_7 = 0$. Then by Theorem 4.1, we have $f_{E_2}(0) \sim 7^2$. Since $(f_{E_1}(T))$ and $(f_{E_2}(T))$ can differ only by multiplication by a power of 7, it is clear that $f_{E_2}(T) = 7f_{E_1}(T)$, up to a factor in $\Lambda^\times$. Thus, $\mu_{E_2} \geq 1$, which also follows from proposition 5.7 because $E_2[7]$ contains the odd, ramified $G_{\mathbb{Q}}$-submodule μ_7.

Conductor = 147. Consider 147(B1, B2), which we denote by E_1 and E_2, respectively. They are related by isogenies of degree 13. For E_1, one has $c_3 = c_7 = 1$, $a_{13} = 1$, $E_1(\mathbb{Q}) = 0$, and $\mathrm{Sel}_{E_1}(\mathbb{Q}) = 0$. Take $p = 13$. By theorem 4.1, $f_{E_1}(0) \sim 13^2$. For E_2, one has $c_3 = 13$, $c_7 = 1$, $a_{13} = 1$, $E_2(\mathbb{Q}) = 0$, and $\mathrm{Sel}_{E_2}(\mathbb{Q}) = 0$. Thus, $f_{E_2}(0) \sim 13^3$. Since an isogeny $E_1 \to E_2$ of degree

13 induces a homomorphism $\mathrm{Sel}_{E_1}(\mathbb{Q}_\infty)_{13} \to \mathrm{Sel}_{E_2}(\mathbb{Q}_\infty)_{13}$ with kernel and cokernel of exponent 13, it is clear that $f_{E_2}(T) = 13 f_{E_1}(T)$, up to a factor in $\Lambda^\times$. Conjecturally, $\mu_{E_1} = 0$ and hence $\mu_{E_2} = 1$. Let ξ be the quadratic character of conductor 7, which is odd. Then E_1^ξ and E_2^ξ are the curves 147(C1, C2). Proposition 5.10 implies that $\mathrm{Sel}_{E_1^\xi}(\mathbb{Q}_\infty)_{13}$ and $\mathrm{Sel}_{E_2^\xi}(\mathbb{Q}_\infty)_{13}$ have μ-invariant equal to zero. In fact, for both E_1^ξ and E_2^ξ, we in fact have $c_3 = c_7 = 1$, $a_{13} = -1$, $E_1^\xi(\mathbb{Q}) = E_2^\xi(\mathbb{Q}) = 0$, $\mathrm{Sel}_{E_1^\xi}(\mathbb{Q})_{13} = \mathrm{Sel}_{E_2^\xi}(\mathbb{Q})_{13} = 0$. By proposition 3.8, we have $\mathrm{Sel}_{E_1^\xi}(\mathbb{Q}_\infty)_{13} = \mathrm{Sel}_{E_2^\xi}(\mathbb{Q}_\infty)_{13} = 0$.

Conductor = 1225. Consider now $E_1 : y^2 + xy + y = x^3 + x^2 - 8x + 6$ and also $E_2 : y^2 + xy + y = x^3 + x^2 - 208083x - 36621194$. These curves have conductor 1225 and are related by a $\mathbb{Q}$-isogeny of degree 37. They have additive reduction at 5 and 7. Hence the Tamagawa factors are at most 4. The j-invariants are in $\mathbb{Z}$ and so these curves have potentially good reduction at 5 and 7. We take $p = 37$. Since $a_{37} = 8$, E_1 and E_2 have good, ordinary reduction at p. Let Φ be the $G_\mathbb{Q}$-invariant subgroup of $E_2[37]$ and let $\Psi = E_2[37]/\Phi$. Thus, Φ is the kernel of the isogeny from E_2 to E_1. The real periods Ω_1, Ω_2 of E_1, E_2 are given by: $\Omega_1 = 4.1353\ldots$, $\Omega_2 = .11176\ldots$. Since $\Omega_1 = 37\Omega_2$, one finds that Φ must be odd. Let φ, ψ be the $(\mathbb{Z}/37\mathbb{Z})^\times$-valued characters which describe the action of $G_\mathbb{Q}$ on Φ and Ψ. We can regard them as Dirichlet characters. They have conductor dividing $5 \cdot 7 \cdot 37$ and one of them (which we denote by θ) is unramified at 37. By examining the Fourier coefficients of the corresponding modular form, one finds that θ is characterized by $\theta(2) = 8 + 37\mathbb{Z}$, $\theta(13) = 6 + 37\mathbb{Z}$. The character θ is even and has order 12 and conductor 35. But since φ is odd, we must have $\theta = \psi$. Thus, Φ is odd and ramified at 37. Therefore, by proposition 5.7, we have $\mu_{E_2} \geq 1$. By using the result given in [Pe2] or [Sch3], one finds that $\mu_{E_2} = \mu_{E_1} + 1$. Conjecturally, $\mu_{E_2} = 1$, $\mu_{E_1} = 0$. In any case, we have $(f_{E_2}(T)) = (37 f_{E_1}(T))$. Now $E_1(\mathbb{Q})$ and $E_2(\mathbb{Q})$ have rank 1. It is interesting to note that the fact that $\mathrm{Sel}_{E_1}(\mathbb{Q})_{37}$, $\mathrm{Sel}_{E_2}(\mathbb{Q})_{37}$ are infinite can be deduced from Theorem 4.1. For if one of these Selmer groups were finite, then so would the other. One would then see that both $f_{E_1}(0)$ and $f_{E_2}(0)$ would have even valuation. This follows from Cassels' theorem that $|\mathrm{Sel}_{E_i}(\mathbb{Q})|$ is a perfect square for $i = 1, 2$ together with the fact that the Tamagawa factors for E_i at 5 and 7 cannot be divisible by 37. But $f_{E_2}(0) \sim 37 f_{3_1}(0)$, which gives a contradiction. Similar remarks apply to even quadratic twists of E_1 and E_2.

Now we will state and prove the analogues of propositions 5.7 and 5.10 for $p = 2$. It is necessary to define the terms "ramified" and "odd" somewhat more carefully. Assume that E is an elliptic curve/$\mathbb{Q}$ with good, ordinary or multiplicative reduction at 2. Suppose that Φ is a cyclic $G_\mathbb{Q}$-invariant subgroup of $E[2^\infty]$. We say that Φ is "ramified at 2" if $\Phi \subseteq C_2$, where C_2 is the subgroup of $E[2^\infty]$ which occurs in the description of the image of the local Kummer map for E over $\mathbb{Q}_2$ given in section 2. (It is characterized by

$C_2 \cong \mu_{2^\infty}$ for the action of $I_{\mathbb{Q}_2}$. Here $I_{\mathbb{Q}_2}$ is the inertia subgroup of $G_{\mathbb{Q}_2}$, identified with a subgroup of $G_{\mathbb{Q}}$ by choosing a prime of $\overline{\mathbb{Q}}$ lying over 2. Then $D_2 = E[2^\infty]/C_2$ is an unramified $G_{\mathbb{Q}_2}$-module.) We say that Φ is "odd" if $\Phi \subseteq C_\infty$, where C_∞ denotes the maximal divisible subgroup of $E[2^\infty]$ on which $\mathrm{Gal}(\mathbb{C}/\mathbb{R})$ acts by -1 : $C_\infty = (E[2^\infty]^-)_{\mathrm{div}}$. Then $C_\infty \cong \mathbb{Q}_2/\mathbb{Z}_2$ as a group. Here we identify $\mathrm{Gal}(\mathbb{C}/\mathbb{R})$ with a subgroup of $G_{\mathbb{Q}}$ by choosing an infinite prime of $\overline{\mathbb{Q}}$. (We remark that $C_\infty \cong \mu_{2^\infty}$ as $\mathrm{Gal}(\mathbb{C}/\mathbb{R})$-modules and that $\mathrm{Gal}(\mathbb{C}/\mathbb{R})$ acts trivially on $D_\infty = E[2^\infty]/C_\infty$.) Since Φ is $G_{\mathbb{Q}}$-invariant, these definitions are easily seen to be independent of the choice of primes of $\overline{\mathbb{Q}}$ lying over 2 and over ∞. We now prove the analogue of proposition 5.7.

Proposition 5.13. *Suppose that E is an elliptic curve/$\mathbb{Q}$ with good, ordinary or multiplicative reduction at 2. Suppose also that $E[2^\infty]$ contains a $G_{\mathbb{Q}}$-invariant subgroup Φ of order 2^m which is ramified at 2 and odd. Then the μ-invariant of $\mathrm{Sel}_E(\mathbb{Q}_\infty)_2$ is at least m.*

Proof. The argument is virtually the same as that for proposition 5.7. We consider $\mathrm{Im}(\epsilon)$ where ϵ is the map

$$\epsilon : H^1(\mathbb{Q}_\Sigma/\mathbb{Q}_\infty, \Phi) \to H^1(\mathbb{Q}_\Sigma/\mathbb{Q}_\infty, E[2^\infty]).$$

The kernel is finite. Since $\Phi \subseteq C_2$, the elements of $\mathrm{Im}(\epsilon)$ satisfy the local conditions defining $\mathrm{Sel}_E(\mathbb{Q}_\infty)_2$ at the prime of $\mathbb{Q}_\infty$ lying over 2. Also, just as previously, a subgroup of finite index in $\mathrm{Im}(\epsilon)$ satisfies the local conditions for all other nonarchimedean primes of $\mathbb{Q}_\infty$. Now we consider the archimedean primes of $\mathbb{Q}_\infty$. Note that $H^1(\mathbb{R}, C_\infty) = 0$. Since $\Phi \subseteq C_\infty$, it is clear that elements in $\mathrm{Im}(\epsilon)$ are locally trivial in $H^1((\mathbb{Q}_\infty)_\eta, E[2^\infty])$ for every infinite prime η of $\mathbb{Q}_\infty$. Therefore, $\mathrm{Im}(\epsilon) \cap \mathrm{Sel}_E(\mathbb{Q}_\infty)_2$ has finite index in $\mathrm{Im}(\epsilon)$.

It remains to show that the Λ-module $H^1(\mathbb{Q}_\Sigma/\mathbb{Q}_\infty, \Phi)$ has μ-invariant equal to m. Since the $G_{\mathbb{Q}}$-composition factors for Φ are isomorphic to $\mathbb{Z}/2\mathbb{Z}$, lemma 5.9 implies that the μ-invariant for $H^1(\mathbb{Q}_\Sigma/\mathbb{Q}_\infty, \Phi)$ is at most m. On the other hand, the Euler characteristic of the $\mathrm{Gal}(\mathbb{Q}_\Sigma/\mathbb{Q}_n)$-module Φ is $\prod_{v|\infty} |\Phi/\Phi^{D_v}|^{-1}$, where v runs over the infinite primes of $\mathbb{Q}_n$ and $D_v = \mathrm{Gal}(\mathbb{C}/\mathbb{R})$ is a corresponding decomposition group. Assume that $m \geq 1$. Then $|\Phi^{D_v}| = 2$ and so this Euler characteristic is $2^{-(m-1)2^n}$ for all $n \geq 0$. Now $H^0(\mathbb{Q}_\Sigma/\mathbb{Q}_n, \Phi)$ just has order 2. As for $H^2(\mathbb{Q}_\Sigma/\mathbb{Q}_n, \Phi)$, it is known that the map

$$H^2(\mathbb{Q}_\Sigma/\mathbb{Q}_n, \Phi) \to \prod_{v|\infty} H^2((\mathbb{Q}_n)_v, \Phi)$$

is surjective. (This is corollary 4.16 in [Mi].) Since $H^2(D_v, \Phi)$ has order 2, it follows that $|H^2(\mathbb{Q}_\Sigma/\mathbb{Q}_n, \Phi)| \geq 2^{2^n}$. Therefore,

$$|H^1(\mathbb{Q}_\Sigma/\mathbb{Q}_n, \Phi)| \geq 2^{m2^n+1}.$$

The restriction map $H^1(\mathbb{Q}_\Sigma/\mathbb{Q}_n,\Phi) \to H^1(\mathbb{Q}_\Sigma/\mathbb{Q}_\infty,\Phi)^{\Gamma_n}$ is surjective and has kernel $H^1(\Gamma_n,\mathbb{Z}/2\mathbb{Z})$, which has order 2. Thus,

$$|H^1(\mathbb{Q}_\Sigma/\mathbb{Q}_n,\Phi)^{\Gamma_n}| \geq 2^{m2^n}$$

for all n. This implies that $H^1(\mathbb{Q}_\Sigma/\mathbb{Q}_\infty,\Phi)$ has μ-invariant at least m. Therefore, the μ-invariant of $H^1(\mathbb{Q}_\Sigma/\mathbb{Q}_\infty,\Phi)$ and hence of $\mathrm{Im}(\epsilon)$ is exactly m, proving proposition 5.13. ∎

Remark. As we mentioned before (for any p), if E admits a $\mathbb{Q}$-isogeny of degree 2 and if $\mathrm{Sel}_E(\mathbb{Q}_\infty)_2$ is Λ-cotorsion, then $\mathrm{Sel}_E(\mathbb{Q}_\infty)_2$ contains a Λ-submodule pseudo-isomorphic to $\widehat{\Lambda}[2^{\mu_E}]$. It is known that there are infinitely many elliptic curves/$\mathbb{Q}$ admitting a cyclic $\mathbb{Q}$-isogeny of degree 16, but none with such an isogeny of degree 32. We will give examples below where the assumptions in proposition 5.13 are satisfied and $|\Phi| = 2^m$ with $m = 0,1,2,3$, or 4. For any elliptic curve $E/\mathbb{Q}$, there is a maximal $G_\mathbb{Q}$-invariant subgroup Φ which is ramified and odd. Define m_E by $|\Phi| = 2^{m_E}$. Conjecturally, $\mu_E = m_E$. Thus the possible values of μ_E as E varies over elliptic curves/$\mathbb{Q}$ with good, ordinary or multiplicative reduction at 2 should be 0, 1, 2, 3, or 4. Examples where $\mu_E > 0$ are abundant. It suffices to have a point $P \in E(\mathbb{Q})$ of order 2 such that $P \in C_2$ and $P \in C_\infty$, using the notation introduced earlier. If the discriminant of a Weierstrass equation for E is negative, then $E(\mathbb{R})$ has just one component. In this case, $C_\infty[2] = E(\mathbb{R})[2]$ and so if $P \in E(\mathbb{Q})$ has order 2, then $\Phi = \langle P \rangle$ is automatically odd. (Note that in this case $H^1(\mathbb{R},E[2^\infty]) = 0$ and so the local conditions at the infinite primes of $\mathbb{Q}_\infty$ occurring in the definition of $\mathrm{Sel}_E(\mathbb{Q}_\infty)_2$ are trivially satisfied anyway.) Similarly, if this discriminant is not a square in $\mathbb{Q}_2^\times$, then $\Phi = \langle P \rangle$ is automatically ramified since then $C_2[2] = E(\mathbb{Q}_2)[2]$.

We now prove the analogue of proposition 5.10, which gives a sufficient condition for $\mu_E = 0$ in case $p = 2$.

Proposition 5.14. *Suppose that E is an elliptic curve/$\mathbb{Q}$ with good, ordinary or multiplicative reduction at 2. Suppose also that $E(\mathbb{Q})$ contains an element P of order 2 and that $\Phi = \langle P \rangle$ is either ramified at 2 but not odd or odd but not ramified at 2. Then $\mathrm{Sel}_E(\mathbb{Q}_\infty)_2$ is Λ-cotorsion and $\mu_E = 0$.*

Proof. We must show that $\mathrm{Sel}_E(\mathbb{Q}_\infty)_2[2]$ is finite. Consider the map

$$\alpha_E : H^1(\mathbb{Q}_\Sigma/\mathbb{Q}_\infty, E[2^\infty]) \to \mathcal{P}^{(\infty)}_{E[2^\infty]}(\mathbb{Q}_\infty)$$

which occurred in the proof of proposition 5.8. By definition we have

$$\mathrm{Sel}_E(\mathbb{Q}_\infty)_2 \subseteq \ker(\alpha_E).$$

Under the hypothesis that E admits a $\mathbb{Q}$-rational isogeny of degree 2 (i.e., that $E(\mathbb{Q})$ has an element of order 2), we showed earlier that $\ker(\alpha_E)[2]$ has $(\Lambda/2\Lambda)$-corank equal to 1. Consider the map

$$\epsilon : H^2(\mathbb{Q}_\Sigma/\mathbb{Q}_\infty,\Phi) \to H^1(\mathbb{Q}_\Sigma/\mathbb{Q}_\infty, E[2^\infty]).$$

Then $\ker(\epsilon)$ is finite and so, by lemma 5.9, $\operatorname{Im}(\epsilon)$ also has $(\Lambda/2\Lambda)$-corank equal to 1.

Assume first that Φ is odd but not ramified at 2. Then $\Phi \subseteq C_\infty$. Since we have $H^1(\mathbb{R}, C_\infty) = 0$, it is clear that $\operatorname{Im}(\epsilon) \subseteq \ker(\alpha_E)$. It follows that $\operatorname{Im}(\epsilon)$ has finite index in $\ker(\alpha_E)[2]$. Thus, it suffices to prove that $\operatorname{Im}(\epsilon) \cap \mathrm{Sel}_E(\mathbb{Q}_\infty)_2$ is finite. To do this, consider the composite map β defined by the commutative diagram

$$\begin{array}{ccccc} H^1(\mathbb{Q}_\Sigma/\mathbb{Q}_\infty, \Phi) & \xrightarrow{\epsilon} & H^1(\mathbb{Q}_\Sigma/\mathbb{Q}_\infty, E[2^\infty]) & \longrightarrow & H^1(I_\pi, E[2^\infty]) \\ & \searrow^{\beta} & & & \downarrow \\ & & & & H^1(I_\pi, D_2) \end{array}$$

where π is the unique prime of $\mathbb{Q}_\infty$ lying above 2 and I_π is the inertia subgroup of $G_{(\mathbb{Q}_\infty)_\pi}$. Let $B = \ker(\beta)$. If $[\sigma] \in H^1(\mathbb{Q}_\Sigma/\mathbb{Q}_\infty, \Phi)$, then the local condition defining $\mathrm{Sel}_E(\mathbb{Q}_\infty)_2$ at the prime π is satisfied by $\epsilon([\sigma])$ precisely when $[\sigma] \in B$. Since $\Phi \not\subseteq C_2$, the map $E[2^\infty] \to D_2$ induces an isomorphism of Φ to $D_2[2]$. Also, since I_π acts trivially on D_2, the map $H^1(I_\pi, D_2[2]) \to H^1(I_\pi, D_2)$ is injective. Hence

$$B = \ker(H^1(\mathbb{Q}_\Sigma/\mathbb{Q}_\infty, \Phi) \to H^1(I_\pi, \Phi)).$$

If we let $H^1_{\mathrm{unr}}(\mathbb{Q}_\infty, \Phi)$ denote the subgroup of $H^1(\mathbb{Q}_\infty, \Phi)$ consisting of elements which are unramified at all *nonarchimedean* primes of $\mathbb{Q}_\infty$, then $H^1_{\mathrm{unr}}(\mathbb{Q}_\infty, \Phi)$ is a subgroup of B and the index is easily seen to be finite. (Only finitely many nonarchimedean primes η of $\mathbb{Q}_\infty$ exist lying over primes in Σ. $H^1((\mathbb{Q}_\infty)_\eta, \Phi)$ is finite if $\eta \nmid 2$.) Now $G_{\mathbb{Q}_\infty}$ acts trivially on Φ. Let L^*_∞ denote the maximal abelian pro-2 extension of $\mathbb{Q}_\infty$ which is unramified at all nonarchimedean primes of $\mathbb{Q}_\infty$. Then

$$H^1_{\mathrm{unr}}(\mathbb{Q}_\infty, \Phi) = \mathrm{Hom}(\mathrm{Gal}(L^*_\infty/\mathbb{Q}_\infty), \Phi).$$

But it is easy to verify that $L^*_\infty = \mathbb{Q}_\infty$. (For example, one can note that $L^*_\infty\mathbb{Q}_\infty(i)/\mathbb{Q}_\infty(i)$ is *everywhere* unramified. But $\mathbb{Q}_\infty(i) = \mathbb{Q}(\mu_{2^\infty})$. It is known that $\mathbb{Q}(\mu_{2^n})$ has odd class number for all $n \geq 0$. Thus, $\mathbb{Q}_\infty \subseteq L^*_\infty \subseteq \mathbb{Q}_\infty(i)$, from which $L^*_\infty = \mathbb{Q}_\infty$ follows.) Therefore B is finite. Therefore $\operatorname{Im}(\epsilon) \cap \mathrm{Sel}_E(\mathbb{Q}_\infty)_2$ is indeed finite.

Now assume that Φ is ramified at 2 but not odd. Let ϵ be as above. Since Φ is not odd, it follows that $E(\mathbb{R})$ must have two connected components. Hence, by proposition 5.8, $H^1(\mathbb{Q}_\Sigma/\mathbb{Q}_\infty, E[2^\infty])[2]$ has $(\Lambda/2\Lambda)$-corank equal to 2. This implies that the $(\Lambda/2\Lambda)$-corank of $H^1(\mathbb{Q}_\Sigma/\mathbb{Q}_\infty, E[2])$ is 2. On the other hand, $H^1((\mathbb{Q}_\infty)_\pi, E[2])$ also has $(\Lambda/2\Lambda)$-corank equal to 2. Consider the map

$$a : H^1(\mathbb{Q}_\Sigma/\mathbb{Q}_\infty, E[2]) \to H^1((\mathbb{Q}_\infty)_\pi, E[2]).$$

We will show that the kernel is finite. It follows from this that the cokernel is also finite. We have an exact sequence

$$0 \to \Phi \to E[2] \to \Psi \to 0$$

of $G_{\mathbb{Q}_\infty}$-modules, where $\Phi \cong \Psi \cong \mathbb{Z}/2\mathbb{Z}$. The finiteness of the group B introduced earlier in this proof, and the corresponding fact for Ψ, implies rather easily that $\ker(a)$ is indeed finite. Consider the map

$$b : H^1((\mathbb{Q}_\infty)_\pi, E[2]) \to H^1((\mathbb{Q}_\infty)_\pi, D_2[2])$$

induced by the map $E[2^\infty] \to D_2$. Since $(\mathbb{Q}_\infty)_\pi$ has 2-cohomological dimension 0, b is surjective. It follows that $\operatorname{coker}(b \circ a)$ is finite. Using the fact that $H^1((\mathbb{Q}_\infty)_\pi, D_2[2])$ has $(\Lambda/2\Lambda)$-corank 1, we see that $\ker(b \circ a)$ also has $(\Lambda/2\Lambda)$-corank 1. Consider the map

$$\gamma_E : H^1(\mathbb{Q}_\Sigma/\mathbb{Q}_\infty, E[2^\infty]) \to H^1((\mathbb{Q}_\infty)_\pi, D_2).$$

The above remarks imply easily that $\ker(\gamma_E)[2]$ has $(\Lambda/2\Lambda)$-corank equal to 1.

The rest of the argument is now rather similar to that for the first case. It is clear that $\operatorname{Im}(\epsilon) \subseteq \ker(\gamma_E)$ since $\Phi \subseteq C_2$. Thus, $\operatorname{Im}(\epsilon)$ has finite index in $\ker(\gamma_E)[2]$. Also, by definition, we have

$$\operatorname{Sel}_E(\mathbb{Q}_\infty)_2 \subseteq \ker(\gamma_E).$$

It then suffices to show that $\operatorname{Im}(\epsilon) \cap \operatorname{Sel}_E(\mathbb{Q}_\infty)_2$ is finite. To do this, we consider the composite map δ_η defined by the following commutative diagram.

$$\begin{array}{ccccc}
H^1(\mathbb{Q}_\Sigma/\mathbb{Q}_\infty, \Phi) & \xrightarrow{\ \epsilon\ } & H^1(\mathbb{Q}_\Sigma/\mathbb{Q}_\infty, E[2^\infty]) & \longrightarrow & H^1((\mathbb{Q}_\infty), E[2^\infty]) \\
 & \searrow^{\delta_\eta} & & & \downarrow \\
 & & & & H^1((\mathbb{Q}_\infty)_\eta, D_\infty)
\end{array}$$

where η is any infinite prime of $\mathbb{Q}_\infty$. If $[\sigma] \in H^1(\mathbb{Q}_\Sigma/\mathbb{Q}_\infty, \Phi)$, then the local condition defining $\operatorname{Sel}_E(\mathbb{Q}_\infty)_2$ at η for the element $\epsilon([\sigma])$ would imply that $\delta_\eta([\sigma]) = 0$. But since $\Phi \not\subseteq C_\infty$, Φ is identified with $D_\infty[2]$. The map $H^1(\mathbb{R}, D_\infty[2]) \to H^1(\mathbb{R}, D_\infty)$ is injective since $\operatorname{Gal}(\mathbb{C}/\mathbb{R})$ acts trivially on D_∞. Hence

$$\ker(\delta_\eta) = \ker\big(H^1(\mathbb{Q}_\Sigma/\mathbb{Q}_\infty, \Phi) \to H^1((\mathbb{Q}_\infty)_\eta, \Phi)\big).$$

By lemma 5.9, we know that $\bigcap_\eta \ker(\delta_\eta)$ is finite, where η varies over all the infinite primes of $\mathbb{Q}_\infty$. It follows from this that $\operatorname{Im}(\epsilon) \cap \operatorname{Sel}_E(\mathbb{Q}_\infty)_2$ is also finite. This implies that $\operatorname{Sel}_E(\mathbb{Q}_\infty)_2[2]$ is finite, finishing the proof of proposition 5.14. ∎

We now consider various examples.

Conductor = 15. There are eight curves of conductor 15, all related by $\mathbb{Q}$-isogenies whose degrees are powers of 2. We will let E_i denote the curve labeled A_i in [Cre] for $1 \le i \le 8$. The following table summarizes the situation for $p = 2$.

		E_1	E_2	E_3	E_4	E_5	E_6	E_7	E_8
$\lvert Ш \rvert$	$=$	1	1	1	1	1	1	1	1
$\lvert T \rvert$	$=$	8	4	8	8	2	2	4	4
c_3, c_5	$=$	2,4	2,2	2,2	2,8	2,1	2,1	1,1	1,1
$f_{E_i}(0)$	$\sim$	2	4	1	4	8	8	1	1
μ_{E_i}	$=$	1	2	0	2	3	3	0	0

For conductor 15, the Selmer group $\mathrm{Sel}_{E_i}(\mathbb{Q}_\infty)_2$ has λ-invariant equal to 0 and the μ-invariant varies from 0 to 3. Now $Ш = \mathrm{Sel}_E(\mathbb{Q})$. Its order was computed under the assumption of the Birch and Swinnerton-Dyer conjecture by evaluating $L(E_i/\mathbb{Q}, 1)/\Omega_{E_i}$. The real period Ω_{E_i} was computed using PARI. $|T|$, c_3, and c_5 are as listed in [Cre]. Using the fact that $a_2 = -1$, and hence $|\widetilde{E}_i(\mathbb{F}_2)| = 4$ for each i, the fourth row is a consequence of theorem 4.1. In particular, it is clear that $f_{E_3}(T) \in \Lambda^\times$. Hence $\mu_{E_3} = \lambda_{E_3} = 0$. The λ-invariant of $\mathrm{Sel}_{E_i}(\mathbb{Q}_\infty)_2$ is unchanged by a $\mathbb{Q}$-isogeny. Hence $\lambda_{E_i} = 0$ for all i. It is then obvious that $f_{E_i}(0) \sim 2^{\mu_{E_i}}$, which gives the final row.

It is not difficult to reconcile these results with propositions 5.13 and 5.14. For example, consider $E_3 : y^2 + xy + y = x^3 + x^2 - 5x + 2$. We have $E_3[2] \cong (\mathbb{Z}/2\mathbb{Z})^2$. The points of order 2 are $(1, -1)$, $(\frac{3}{4}, -\frac{7}{8})$, and $(-3, 1)$. The second point generates $C_2[2]$; the third point generates $C_\infty[2]$. (Remark: It is not hard to find the generator of $C_\infty[2]$. It is the point in $E(\mathbb{R})[2]$ whose x-coordinate is minimal.) Thus, $E_3(\mathbb{Q})[2]$ contains a subgroup which is ramified at 2 but not odd and another subgroup which is odd but not ramified at 2. Proposition 5.14 implies that $\mu_{E_3} = 0$. Similarly, one can verify that $\mu_{E_7} = \mu_{E_8} = 0$. Both $E_7(\mathbb{Q})[2]$ and $E_8(\mathbb{Q})[2]$ have order 2. $E_7(\mathbb{Q})[2]$ is ramified at 2 but not odd. $E_8(\mathbb{Q})[2]$ is odd but not ramified at 2. Proposition 5.14 again applies.

The μ-invariants listed above turn out to be just as predicted by proposition 5.13. One can deduce this from the isogeny data given in [Cre]. One uses the following observation. *Suppose that $\varphi : E \to E'$ is a $\mathbb{Q}$-isogeny such that $\Phi = \ker(\varphi)$ is ramified and odd. Suppose also that $E'[2^\infty]$ contains a $\mathbb{Q}$-rational subgroup Φ' which is ramified and odd. Then $\varphi^{-1}(\Phi')$ is ramified and odd too. Its order is $|\Phi| \cdot |\Phi'|$.* For example, $E_2(\mathbb{Q})$ has three subgroups of order 2, one of which is ramified and odd. There is a $\mathbb{Q}$-isogeny of degree 2 from E_2 to E_1, E_5, and E_6. One can verify that $E_1(\mathbb{Q})$, $E_5(\mathbb{Q})$, and $E_6(\mathbb{Q})$ each has a subgroup of order 2 which is ramified and odd. Thus, $E_2[2^\infty]$ must have a subgroup Φ' of order 4 which is ramified and odd. Now $\Phi = E_5(\mathbb{Q})[2]$

is of order 2, generated by $(-\frac{109}{4}, \frac{105}{8})$. This Φ is ramified and odd. Since $E_5/\Phi \cong E_2$, it follows that $E_5[2^\infty]$ has a ramified and odd $\mathbb{Q}$-rational subgroup of order 8. Thus, proposition 5.13 implies that $\mu_{E_5} \geq 3$.

Conductor = 69. There are two such elliptic curves $E/\mathbb{Q}$. Both should have $|Ш_E(\mathbb{Q})| = 1$ and $|E(\mathbb{Q})| = 2$. For one of them, we have $c_3 = 1, c_{23} = 2$. For the other, $c_3 = 2$, $c_{23} = 1$. We have $a_2 = 1$ and so $|\widetilde{E}(\mathbb{F}_2)| = 2$. By theorem 4.1, we have $f_E(0) \sim 2$. Hence $f_E(T)$ is an irreducible element of Λ. Now let $\Phi = E(\mathbb{Q}) \cong \mathbb{Z}/2\mathbb{Z}$. For one of these curves, Φ is ramified at 2 but not odd. For the other, Φ is odd but not ramified at 2. Hence proposition 5.14 implies that $\mu_E = 0$. Since $f_E(T) \notin \Lambda^\times$, it follows that $\lambda_E \geq 1$. In fact, it turns out that $\lambda_E = \lambda_E^{M\text{-}W} = 1$, $\lambda_E^{Ш} = 0$, and $f_E(T) = T+2$, up to a factor in $\Lambda^\times$. To see this, consider the quadratic twist E^ξ, where ξ is the quadratic character corresponding to $\mathbb{Q}(\sqrt{2})$. Now $E^\xi(\mathbb{Q})$ has rank 1. Therefore, $E(\mathbb{Q}(\sqrt{2}))$ has rank 1. But $\mathbb{Q}(\sqrt{2})$ is the first layer in the cyclotomic $\mathbb{Z}_2$-extension $\mathbb{Q}_\infty/\mathbb{Q}$. Therefore, $\mathrm{Sel}_E(\mathbb{Q}_\infty)_2$ contains the image of $E(\mathbb{Q}(\sqrt{2})) \otimes (\mathbb{Q}_2/\mathbb{Z}_2)$ under restriction as a Λ-submodule. Its characteristic ideal is $(T+2)$. The assertions made above follow easily.

Conductor = 195. We will discuss the isogeny class consisting of A1–A8 in [Cre]. Some of the details below were worked out by Karl Rubin and myself with the help of PARI. We denote these curves by $E_1, \ldots, E_8$, respectively. We will show that $\lambda_{E_i} = \lambda_{E_i}^{M\text{-}W} = 3$, $\lambda_{E_i}^{Ш} = 0$, and that μ_{E_i} varies from 0 to 4 for $1 \leq i \leq 8$. Here is a table of the basic data.

	E_1	E_2	E_3	E_4	E_5	E_6	E_7	E_8
$\lvert Ш\rvert =$	1	1	1	1	1	1	4	1
$\lvert T\rvert =$	4	8	8	4	4	4	2	2
$c_3, c_5, c_{13} =$	4,1,1	8,2,2	4,4,4	16,1,1	2,8,2	2,2,8	1,4,1	1,16,1
$f_{E_i}(0) \sim$	4	8	16	16	32	32	64	64
$\mu_{E_i} =$	0	1	2	2	3	3	4	4

As before, we evaluated $|Ш|$ by assuming the Birch and Swinnerton-Dyer conjecture. But one could confirm directly that $|Ш_2|$ is as listed, which would be sufficient for us. [Cre] gives $|T|$ and the Tamagawa factors c_3, c_5, and c_{13}. The fourth row is a consequence of theorem 4.1. Since the λ_{E_i}'s are equal, clearly the μ-invariants must vary. The last row becomes clear if we can show that $\mu_{E_1} = 0$. Unfortunately, this does not follow from proposition 5.14. The problem is that $\Phi = E(\mathbb{Q})[2]$ is of order 2, but is neither ramified at 2 nor odd. In fact, Φ is generated by $(6, -3)$, which is clearly not in the kernel of reduction modulo 2 and so is not in C_2. Also, $E(\mathbb{R})[2]$ has order 4 and $(6, -3)$ is in the connected component of O_E. This implies that $(6, -3) \notin C_\infty$.

We will verify that $\mu_{E_1} = 0$ by showing that $f_{E_1}(T)$ is divisible by $g(T) = (T+2)(T^2+2T+2)$ in Λ. Since the characteristic ideals of $\mathrm{Sel}_{E_i}(\mathbb{Q}_\infty)_2$ differ only by multiplication by a power of 2, it is equivalent to show that

$g(T)$ divides $f_{E_i}(T)$ for any i. It then follows that $(f_{E_1}(T)) = (g(T))$ since $f_{E_1}(0)$ and $g(0)$ have the same valuation. Therefore, μ_{E_1} must indeed be zero. Let F and K denote the first and second layers in the cyclotomic $\mathbb{Z}_2$-extension $\mathbb{Q}_\infty/\mathbb{Q}$. Thus $\mathrm{Gal}(K/\mathbb{Q})$ is cyclic of degree 4 and F is the unique quadratic subfield of K. In fact, $F = \mathbb{Q}(\sqrt{2})$, $K = F(\sqrt{2+\sqrt{2}})$. We will show that $E_2(K)\otimes\mathbb{Q}$, considered as a $\mathbb{Q}$-representation of $\mathrm{Gal}(K/\mathbb{Q})$ contains the two nontrivial, $\mathbb{Q}$-irreducible representations of $\mathrm{Gal}(K/\mathbb{Q})$. One of them has degree 1 and factors through $\mathrm{Gal}(F/\mathbb{Q})$. The other has degree 2 and is faithful. The fact that $g(T)$ divides $f_{E_2}(T)$, and hence $f_{E_1}(T)$, follows easily.

The equation $y^2+xy = x^3-115x+392$ is the minimal Weierstrass equation defining E_2. It is slightly more convenient to calculate with the nonminimal equation $y^2 = (x-1)(x-2)(16x+49)$, obtained by a simple change of variables. We single out the following two points satisfying this equation:

$$P = \left(0,\, 7\sqrt{2}\right), \qquad Q = \left(10+9\sqrt{2},\, (123+78\sqrt{2})\sqrt{2+\sqrt{2}}\right).$$

Now P is rational over F, Q is rational over K. To study $E_2(K)$, it is useful to first determine its torsion subgroup. In fact, we have

$$E_2(\mathbb{Q}_\infty)_{\mathrm{tors}} = E_2(\mathbb{Q})_{\mathrm{tors}} \cong (\mathbb{Z}/2\mathbb{Z}) \times (\mathbb{Z}/4\mathbb{Z}).$$

The structure of $E_2(\mathbb{Q})$ is given in [Cre]. It is easy to see that $E_2(\mathbb{Q}_\infty)_{\mathrm{tors}}$ is a 2-primary group since E_2 has good reduction at 2, 2 is totally ramified in $\mathbb{Q}_\infty/\mathbb{Q}$, and $|\widetilde{E}_2(\mathbb{F}_2)| = 4$. Now $\mathbb{Q}_\infty$ is totally real and so $E_2(\mathbb{Q}_\infty)_{\mathrm{tors}} \cong \mathbb{Z}/2\mathbb{Z} \times \mathbb{Z}/2^t\mathbb{Z}$ where $t \geq 2$. Assume $t \geq 3$. Then $E_2(\mathbb{Q}_\infty)_{\mathrm{tors}}$ would have 8 elements of order 8. Since their squares are in $E_2(\mathbb{Q})$, the orbit under $\Gamma = \mathrm{Gal}(\mathbb{Q}_\infty/\mathbb{Q})$ of an element of order 8 has cardinality at most 4. Hence such an element would be rational over K. We can rule out this possibility by noting that E_2 has good reduction at 31, 31 splits completely in $K/\mathbb{Q}$, and $|\widetilde{E}_2(\mathbb{F}_{31})| = 40$, which is not divisible by 16.

It is now clear that P and Q have infinite order. Also, $\mathrm{Gal}(F/\mathbb{Q})$ acts on $\langle P\rangle$ by -1 since $(0, -7\sqrt{2}) = -P$. Thus, $\langle P\rangle \otimes \mathbb{Q}$ is a $\mathrm{Gal}(K/\mathbb{Q})$-invariant subspace of $E_2(K)\otimes\mathbb{Q}$ giving the degree 1, nontrivial representation of $\mathrm{Gal}(K/\mathbb{Q})$. Similarly, Q belongs to $\ker(\mathrm{Tr}_{K/F})$, the kernel of the trace map from $E_2(K)$ to $E_2(F)$. Thus, $\ker(\mathrm{Tr}_{K/F})\otimes\mathbb{Q}$ is nonzero and provides at least one copy of the 2-dimensional, irreducible $\mathbb{Q}$-representation of $\mathrm{Gal}(K/\mathbb{Q})$. Therefore, $\mathrm{rank}(E_2(K)) \geq 3$. Considering the action of $\gamma = 1+T$ on the image of $E_2(K)\otimes(\mathbb{Q}_2/\mathbb{Z}_2)$ in $\mathrm{Sel}_{E_2}(\mathbb{Q}_\infty)_2$ makes it clear that $g(T)$ does indeed divide $f_{E_2}(T)$, as claimed. As noted above, it now follows that $(f_{E_1}(T)) = (g(T))$. This implies that $\lambda_{E_1} = 3$, $\mu_{E_1} = 0$. More precisely, it is clear that $\lambda_{E_1}^{M\text{-}W} = 3$, $\lambda_{E_1}^{Ш} = 0$. For the $\mathbb{Q}$-isogenous curves E_i, $1 \leq i \leq 8$, we also have $\lambda_{E_i}^{M\text{-}W} = 3$, $\lambda_{E_i}^{Ш} = 0$, but $(f_{E_i}(T)) = (2^{\mu_{E_i}} g(T))$.

One can verify that in this example $\mu_{E_i} = m_{E_i}$. It is not hard to prove the existence of a $G_{\mathbb{Q}}$-invariant subgroup Φ_i of $E_i[2^\infty]$ with the expected order satisfying the hypotheses of proposition 5.13. Just as for conductor 15, one uses the $\mathbb{Q}$-isogenies between the E_i's. By direct verification, one finds that $E_i[2]$ contains a ramified, odd $G_{\mathbb{Q}}$-invariant subgroup for $i = 2, \ldots, 8$. The listed isogenies then imply that $E_i[4]$ has a ramified, odd $G_{\mathbb{Q}}$-invariant subgroup of order 4 for $i = 3, \ldots, 8$. Then one sees that $E_i[8]$ contains such a subgroup of order 8 for $i = 5, \ldots, 8$. Finally, both E_7 and E_8 admit $\mathbb{Q}$-isogenies of degree 2 to E_5.The kernels of these $\mathbb{Q}$-isogenies are ramified and odd. The inverse image of the ramified, odd, $G_{\mathbb{Q}}$-invariant subgroup Φ_5 of $E_5[8]$ will be the ramified, odd, $G_{\mathbb{Q}}$-invariant subgroup Φ_i of $E_i[16]$ of order 16, for $i = 7$ or 8.

Ken Kramer has found a description of the family of elliptic curves/$\mathbb{Q}$ which satisfy the hypotheses of proposition 5.13 for $m = 1, 2, 3$, and 4. Here we will give his description for $m = 1$ and $m = 4$, with the additional condition that E have square-free conductor. For $m = 1$, his family is

$$E : y^2 + xy = x^3 - ax^2 - 4bx + (4a-1)b, \qquad \Phi = \left\langle \left(\frac{4a-1}{4}, \frac{1-4a}{8} \right) \right\rangle$$

where $a, b \in \mathbb{Z}$, $\gcd(4a-1, b) = 1$, $(4a-1)^2 > 64b$, and either a or b is negative. The last conditions assure that $\Phi \subseteq C_\infty$. (If $b < 0$, then $E(\mathbb{R})$ has only one connected component. Then Φ is automatically contained in C_∞. If $b > 0$ and $a < 0$, then the inequality $1 - 4a > 8\sqrt{b}$ implies that the above generator of Φ is the element of $E(\mathbb{R})[2]$ with minimal x-coordinate.) The discriminant of this equation, which is minimal, is $b((4a-1)^2 - 64b)^2$. If b is odd, then E has good, ordinary reduction at 2. Conjecturally, this family should give all elliptic curves/$\mathbb{Q}$ with good, ordinary or multiplicative reduction at 2 and square-free conductor such that $\mathrm{Sel}_E(\mathbb{Q}_\infty)_2$ has positive μ-invariant. Kramer describes the elliptic curves with square-free conductor having a subgroup Φ of order 16 which is ramified and odd by the following equation:

$$E : y^2 = (x + 2c^4 - d^4)(x^2 + 4(cd)^4 - 4c^8)$$

where c, d are distinct odd, positive integers, $c \equiv d \pmod 4$, and $\gcd(c, d) = 1$. This equation is not minimal, but the discriminant of a minimal Weierstrass equation for E is $(c^4 - d^4)c^4d^{16}/16$. Interchanging c and d gives a second elliptic curve, $\mathbb{Q}$-isogenous to E, but with discriminant of opposite sign. Thus, there are an even number of such elliptic curves in a $\mathbb{Q}$-isogeny class. A similar statement is true for $m = 2$ or 3, as Kramer shows. We refer to [K] for a more complete discussion.

We will end this article by returning to some of our earlier examples and discussing a few other examples, but now using Kato's theorem in conjunction with some calculations recently carried out by Ted McCabe. Assume that E is a modular elliptic curve/$\mathbb{Q}$ and that p is a prime where E has good, ordinary reduction. Kato's theorem asserts that $f_E(T)$ divides $p^m f_E^{\mathrm{anal}}(T)$ in Λ for some $m \geq 0$. Let $\lambda_E^{\mathrm{anal}}$ and μ_E^{anal} denote $\lambda(f_E^{\mathrm{anal}})$ and $\mu(f_E^{\mathrm{anal}})$. Kato's theorem implies that $\lambda_E \leq \lambda_E^{\mathrm{anal}}$. McCabe has calculated approximations to the first few coefficients when $f_E^{\text{anal}}(T)$ is written as a power series in T, enough to verify that $\mu_E^{\text{anal}} = 0$ and to determine λ_E^{anal} for the examples he considers. These calculations allow us to justify several statements that were made earlier. As previously, we will use the value of $|\mathrm{Sel}_E(\mathbb{Q})_p|$ which is predicted by the Birch and Swinnerton-Dyer conjecture. In [M-SwD], one finds the results of calculations of the p-adic L-functions for the elliptic curves of conductors 11 and 17 and all primes < 100.

Kato's theorem reduces the verification of conjecture 1.13 to showing that $\lambda_E = \lambda_E^{\text{anal}}$ and $\mu_E = \mu_E^{\text{anal}}$. In a number of the following examples, these equalities can be shown. Before discussing the examples, we want to mention two situations which occur rather frequently.

$\boldsymbol{\lambda_E^{\text{anal}} = \mu_E^{\text{anal}} = 0}$. This means that $f_E^{\text{anal}}(T) \in \Lambda^\times$. By Kato's theorem, it follows that $\lambda_E = 0$. Also, $f_E^{\text{anal}}(0) = (1-\beta_p p^{-1})^2 L(E/\mathbb{Q},1)/\Omega_E$ is a p-adic unit. Kolyvagin's theorem can then be used to verify the Birch and Swinnerton-Dyer conjecture, i.e., that $\mathrm{Sel}_E(\mathbb{Q})_p$ has the predicted order. Then by theorem 4.1, one would obtain that $f_E(0) \in \mathbb{Z}_p^\times$ too. That is, $f_E(T) \in \Lambda^\times$ and hence $\mu_E = 0$ and conjecture 1.13 is valid for E and p.

$\boldsymbol{\lambda_E^{\text{anal}} = 1, \mu_E^{\text{anal}} = 0}$. We will also assume that p is odd. Since $\lambda_E^{\text{anal}} = 1$, $f_E^{\text{anal}}(T)$ has exactly one root: $T = a$, where $a \in p\mathbb{Z}_p$. We mentioned in the introduction that $f_E^{\text{anal}}(T^\iota)/f_E^{\text{anal}}(T) \in \Lambda^\times$, where $T^\iota = (1+T)^{-1} - 1$. Thus $(1+a)^{-1} - 1$ is also a root of $f_E^{\text{anal}}(T)$. It follows that $(1+a)^2 = 1$ and, since p is odd, $a = 0$. (For $p = 2$, we would have another possibility: $a = -2$.) Hence $f_E^{\text{anal}}(0) = 0$ and so $T | f_E^{\text{anal}}(T)$. We then must have $f_E^{\text{anal}}(T)/T \in \Lambda^\times$. The p-adic L-function $L_p(E/\mathbb{Q}, s)$ would have a simple zero at $s = 1$. Assuming that E has good, ordinary reduction at p, the complex L-function $L(E/\mathbb{Q}, s)$ would have an odd order zero at $s = 1$. (The "signs" in the functional equations for $L_p(E/\mathbb{Q}, s)$ and $L(E/\mathbb{Q}, s)$ are the same. See [M-T-T].) Perrin-Riou's analogue of the Gross-Zagier formula implies that $L'(E/\mathbb{Q}, 1) \neq 0$. Hence $\mathrm{rank}(E(\mathbb{Q})) = 1$. Consequently, $f_E(0) = 0$, $\lambda_E = 1$, and $f_E(T) = p^{\mu_E} T$, up to a factor in $\Lambda^\times$. Furthermore, Perrin-Riou's formula also shows that $h_p(P) \neq 0$, where P is a generator of $E(\mathbb{Q})/E(\mathbb{Q})_{\text{tors}}$, and that

$$h_p(P) \sim p(1-\beta_p p^{-1})^{-2}(L'(E/\mathbb{Q},1)/\Omega_E h_\infty(P))^{-1}.$$

Kolyvagin's theorem should allow one to verify that $L'(E/\mathbb{Q},1)/\Omega_E h_\infty(P) \sim (\prod_{v \text{ bad}} c_v^{(p)}) |\text{Ш}_E(\mathbb{Q})_p| / |E(\mathbb{Q})_p|^2$. If one then uses Schneider's result (for the case $F = \mathbb{Q}$, $r = 1$), one would obtain that $f_E(T)/T \in \Lambda^\times$, thus verifying that $\lambda_E = 1$, $\mu_E = 0$, and that conjecture 1.13 is valid for E and p.

Conductor = 67. We consider $p = 3$. As expected, $\mu_E^{\text{anal}} = 0$. We can't verify that $\mu_E = 0$, as conjecture 1.11 predicts. ($E[3]$ is irreducible as a $G_{\mathbb{Q}}$-module.) McCabe finds that $\lambda_E^{\text{anal}} = 2$. As pointed out earlier, $\mathrm{Sel}_E(\mathbb{Q}_\infty)_3$ is infinite. Hence, assuming that $\mu_E = 0$, we have $\lambda_E > 0$. By proposition 3.10, λ_E must be even. Thus, $\lambda_E = 2$. Hence, if $\mu_E = 0$, it is clear from

Kato's theorem, that $(f_E(T)) = (f_E^{\mathrm{anal}}(T))$, i.e., conjecture 1.13 holds for E and $p = 3$. In fact, we would have $\lambda_E^{M\text{-}W} = 0$ and $\lambda_E^{Ш} = 2$. To see this, suppose that $\lambda_E^{M\text{-}W} > 0$. Now Γ acts in the finite-dimensional $\mathbb{Q}$-vector space $E(\mathbb{Q}_\infty) \otimes \mathbb{Q}$. The irreducible $\mathbb{Q}$-representations of Γ have degrees $1, 2, 6, \dots,$ $2 \cdot 3^{n-1}$ for $n \geq 1$. Since $E(\mathbb{Q})$ is finite and $\lambda_E = 2$, we would have $\lambda_E^{M\text{-}W} = 2$ and $\mathrm{rank}(E(\mathbb{Q}_1)) = 2$, where $\mathbb{Q}_1$ is the first layer in $\mathbb{Q}_\infty/\mathbb{Q}$. This would imply that $g(T) = T^2 + 3T + 3$ divides $f_E(T)$. Hence $g(T)$ and $f_E(T)$ would differ by a factor in $\Lambda^\times$, which is impossible since $f_E(0) \sim 3^2$, $g(0) \sim 3$.

Conductor = 915. We consider again the elliptic curve E corresponding to 915(A1) in [Cre]. We take $p = 7$ or $p = 43$. In both cases, McCabe finds that $\lambda_E^{\mathrm{anal}} = 2$. Thus, $\lambda_E \leq 2$. It is then clear that $\lambda_E^{M\text{-}W} = 0$. For the only irreducible $\mathbb{Q}$-representation of Γ with degree ≤ 2 is the trivial representation. (The nontrivial irreducible $\mathbb{Q}$-representations have degree divisible by $p-1$.) But $E(\mathbb{Q})$ is finite in this case. Hence, assuming that $\mu_E = 0$, we have $\lambda_E^{Ш} = 2$ for both $p = 7$ and $p = 43$. Also, just as for the preceding example, conjecture 1.13 would hold if $\mu_E = 0$.

Conductor = 34. We considered before the elliptic curve E corresponding to 34(A1) in [Cre] and found that $\lambda_E^{M\text{-}W} = 2$, $\lambda_E^{Ш} = 0$, and $\mu_E = 0$ for $p = 3$. In this case, McCabe finds that $\lambda_E^{\mathrm{anal}} = 2$ and $\mu_E^{\mathrm{anal}} = 0$. Thus, Kato's theorem again implies conjecture 1.13: $(f_E(T)) = (f_E^{\mathrm{anal}}(T))$ for $p = 3$. There are four elliptic curves of conductor 34, all $\mathbb{Q}$-isogenous. In general, conjecture 1.13 is preserved by $\mathbb{Q}$-isogeny. The power of p dividing $f_E(T)$ changes in a way predicted by the result of [Sch3] or [Pe2]. The power of p dividing $f_E^{\mathrm{anal}}(T)$ changes in a compatible way, determined just by the change in Ω_E. (Ω_E is the only thing that changes in the definition of $f_E^{\mathrm{anal}}(T)$.) One can verify all of this directly. For E, PARI gives $\Omega_E = 4.4956\dots$. Let E' be 34(A3) in [Cre], which is related to E by a $\mathbb{Q}$-isogeny of degree 3. Using the fact that $\mu_E = 0$, one finds that $\mu_{E'} = 1$. Therefore, $f_{E'}(T) = 3f_E(T)$. But PARI gives $\Omega_{E'} = 1.4985\dots = \Omega_E/3$. (This must be exact.) Thus, one sees that $f_{E'}^{\mathrm{anal}}(T) = 3f_E^{\mathrm{anal}}(T)$. Conjecture 1.13 is valid for E' too.

Conductor = 26. We take $p = 7$. For 26(B1), which we previously denoted by E_1, McCabe finds that $\mu_{E_1}^{\mathrm{anal}} = 0$, $\lambda_{E_1}^{\mathrm{anal}} = 4$, and $f_{E_1}^{\mathrm{anal}}(0) \sim 7$. Thus, $f_{E_1}^{\mathrm{anal}}(T)$ is an irreducible element of Λ. If $\mu_{E_1} = 0$, as conjecturally should be true, then Kato's theorem implies that $f_{E_1}(T) = f_{E_1}^{\mathrm{anal}}(T)$, up to a factor in $\Lambda^\times$. Conjecture 1.13 would then be valid for E_1 (and for E_2 too). Thus, in this example, if $\mu_{E_1} = 0$, then $\lambda_{E_1} = 4$. Note that proposition 3.10 would tell us only that λ_{E_1} is even. Also, just as in the example of conductor 915, we would have $\lambda_{E_1}^{M\text{-}W} = 0$, $\lambda_{E_1}^{Ш} = 4$.

Conductor = 147. Let $p = 13$. We will denote 147(B1, B2) by E_1 and E_2 as earlier. McCabe's calculation for 147B1 gives $\mu_{E_1}^{\mathrm{anal}} = 0$, $\lambda_{E_1}^{\mathrm{anal}} = 2$. Proposition 3.10 shows that λ_{E_1} is even. If $\mu_{E_1} = 0$, as conjecture 1.11 predicts, then $\lambda_{E_1} > 0$. Hence $\lambda_{E_1} = 2$ and conjecture 1.13 would again follow from Kato's theorem. As in previous examples, we would have $\lambda_{E_1}^{M\text{-}W} = 0$, $\lambda_{E_1}^{Ш} = 2$.

Conductor = 1225. We consider again the two curves E_1 and E_2 of conductor 1225 discussed earlier. We take $p = 37$. McCabe finds that $\lambda_{E_1}^{\text{anal}} = 1$, $\mu_{E_1}^{\text{anal}} = 0$. Since $L(E_1/\mathbb{Q}, 1) = 0$, it follows that $f_{E_1}^{\text{anal}}(0) = 0$ and that $f_{E_1}^{\text{anal}}(T)/T \in \Lambda^\times$. As remarked earlier, it then should follow that $\lambda_{E_1} = 1$, $\mu_{E_1} = 0$ and that conjecture 1.13 holds. For E_2, we have $\lambda_{E_2} = 1$, $\mu_{E_2} = 1$. Conjecture 1.13 holds for E_2 too.

Conductor = 58. We consider $E' : y^2 + xy = x^3 - x^2 - x + 1$ and $p = 5$. In this case, $E'(\mathbb{Q}) \cong \mathbb{Z}$ and the predicted order of $Ш_{E'}(\mathbb{Q})$ is 1. McCabe finds that $\lambda_{E'}^{\text{anal}} = 1$, $\mu_{E'}^{\text{anal}} = 0$. It then follows that $\lambda_{E'} = 1$, $\mu_{E'} = 0$.

Conductor = 406. Consider $E : y^2 + xy = x^3 + x^2 - 2124x - 60592$. This is 406(D1) in [Cre]. We take $p = 5$. We have $c_2 = c_{29} = 2$, $c_7 = 5$, $|E(\mathbb{Q})| = 2$, and $\mathrm{Sel}_E(\mathbb{Q})$ is predicted to have order 1. Thus, by theorem 4.1, $f_E(0) \sim 5$. Now it turns out that $E[5] \cong E'[5]$ as $G_{\mathbb{Q}}$-modules, where E' is the elliptic curve of conductor 58 considered above. One verifies this by comparing the q-expansions of the modular forms corresponding to these curves. Since $\mu_{E'} = 0$, it follows that $\mu_E = 0$. Therefore, $\lambda_E \geq 1$. By proposition 3.10 λ_E must be even. However, 7 splits completely in $\mathbb{Q}_1/\mathbb{Q}$, where $\mathbb{Q}_1$ denotes the first layer of the cyclotomic $\mathbb{Z}_5$-extension $\mathbb{Q}_\infty$ of $\mathbb{Q}$. (This is because $7^4 \equiv 1 \pmod{5^2}$.) Thus, there are 5 primes of $\mathbb{Q}_1$ lying over 7, each with Tamagawa factor equal to 5. The proof of corollary 5.6 can be used to show that $\lambda_E \geq 5$ and hence, since it is even, we must have $\lambda_E \geq 6$. McCabe finds that $\lambda_E^{\text{anal}} = 6$, $\mu_E^{\text{anal}} = 0$. Therefore, it follows that $\lambda_E = 6$, $\mu_E = 0$, and conjecture 1.13 holds for E and $p = 5$. We also can conclude that $\lambda_E^{M\text{-}W} = 0$. This is so because $E(\mathbb{Q})$ is finite, $E(\mathbb{Q}_\infty) \otimes \mathbb{Q}$ is a finite dimensional $\mathbb{Q}$-representation of Γ, and the nontrivial irreducible $\mathbb{Q}$-representations of Γ have degree divisible by 4. Hence $\mathrm{Sel}_E(\mathbb{Q}_\infty)_5 = Ш_E(\mathbb{Q}_\infty)_5$ and $\lambda_E^{Ш} = 6$.

References

[B-D-G-P] K. Barré-Sirieix, G. Diaz, F. Gramain, G. Philibert, Une preuve de la conjecture de Mahler-Manin, *Invent. Math.* **124** (1996), 1–9.

[Be] M. Bertolini, Selmer groups and Heegner points in anticyclotomic $\mathbb{Z}_p$-extensions, *Compositio Math.* **99** (1995), 153–182.

[BeDa1] M. Bertolini, H. Darmon, Heegner points on Mumford-Tate curves, *Invent. Math.* **126** (1996), 413–456.

[BeDa2] M. Bertolini, H. Darmon, Nontriviality of families of Heegner points and ranks of Selmer groups over anticyclotomic towers, *Journal of the Ramanujan Math. Society* **13** (1998), 15–25.

[B-G-S] D. Bernardi, C. Goldstein, N. Stephens, Notes p-adiques sur les courbes elliptiques, *J. reine angew. Math.* **351** (1984), 129–170.

[CoMc] J. Coates, G. McConnell, Iwasawa theory of modular elliptic curves of analytic rank at most 1, *J. London Math. Soc.* **50** (1994), 243–269.

[CoGr] J. Coates, R. Greenberg, Kummer theory for abelian varieties over local fields, *Invent. Math.* **124** (1996), 129–174.

[CoSc] J. Coates, C.-G. Schmidt, Iwasawa theory for the symmetric square of an elliptic curve, *J. reine angew. Math.* **375/376** (1987), 104–156.

[Cre] J.E. Cremona, *Algorithms for Modular Elliptic Curves*, Cambridge University Press (1992).

[D] F. Diamond, On deformation rings and Hecke rings, *Ann. of Math.* **144** (1996), 137–166.

[F] E.C. Friedman, Ideal class groups in basic $\mathbb{Z}_{p_1} \times \cdots \times \mathbb{Z}_{p_s}$-extensions of abelian number fields, *Invent. Math.* **65** (1982), 425–440.

[FeWa] B. Ferrero, L.C. Washington, The Iwasawa invariant μ_p vanishes for abelian number fields, *Ann. of Math.* **109** (1979), 377–395.

[Gr1] R. Greenberg, On a certain l-adic representation, *Invent. Math.* **21** (1973), 117–124.

[Gr2] R. Greenberg, Iwasawa theory for p-adic representations, *Advanced Studies in Pure Mathematics* **17** (1989), 97–137.

[Gr3] R. Greenberg, Iwasawa theory for motives, *LMS Lecture Notes Series* **153** (1991), 211–233.

[Gr4] R. Greenberg, Trivial zeroes of p-adic L-functions, *Contemporary Math.* **165** (1994), 149–174.

[Gr5] R. Greenberg, The structure of Selmer groups, *Proc. Nat. Acad. Sci.* **94** (1997), 11125–11128.

[Gr6] R. Greenberg, Iwasawa theory for p-adic representations II, in preparation.

[GrVa] R. Greenberg, V. Vatsal, On the Iwasawa invariants of elliptic curves, in preparation.

[Gu1] L. Guo, On a generalization of Tate dualities with application to Iwasawa theory, *Compositio Math.* **85** (1993), 125–161.

[Gu2] L. Guo, General Selmer groups and critical values of Hecke L-functions, *Math. Ann.* **297** (1993), 221–233.

[HaMa] Y. Hachimori, K. Matsuno, On finite Λ-submodules of Selmer groups of elliptic curves, to appear in *Proc. Amer. Math. Soc.*

[Im] H. Imai, A remark on the rational points of abelian varieties with values in cyclotomic $\mathbb{Z}_p$-extensions, *Proc. Japan Acad.* **51** (1975), 12–16.

[Jo] J.W. Jones, Iwasawa L-functions for multiplicative abelian varieties, *Duke Math. J.* **59** (1989), 399–420.

[K] K. Kramer, Elliptic curves with non-trivial 2-adic Iwasawa μ-invariant, to appear in *Acta Arithmetica.*

[Man] Yu.I. Manin, Cyclotomic fields and modular curves, *Russian Math. Surveys* **26** no. 6 (1971), 7–78.

[Maz1] B. Mazur, Rational points of abelian varieties with values in towers of number fields, *Invent. Math.* **18** (1972), 183–266.

[Maz2] B. Mazur, Rational isogenies of prime degree, *Invent. Math.* **44** (1978), 129–162.

[Maz3] B. Mazur, On the arithmetic of special values of L-functions, *Invent. Math.* **55** (1979), 207–240.

[Maz4] B. Mazur, Modular curves and arithmetic, Proceedings of the International Congress of Mathematicians, Warszawa (1983), 185–211.

[M-SwD] B. Mazur, P. Swinnerton-Dyer, Arithmetic of Weil curves, *Invent. Math.* **25** (1974), 1–61.

[M-T-T] B. Mazur, J. Tate, J. Teitelbaum, On p-adic analogues of the conjectures of Birch and Swinnerton-Dyer, *Invent. Math.* **84** (1986), 1–48.

[Mi] J. S. Milne, *Arithmetic Duality Theorems*, Academic Press (1986).

[Mo] P. Monsky, Generalizing the Birch-Stephens theorem, I. Modular curves, *Math. Zeit.* **221** (1996), 415–420.

[Pe1] B. Perrin-Riou, Arithmétique des courbes elliptiques et théorie d'Iwasawa, *Mémoire Soc. Math. France* **17** (1984).

[Pe2] B. Perrin-Riou, Variation de la fonction L p-adique par isogénie, *Advanced Studies in Pure Mathematics* **17** (1989), 347–358.

[Pe3] B. Perrin-Riou, Points de Heegner et dérivées de fonctions L p-adiques, *Invent. Math.* **89** (1987), 455–510.

[Pe4] B. Perrin-Riou, Théorie d'Iwasawa p-adique locale et globale, *Invent. Math.* **99** (1990), 247–292.

[Ri] K. A. Ribet, Torsion points of abelian varieties in cyclotomic extensions, *Enseign. Math.* **27** (1981), 315–319.

[Ro] D. E. Rohrlich, On L-functions of elliptic curves and cyclotomic towers, *Invent. Math.* **75** (1984), 409–423.

[Ru1] K. Rubin, On the main conjecture of Iwasawa theory for imaginary quadratic fields, *Invent. Math.* **93** (1988), 701–713.

[Ru2] K. Rubin, The "main conjectures" of Iwasawa theory for imaginary quadratic fields, *Invent. Math.* **103** (1991), 25–68.

[R-W] K. Rubin, A. Wiles, Mordell-Weil groups of elliptic curves over cyclotomic fields, *Progress in Mathematics* **26** (1982), 237–254.

[Sch1] P. Schneider, Iwasawa L-functions of varieties over algebraic number fields, A first approach, *Invent. Math.* **71** (1983), 251–293.

[Sch2] P. Schneider, p-adic height pairings II, *Invent. Math.* **79** (1985), 329–374.

[Sch3] P. Schneider, The μ-invariant of isogenies, *Journal of the Indian Math. Soc.* **52** (1987), 159–170.

[Se1] J.-P. Serre, *Abelian l-adic Representations and Elliptic Curves*, W. A. Benjamin (1968).

[Se2] J.-P. Serre, Cohomologie Galoisienne, *Lecture Notes in Mathematics* **5**, Springer-Verlag (1964).

[Si] J. Silverman, The Arithmetic of Elliptic Curves, *Grad. Texts in Math.* **106**, Springer-Verlag (1986).

[St] G. Stevens, Stickelberger elements and modular parametrizations of elliptic curves, *Invent. Math.* **98** (1989), 75–106.

[Ta] J. Tate, The arithmetic of elliptic curves, *Invent. Math.* **23** (1974), 179–206.

[Wa1] L. C. Washington, The non-p-part of the class number in a cyclotomic $\mathbb{Z}_p$-extension, *Invent. Math.* **49** (1978), 87–97.

[Wa2] L. C. Washington, Introduction to Cyclotomic Fields, *Grad. Texts in Math.* **83**, Springer-Verlag (1982).

Torsion Points on $J_0(N)$ and Galois Representations

Kenneth A. Ribet

University of California, Berkeley

To Barry Mazur, for his 60th *birthday*

Suppose that N is a prime number greater than 19 and that P is a point on the modular curve $X_0(N)$ whose image in $J_0(N)$ (under the standard embedding $\iota\colon X_0(N) \hookrightarrow J_0(N)$) has finite order. In [2], Coleman-Kaskel-Ribet conjecture that either P is a hyperelliptic branch point of $X_0(N)$ (so that $N \in \{23, 29, 31, 41, 47, 59, 71\}$) or else that $\iota(P)$ lies in the cuspidal subgroup C of $J_0(N)$. That article suggests a strategy for the proof: assuming that P is not a hyperelliptic branch point of $X_0(N)$, one should show for each prime number ℓ that the ℓ-primary part of $\iota(P)$ lies in C. In [2], the strategy is implemented under a variety of hypotheses but little is proved for the primes $\ell = 2$ and $\ell = 3$. Here I prove the desired statement for $\ell = 2$ whenever N is prime to the discriminant of the ring $\operatorname{End} J_0(N)$. This supplementary hypothesis, while annoying, seems to be a mild one; according to W. A. Stein of Berkeley, California, in the range $N < 5021$, it is false only in case $N = 389$.

1 Introduction

At the C.I.M.E. conference on the arithmetic of elliptic curves, I lectured on interrelated questions with a common underlying theme: the action of $\mathrm{Gal}(\overline{\mathbf{Q}}/\mathbf{Q})$ on torsion points of semistable abelian varieties over $\mathbf{Q}$. In this written record of my lectures, I focus on the modular curve $X_0(N)$ and its Jacobian $J_0(N)$ when N is a prime number. In this special case, $X_0(N)$ and $J_0(N)$ were studied intensively by B. Mazur in [9] and [10], so that we have a wealth of arithmetic information at our disposal.

The main theorem of this article complements the results of Coleman-Kaskel-Ribet [2] on the "cuspidal torsion packet" of $X_0(N)$. Recall that $X_0(N)$ has two cusps, customarily denoted 0 and ∞. Selecting the latter cusp as the more "standard" of the two, we use it to map $X_0(N)$ to $J_0(N)$,

---

The author's research was partially supported by National Science Foundation contract #DMS 96 22801. The author thanks M. Baker, J. A. Csirik and H. W. Lenstra, Jr. for helpful conversations and suggestions.

via the Albanese mapping ι which takes a point P of the curve to the class of the divisor $(P)-(\infty)$. This map is injective if the genus of $X_0(N)$ is non-zero.

Let g be the genus of $X_0(N)$. For the remainder of this preliminary discussion, make the hypothesis $g \geq 2$. (This hypothesis is satisfied if and only if $N \geq 23$.) Then ι identifies $X_0(N)$ with a subvariety of $J_0(N)$ of positive codimension. The torsion packet in question is the set Ω of points of $X_0(N)$ whose images in $J_0(N)$ have finite order. According to the Manin-Mumford conjecture, first proved by Raynaud in 1983 [13], Ω is a finite set.

The article [2] introduces a strategy for identifying Ω precisely. Clearly, Ω contains the two cusps 0 and ∞ of $X_0(N)$, whose images under ι have order $n := \operatorname{num}(\frac{N-1}{12})$ and 1, respectively [9, p. 98]. Further, in the special case when $X_0(N)$ is hyperelliptic, we note in [2] that the hyperelliptic branch points of $X_0(N)$ belong to Ω if and only if N is different from 37. (Results of Ogg [11, 12] show that $X_0(N)$ is hyperelliptic if and only if N lies in the set $\{23, 29, 31, 37, 41, 47, 59, 71\}$.) In fact, suppose that $X_0(N)$ is hyperelliptic and that P is a hyperelliptic branch point on $X_0(N)$. Then $2\iota(P) = \iota(0)$ if $N \neq 37$, but P has infinite order when $N = 37$.

In [2], we advance the idea that Ω might contain only the points we have just catalogued:

Guess 1.1. *Suppose that P is a point on $X_0(N)$ whose image in $J_0(N)$ has finite order. Then either P is one of the two cusps of $X_0(N)$, or $X_0(N)$ is a hyperelliptic curve and P is a hyperelliptic branch point of $X_0(N)$.*

In the latter case, (i.e., $X_0(N)$ hyperelliptic and P a hyperelliptic branch point with finite order in $J_0(N)$), it follows automatically that N is different from 37.

A reformulation of Guess 1.1 involves the cuspidal subgroup C of $J_0(N)$, i.e., the group generated by the point $\iota(0)$. As we point out in [2], the results of [10] imply that the intersection of $X_0(N)$ and C (computed in $J_0(N)$) consists of the two cusps 0 and ∞. In words, to prove that a torsion point P of $X_0(N)$ is a cusp is to prove that it lies in the group C. For this, it is useful to decompose P into its primary parts: If P is a torsion point of $J_0(N)$ and ℓ is a prime number, we let P_ℓ be the ℓ-primary part of P. Thus $P = \sum P_\ell$, the sum being extended over all primes, and we have $P \in C$ if and only if $P_\ell \in C$ for all primes ℓ.

Consider the following two statements (in both, we regard $X_0(N)$ as embedded in its Jacobian via ι):

Statement 1.2. *Suppose that P is an element of Ω and that ℓ is an odd prime. Then we have $P_\ell \in C$.*

Statement 1.3. *Suppose that P is an element of Ω and that $P_2 \notin C$. Then P is a hyperelliptic branch point of $X_0(N)$.*

It is clear that Guess 1.1 is equivalent to the conjunction of Statements 1.2 and 1.3. Indeed, suppose first that (1.1) is correct and that P is an element of Ω. If P is a cuspidal point (i.e., one of 0, ∞), then one has $P_\ell \in C$ for all primes ℓ. If P is not a cuspidal point, then P is a hyperelliptic branch point and $N \neq 37$; we then have $2P \in C$, so that $P_\ell \in C$ for all $\ell > 2$. Conversely, suppose that Statements 1.2 and 1.3 are true and that P is an element of Ω. If P_2 is not in C, then P is a hyperelliptic branch point (and is thus accounted for by the guess). If P_2 lies in C, then P_ℓ is in C for all primes ℓ, so that P is a point of C. As was mentioned above, this implies that P is one of the two cuspidal points on $X_0(N)$.

Our article [2] proves a number of results in the spirit of (1.2). For example, suppose that P is an element of Ω and ℓ is an odd prime different from N. Let g again be the genus of $X_0(N)$. Then $P_\ell \in C$ if ℓ is greater than $2g$ or if ℓ satisfies $5 \leq \ell < 2g$ and at least one of a number of supplementary conditions.

These notes prove a theorem in the direction of (1.3). This theorem requires an auxiliary hypothesis concerning the discriminant of the subring $\mathbf{T}$ of $\operatorname{End} J_0(N)$ which is generated by the Hecke operators T_m (with $m \geq 1$) on $J_0(N)$. (Many authors write the Hecke operator T_N as U_N.) According to [9, Prop. 9.5, p. 95], the Hecke ring $\mathbf{T}$ is in fact the full endomorphism ring of $J_0(N)$. Concerning the structure of $\mathbf{T}$, it is known that $\mathbf{T}$ is an order in a product $E = \prod E_i$ of totally real number fields. The discriminant $\operatorname{disc}(\mathbf{T})$ is the product of the discriminants of the number fields E_i, multiplied by the square of the index of $\mathbf{T}$ in its normalization. Our auxiliary hypothesis is the following statement:

Hypothesis 1.4. *The discriminant of* $\mathbf{T}$ *is prime to* N.

According to William Arthur Stein of Berkeley, California, Hypothesis 1.4 is false when $N = 389$ and true for all other primes $N \leq 5011$.

Theorem 1.5. *Suppose that* P *lies in* Ω *and that* P_2 *does not belong to* C. *In addition, suppose either that the order of* P *is prime to* N *or that Hypothesis* 1.4 *holds. Then* $X_0(N)$ *is hyperelliptic, and* P *is a hyperelliptic branch point of* $X_0(N)$.

Theorem 1.5 is a direct consequence of a Galois-theoretic statement which we prove in §7. Since this latter theorem is the main technical result of these notes, we state it now and then show how it implies Theorem 1.5.

Theorem 1.6. *Let* N *be a prime number, and let* $J = J_0(N)$. *Let* ℓ *be a prime different from* N. *Suppose that* P *is a point of finite order on* $J_0(N)$ *whose* ℓ*-primary component* P_ℓ *is not defined over* $\mathbf{Q}$. *Assume that at least one of the following hypotheses holds:* (1) N *is prime to the order of* P; (2) ℓ *is prime to* $N-1$; (3) N *is prime to the discriminant of* $\mathbf{T}$. *Then there is a* $\sigma \in \operatorname{Gal}(\overline{\mathbf{Q}}/\mathbf{Q})$ *such that* $\sigma P - P$ *has order* ℓ.

Note that the hypothesis $g \geq 2$ is not needed for Theorem 1.6.

Proof that (1.6) *implies* (1.5). Let P be as in Theorem 1.5. Because P_2 does not lie in C, P_2 is not a rational point of $J_0(N)$ [9, Ch. III, Th. 1.2]. We apply Theorem 1.6 in this situation, taking $\ell = 2$. The theorem shows that there is a $\sigma \in \mathrm{Gal}(\overline{\mathbf{Q}}/\mathbf{Q})$ such that the divisor $(\sigma P) - (P)$ on $X_0(N)$ has order 2 in $J_0(N)$. Accordingly, the points P and σP are distinct, and there is a rational function f on the curve $X_0(N)$ whose divisor is $2\big((\sigma P) - (P)\big)$. The function f has a double zero at σP, a double pole at P, and no other zeros or poles. It follows that the covering $X \to \mathbf{P}^1$ defined by f is of degree two and that P is ramified in the covering. Since the genus of X is at least 2, X is hyperelliptic and P is a hyperelliptic branch point. ■

We conclude this discussion with a second statement which will be proved only below. For this statement and for most of what follows, we again allow N be an arbitrary prime; i.e., we have no need of the assumption that $J = J_0(N)$ has dimension > 1. As in [9], we consider the Eisenstein ideal $\mathscr{I} \subseteq \mathbf{T}$ and form the kernel $J[\mathscr{I}] \subseteq J(\overline{\mathbf{Q}})$. Let $K = \mathbf{Q}(J[\mathscr{I}])$ be the field generated by the coordinates of the points in $J[\mathscr{I}]$. Recall that $n = \mathrm{num}\big(\frac{N-1}{12}\big)$. Then we have:

Theorem 1.7. *The field K is the field of $2n^{\mathrm{th}}$ roots of unity.*

Theorem 1.7 is an essential ingredient in our proof of Theorem 1.6 in the crucial case where $\ell = 2$. Readers who are familiar with Mazur's article [9] will recognize that Theorem 1.7 follows directly from the results of that article if n is not divisible by 4. Moreover, as H. W. Lenstra, Jr. has pointed out, Theorem 1.7 may be proved rather easily by elementary arguments if n is divisible by 8. The most difficult case is therefore that for which n is divisible by 4 but not by 8; this case occurs precisely when $N \equiv 17 \bmod 32$. We will discuss Lenstra's observations in §4 and then prove Theorem 1.7 in the general case in §5 by exploiting Mazur's "congruence formula for the modular symbol" [9, Ch. II, §18]. An alternative proof of Theorem 1.7 was given recently by J. A. Csirik [3]. Csirik provides a complete concrete description of $J_0(N)[\mathscr{I}]$ which yields Theorem 1.7 as a corollary.

2 A local study at N

For the rest of this article, we take N to be a prime number and let $J = J_0(N)$. The assumption of §1 concerning the genus of $X_0(N)$ is no longer required.

We remind the reader that the results of Deligne and Rapoport [4] imply that J has purely multiplicative reduction at N. As explained in the Mazur-Rapoport appendix to [9], the fiber over $\mathbf{F}_N$ of the Néron model of J is the product of a cyclic component group Φ and a torus $J^0_{/\mathbf{F}_N}$.

The character group of this torus,

$$\mathscr{X} := \operatorname{Hom}_{\overline{\mathbf{F}}_N}\left(J^0_{/\mathbf{F}_N}, \mathbf{G}_{\mathrm{m}}\right),$$

is a free $\mathbf{Z}$-module of rank $\dim J$ which is furnished with compatible actions of $\mathbf{T}$ and the Galois group $\operatorname{Gal}(\overline{\mathbf{F}}_N/\mathbf{F}_N)$. Here, $\overline{\mathbf{F}}_N$ is of course an algebraic closure of the prime field $\mathbf{F}_N$. It will be convenient to choose a prime dividing N in $\overline{\mathbf{Q}}$ and to let $\overline{\mathbf{F}}_N$ be the residue field of this prime. Then if $D \subset \operatorname{Gal}(\overline{\mathbf{Q}}/\mathbf{Q})$ is the decomposition group corresponding to the chosen prime, $\operatorname{Gal}(\overline{\mathbf{F}}_N/\mathbf{F}_N)$ is the quotient of D by its inertia subgroup I. Using the quotient map $D \to \operatorname{Gal}(\overline{\mathbf{F}}_N/\mathbf{F}_N)$, we view $\mathscr{X}$ as an unramified representation of D. As one knows, this action is "nearly" trivial: the generator $x \mapsto x^N$ of $\operatorname{Gal}(\overline{\mathbf{F}}_N/\mathbf{F}_N)$ acts on $\mathscr{X}$ as an automorphism of order 1 or 2, so that the group $\operatorname{Gal}(\overline{\mathbf{F}}_N/\mathbf{F}_{N^2})$ acts trivially on $\mathscr{X}$. (The group $\mathscr{X}$ is discussed in [14, §3] in the more general case where N is replaced by the product of a prime q and a positive integer which is prime to q.)

As far as the Hecke action goes, the group $\mathscr{X}$ is a free $\mathbf{Z}$-module whose rank is the same as that of $\mathbf{T}$, namely the dimension of J. Because $\mathbf{T}$ acts faithfully on $\mathscr{X}$, it is clear that $\mathscr{X}\otimes\mathbf{Q}$ is free of rank 1 over $\mathbf{T}\otimes\mathbf{Q}$. Thus $\mathscr{X}$ is a "$\mathbf{T}$-module of rank 1" in the sense of [9, Ch. II, §8]. (In [9, Ch. II, Prop. 8.3], Mazur notes in effect that $\mathscr{X}\otimes\mathbf{Q}_p$ is free of rank 1 over $\mathbf{T}\otimes\mathbf{Q}_p$ for each prime $p \neq N$.) It is natural to ask whether $\mathscr{X}$ is locally free of rank 1 over $\mathbf{T}$. In this section, we will answer the question affirmatively, except perhaps for certain primes (meaning: maximal ideals) of $\mathbf{T}$ which divide 2.

In what follows, we consider a maximal ideal $\mathfrak{m}$ of $\mathbf{T}$. Let p be the characteristic of the finite field $\mathbf{T}/\mathfrak{m}$. As in [9, Ch. II, §7], we let $\mathbf{T}_\mathfrak{m} = \varprojlim_\nu \mathbf{T}/\mathfrak{m}^\nu$ be the completion of $\mathbf{T}$ at $\mathfrak{m}$. As usual, we say that $\mathfrak{m}$ is ordinary if T_p is non-zero mod $\mathfrak{m}$ and supersingular otherwise.

Also, we recall that $\mathfrak{m}$ is *Eisenstein* if it divides (i.e., contains) the Eisenstein ideal $\mathscr{I}$ of $\mathbf{T}$. This latter ideal is defined (on p. 95 of [9]) as the ideal generated by the difference $T_N - 1$ and by the quantities $\eta_\ell := 1 + \ell - T_\ell$ as ℓ ranges over the set of primes different from N. The natural map $\mathbf{Z} \to \mathbf{T}/\mathscr{I}$ induces an isomorphism $\mathbf{Z}/n\mathbf{Z} \xrightarrow{\sim} \mathbf{T}/\mathscr{I}$, where n is the numerator of $\frac{N-1}{12}$. Thus the Eisenstein primes of $\mathbf{T}$ are in 1-1 correspondence with the prime ideals of $\mathbf{Z}/n\mathbf{Z}$ and therefore with the prime numbers which divide n.

Next, we write $J[\mathfrak{m}]$ for the group of points in $J(\overline{\mathbf{Q}})$ which are killed by all elements of $\mathfrak{m}$ (cf. [9, p. 91]). This group is a $\mathbf{T}/\mathfrak{m}$-vector space which is furnished with an action of $\operatorname{Gal}(\overline{\mathbf{Q}}/\mathbf{Q})$. Recall the following key result of [9]:

Theorem 2.1. *Let $\mathfrak{m}$ be a maximal ideal of $\mathbf{T}$. If $\mathfrak{m}$ divides 2, suppose that $\mathfrak{m}$ is either Eisenstein or supersingular. Then $J[\mathfrak{m}]$ is of dimension two.*

Theorem 2.1 is proved in [9, Ch. II]. Note, however, that the discussions for $\mathfrak{m}$ Eisenstein and $\mathfrak{m}$ non-Eisenstein occur in different sections: one may consult Proposition 14.2 if $\mathfrak{m}$ is non-Eisenstein and (16.3) if $\mathfrak{m}$ is Eisenstein. (See also (17.9) if $\mathfrak{m}$ is Eisenstein and $\mathfrak{m}$ divides 2.)

When $\mathfrak{m}$ is non-Eisenstein, Theorem 2.1 relates $J[\mathfrak{m}]$ and the standard representation $\rho_{\mathfrak{m}}$ of $\mathrm{Gal}(\overline{\mathbf{Q}}/\mathbf{Q})$ which is attached to $\mathfrak{m}$. By definition, $\rho_{\mathfrak{m}}$ is the unique (up to isomorphism) continuous semisimple representation $\mathrm{Gal}(\overline{\mathbf{Q}}/\mathbf{Q}) \to \mathbf{GL}(2, \mathbf{T}/\mathfrak{m})$ satisfying: (i) $\det \rho_{\mathfrak{m}}$ is the mod p cyclotomic character; (ii) for each prime ℓ prime to pN, $\rho_{\mathfrak{m}}$ is unramified at ℓ and $\rho_{\mathfrak{m}}(\mathrm{Frob}_\ell)$ has trace T_ℓ mod $\mathfrak{m}$. (The existence and uniqueness of $\rho_{\mathfrak{m}}$ are discussed, for instance, in [14, §5].) The representation $\rho_{\mathfrak{m}}$ is irreducible if and only if $\mathfrak{m}$ is non-Eisenstein [9, Ch. II, Prop. 14.1 and Prop. 14.2]. The relation between $J[\mathfrak{m}]$ and $\rho_{\mathfrak{m}}$ is that the former representation *is* (i.e., defines or affords) the latter representation whenever $J[\mathfrak{m}]$ is irreducible and 2-dimensional [9, Ch. II, §14]. In particular, if $\mathfrak{m}$ is non-Eisenstein, then $J[\mathfrak{m}]$ affords the representation $\rho_{\mathfrak{m}}$ if either p is odd or $\mathfrak{m}$ is supersingular.

Suppose that $\rho_{\mathfrak{m}}$ is irreducible. Following [18, p. 189], we define $\rho_{\mathfrak{m}}$ to be finite at N if there is a finite flat $\mathbf{T}/\mathfrak{m}$-vector space scheme $\mathscr{V}$ of rank 2 over $\mathbf{Z}_N$ such that the restriction of $\rho_{\mathfrak{m}}$ to $D = \mathrm{Gal}(\overline{\mathbf{Q}}_N/\mathbf{Q}_N)$ is isomorphic to the two-dimensional representation $\mathscr{V}(\overline{\mathbf{Q}}_N)$. The following result is obtained by combining a 1973 theorem of Tate with the author's level-lowering result.

Proposition 2.2. *Let $\mathfrak{m}$ be a non-Eisenstein prime of $\mathbf{T}$. Then the two-dimensional Galois representation $\rho_{\mathfrak{m}}$ is not finite at the prime N.*

Proof. Suppose first that $\mathfrak{m}$ does not divide 2. Assume that $\rho_{\mathfrak{m}}$ is finite at N. Then [14, Th. 1.1] shows that $\rho_{\mathfrak{m}}$ is modular of level 1. (In applying [14, Th. 1.1], we take $N = N$, $p = N$, and $\ell = p$. Note that condition 2 of the theorem is satisfied except when $\mathfrak{m}$ divides N. In this case, however, condition 1 of the theorem holds since we do not have $N \equiv 1 \bmod N$.) This is a contradiction, since there are no non-zero weight-2 cusp forms on $\Gamma_0(1)$.

Assume now that $\mathfrak{m}$ does divide 2. Suppose again that $\rho_{\mathfrak{m}}$ is finite at N. Then $\rho_{\mathfrak{m}}$ is an irreducible mod 2 two-dimensional representation of $\mathrm{Gal}(\overline{\mathbf{Q}}/\mathbf{Q})$ which is unramified outside of the prime 2. An important theorem of Tate [20] proves, however, that there is no such representation. ∎

Note that when p is different from N, $\rho_{\mathfrak{m}}$ is finite at N if and only if $\rho_{\mathfrak{m}}$ is unramified at N. Thus Proposition 2.2 shows, in particular, that $\rho_{\mathfrak{m}}$ is ramified at N for all $\mathfrak{m}$ such that $\rho_{\mathfrak{m}}$ is irreducible.

Theorem 2.3. *Let $\mathfrak{m}$ be a maximal ideal of $\mathbf{T}$. If $\mathfrak{m}$ divides 2, suppose that $\mathfrak{m}$ is either Eisenstein or supersingular. Then $\mathscr{X} \otimes_{\mathbf{T}} \mathbf{T}_{\mathfrak{m}}$ is free of rank 1 over $\mathbf{T}_{\mathfrak{m}}$.*

Proof. Since $\mathscr{X}$ is of rank 1, $\mathscr{X} \otimes \mathbf{T}_{\mathfrak{m}}$ is free of rank 1 if and only if it is cyclic. By Nakayama's lemma, the cyclicity amounts to the statement that $\mathscr{X}/\mathfrak{m}\mathscr{X}$ has dimension ≤ 1 over the field $\mathbf{T}/\mathfrak{m}$.

To prove this latter statement, i.e., the cyclicity of $\mathscr{X}/\mathfrak{m}\mathscr{X}$, we exploit the relation between $\mathscr{X}$ and torsion points of J. In the following discussion, for each integer $m \geq 1$, we let $J[m]$ be the group of points of J with values in $\overline{\mathbf{Q}}$ which have order dividing m. Thus $J[m]$ is a $\mathbf{T}[\mathrm{Gal}(\overline{\mathbf{Q}}/\mathbf{Q})]$-module. Especially, we shall view $J[m]$ locally at N, i.e., as a $\mathbf{T}[D]$-module. One obtains from [6, 11.6.6–11.6.7] a $\mathbf{T}[D]$-equivariant exact sequence

$$(2.4) \qquad 0 \to \mathrm{Hom}(\mathscr{X}/m\mathscr{X}, \mu_m) \to J[m] \to \mathscr{X}/m\mathscr{X} \to 0.$$

(See, e.g., [15, pp. 669–670] for a discussion of this exact sequence when m is a prime number.) Especially, there is a natural identification of $\mathrm{Hom}(\mathscr{X}/m\mathscr{X}, \mu_m)$ with a subgroup of $J[m]$.

In particular, we find an injection

$$j : \mathrm{Hom}(\mathscr{X}/\mathfrak{m}\mathscr{X}, \mu_p) \hookrightarrow J[\mathfrak{m}];$$

here, p is again the residue characteristic of $\mathfrak{m}$. By Theorem 2.1, j is an isomorphism if $\mathscr{X}/\mathfrak{m}\mathscr{X}$ is not cyclic.

On the other hand, it is clear that j cannot be an isomorphism. Indeed, the group $\mathrm{Hom}(\mathscr{X}/\mathfrak{m}\mathscr{X}, \mu_p)$ is finite at N in the sense of [18] (since μ_p is finite), and we have seen in Proposition 2.2 that $J[\mathfrak{m}]$ is not finite at N. ■

3 The kernel of the Eisenstein ideal

We turn now to a study of the action of $\mathrm{Gal}(\overline{\mathbf{Q}}/\mathbf{Q})$ on the Eisenstein kernel in the Jacobian $J = J_0(N)$. Let $\mathscr{I}$ again be the Eisenstein ideal of $\mathbf{T}$, and recall that $n = \mathrm{num}(\frac{N-1}{12})$. By $J[\mathscr{I}]$ we mean the kernel of $\mathscr{I}$ on J, i.e., the group of points in $J(\overline{\mathbf{Q}})$ which are annihilated by all elements of $\mathscr{I}$. The analysis of [9, Ch. II, §§16–18] shows that $J[\mathscr{I}]$ is free of rank two over $\mathbf{T}/\mathscr{I} \approx \mathbf{Z}/n\mathbf{Z}$.

The group $J[\mathscr{I}]$ contains the cuspidal group C, which was mentioned above, and also the Shimura subgroup Σ of J [9, Ch. II, §11]. The two groups C and Σ are $\mathrm{Gal}(\overline{\mathbf{Q}}/\mathbf{Q})$-stable and cyclic of order n. The actions of $\mathrm{Gal}(\overline{\mathbf{Q}}/\mathbf{Q})$ on these two groups are respectively the trivial action and the cyclotomic action ($\Sigma \approx \mu_n$). Accordingly, the intersection of C and Σ is trivial if n is odd; in that case, the inclusions of C and Σ in $J[\mathscr{I}]$ induce an isomorphism of $\mathrm{Gal}(\overline{\mathbf{Q}}/\mathbf{Q})$-modules $C \oplus \Sigma \xrightarrow{\sim} J[\mathscr{I}]$. If n is even, however, $C \cap \Sigma$ has order 2, and the sum $C + \Sigma$ in $J[\mathscr{I}]$ (which is no longer direct) has index 2 in $J[\mathscr{I}]$.

In much of what follows, the reader may wish to assume that n is even; when n is odd, almost everything that we prove may be deduced immediately from the decomposition $J[\mathscr{I}] \approx C \oplus \Sigma$.

Proposition 3.1. *The group $J[\mathscr{I}]$ is unramified at N.*

Proof. We regard $J[\mathscr{I}]$ as a D-module, where $D = \mathrm{Gal}(\overline{\mathbf{Q}}_N/\mathbf{Q}_N)$ as above. We have a natural injection (analogous to the map j above)

$$\mathrm{Hom}(\mathcal{X}/\mathcal{I}\mathcal{X}, \mu_n) \hookrightarrow J[\mathcal{I}],$$

where $\mathcal{X}$ is again the character group associated with the reduction of J mod N. By combining this injection with the inclusion of Σ in $J[\mathcal{I}]$, we obtain a map of D-modules

$$\theta : \Sigma \oplus \mathrm{Hom}(\mathcal{X}/\mathcal{I}\mathcal{X}, \mu_n) \longrightarrow J[\mathcal{I}].$$

This map is again injective, in view of Proposition 11.9 of [9, Ch. II].

Now by Theorem 2.3, $\mathcal{X}$ is free of rank 1 locally at each prime $\mathfrak{m}$ dividing $\mathcal{I}$. Hence $\mathcal{X}/\mathcal{I}\mathcal{X}$, and therefore $\mathrm{Hom}(\mathcal{X}/\mathcal{I}\mathcal{X}, \mu_n)$, has order n. Thus the source of θ has n^2 elements. Since the target of θ has the same cardinality, we conclude that θ is an isomorphism of D-modules. The group $\Sigma \oplus \mathrm{Hom}(\mathcal{X}/\mathcal{I}\mathcal{X}, \mu_n)$, however, is unramified; note that D acts on $\mathcal{X}$ through its quotient $\mathrm{Gal}(\overline{\mathbf{F}}_N/\mathbf{F}_N)$. ■

We continue our study of the action of $\mathrm{Gal}(\overline{\mathbf{Q}}/\mathbf{Q})$ on $J[\mathcal{I}]$:

Proposition 3.2. *The Galois group* $\mathrm{Gal}(\overline{\mathbf{Q}}/\mathbf{Q})$ *acts trivially on* $J[\mathcal{I}]/\Sigma$.

Proof. It is clear that Jordan-Hölder constituents of the $\mathrm{Gal}(\overline{\mathbf{Q}}/\mathbf{Q})$-module $J[\mathcal{I}]$ are all of the form μ_p or $\mathbf{Z}/p\mathbf{Z}$, with p dividing n. Indeed, $J[\mathcal{I}]$ is an extension of a group whose order divides 2 by a quotient of $\Sigma \oplus C$, where the latter group has the indicated property. Because $J[\mathcal{I}]$ is unramified at N, it is finite at N in Serre's sense; it extends to a finite flat group scheme over $\mathbf{Z}$. In the language of Chapter I of [9], $J[\mathcal{I}]$ is thus an admissible group scheme over $\mathrm{Spec}\,\mathbf{Z}[\frac{1}{N}]$ which extends to a finite flat group scheme G over $\mathrm{Spec}\,\mathbf{Z}$.

To analyze G, we follow the proof of Proposition 4.5 in [9, Ch. I]. The last step in the proof of that Proposition uses a result above it (Proposition 4.1) which applies only to groups of odd order. However, Steps 1–3 are perfectly applicable; they show that G is an extension of a constant group scheme by a μ-type group (dual of a constant group) $H \subseteq G$.

In particular, there is a subgroup Σ' of $J[\mathcal{I}]$ with the property that the action of $\mathrm{Gal}(\overline{\mathbf{Q}}/\mathbf{Q})$ on Σ' is cyclotomic, whereas the action of $\mathrm{Gal}(\overline{\mathbf{Q}}/\mathbf{Q})$ on $J[\mathcal{I}]/\Sigma'$ is trivial. By [9, Ch. III, Th. 1.3], Σ' is contained in Σ. Hence the action of $\mathrm{Gal}(\overline{\mathbf{Q}}/\mathbf{Q})$ on the quotient $J[\mathcal{I}]/\Sigma$ is indeed trivial. ■

Before studying further the $\mathrm{Gal}(\overline{\mathbf{Q}}/\mathbf{Q})$-action on $J[\mathcal{I}]$, we pause to establish a converse to Proposition 3.1.

Proposition 3.3. *Let* $P \in J(\overline{\mathbf{Q}})$ *be a torsion point on* J *for which the finite extension* $\mathbf{Q}(P)/\mathbf{Q}$ *is unramified at* N. *Then* P *lies in* $J[\mathcal{I}]$.

Proof. Let M be the smallest $\mathbf{T}[\mathrm{Gal}(\overline{\mathbf{Q}}/\mathbf{Q})]$-submodule of $J(\overline{\mathbf{Q}})$ which contains both P and $J[\mathcal{I}]$. We must prove that M is annihilated by $\mathcal{I}$. Clearly, M is finite; indeed, we have $M \subseteq J[mn]$ if m is the order of P.

Consider the Jordan-Hölder constituents of M, regarded as a $\mathbf{T}[\mathrm{Gal}(\overline{\mathbf{Q}}/\mathbf{Q})]$-module. If V is such a constituent, then the annihilator of V is a maximal ideal $\mathfrak{m}$ of $\mathbf{T}$. It follows from the discussion of [9, Ch. II, §14] that V is 1-dimensional over $\mathbf{T}/\mathfrak{m}$ if and only if $\mathfrak{m}$ is Eisenstein. If $\mathfrak{m}$ is not Eisenstein, then V is isomorphic to the irreducible representation $\rho_{\mathfrak{m}}$. (This follows from the discussion on page 115 of [9]. In fact, the main result of [1] can be used to prove the more precise fact that $J[\mathfrak{m}]$ is a direct sum of copies of $\rho_{\mathfrak{m}}$ when $\mathfrak{m}$ is non-Eisenstein.) However, Proposition 2.2 shows that $\rho_{\mathfrak{m}}$ is ramified at N when $\mathfrak{m}$ is non-Eisenstein. We conclude that all constituents of M belong to Eisenstein primes of $\mathbf{T}$. These constituents therefore have the form μ_p or $\mathbf{Z}/p\mathbf{Z}$, with p dividing n.

Returning to the language of [9, Ch. I], we see that M is an admissible group. As explained in the proof of the proposition above, M must be an extension of a constant group Q by a μ-type group M_0. Since M contains $J[\mathcal{I}]$ and since Σ is the maximal μ-type group in $J(\overline{\mathbf{Q}})$, we have $M_0 = \Sigma$. Next, note that the extension of $\mathbf{T}$-modules

$$0 \to \Sigma \to M \to Q \to 0$$

splits. The splitting is obtained as in the argument on p. 142 of [9] which proves [9, Ch. III, Th. 1.3]. Namely, specialization to characteristic N provides a map $M \to \Phi$, where Φ is the component group of J in characteristic N. We get a splitting because the restriction of this map to Σ is an isomorphism $\Sigma \xrightarrow{\sim} \Phi$. It follows that $\mathcal{I}$ annihilates M if and only if $\mathcal{I}$ annihilates Q.

Since $\mathrm{Gal}(\overline{\mathbf{Q}}/\mathbf{Q})$ acts trivially on Q, the Eichler-Shimura relation shows that Q is annihilated by the differences $\eta_\ell = 1 + \ell - T_\ell$. To deduce from this the apparently stronger fact that Q is annihilated by all of $\mathcal{I}$ (which includes the generator $T_N - 1$), write Q as the direct sum $\bigoplus_{\mathfrak{m}} Q_{\mathfrak{m}}$, where the sum runs over the set of Eisenstein primes of $\mathbf{T}$. Each summand $Q_{\mathfrak{m}}$ is a module over $\mathbf{T}/\mu^\nu$, where ν is a suitable positive integer. It follows from [9, Ch. II, Th. 18.10] that the image of $\mathcal{I}$ in $\mathbf{T}/\mu^\nu$ is generated by a single element of the form η_ℓ. Thus $Q_{\mathfrak{m}}$ is annihilated by $\mathcal{I}$. Since this statement is true for each $\mathfrak{m}$, Q is annihilated by $\mathcal{I}$. ■

Our next goal is to study $J[\mathcal{I}]$ sufficiently closely to permit identification of the field $\mathbf{Q}(J[\mathcal{I}])$, i.e., to prove Theorem 1.7. For an alternative proof of Theorem 1.7, the reader may consult Csirik's forthcoming article [3], which determines $J[\mathcal{I}]$ completely by a method generalizing that of [9, Ch. II, §12–§13].

Recall that the cuspidal group C is provided with a natural generator, namely the image of the cusp 0 in J. We select generators for certain other cyclic groups by making use of the place over N that we have chosen in $\overline{\mathbf{Q}}$. As explained in §11 of [9, Ch. II], reduction to characteristic N induces isomorphisms among C, Σ and the group of components of $J_{/\mathbf{F}_N}$. In particular, we have a distinguished isomorphism $C \approx \Sigma$. Since C is provided with a generator, we obtain a basis of Σ. (See [5] for a comparison of the isomorphism $C \approx \Sigma$ with a second natural one.)

Since Σ and $J[\mathscr{I}]$ are free of ranks 1 and 2 over $\mathbf{Z}/n\mathbf{Z}$, the group $Q := J[\mathscr{I}]/\Sigma$ is cyclic of order n. The intersection $C \cap \Sigma$ has order $\gcd(2,n)$ [9, Ch. II, Prop. 11.11]. The image of C in Q has order $n/\gcd(2,n)$. Choose a generator g of Q such that $2g$ is the image in Q of the chosen generator of C. Finally, note as above that reduction to characteristic N provides us with a splitting of the tautological exact sequence which displays Q as a quotient of $J[\mathscr{I}]$. This splitting writes $J[\mathscr{I}]$ as the direct sum $\Sigma \oplus Q$. (Said differently, $J[\mathscr{I}]$ is the direct sum of Σ and the "toric part" $\mathrm{Hom}(\mathscr{X}/\mathscr{I}\mathscr{X}, \mu_n)$ of $J[\mathscr{I}]$. The natural map $\mathrm{Hom}(\mathscr{X}/\mathscr{I}\mathscr{X}, \mu_n) \to Q$ is an isomorphism.)

Using the chosen generators of Σ and Q, we write $J[\mathscr{I}] = (\mathbf{Z}/n\mathbf{Z})^2$. In this model of $J[\mathscr{I}]$, Σ is the group generated by $(1,0)$ and C is the group generated by $(1,2)$. Since $\mathrm{Gal}(\overline{\mathbf{Q}}/\mathbf{Q})$ preserves Σ and operates on Σ as the mod n cyclotomic character χ, and since $\mathrm{Gal}(\overline{\mathbf{Q}}/\mathbf{Q})$ operates trivially on Q, the action of $\mathrm{Gal}(\overline{\mathbf{Q}}/\mathbf{Q})$ on $J[\mathscr{I}]$ is given in matrix terms by a map

$$\sigma \longmapsto \rho(\sigma) := \begin{pmatrix} \chi & b(\sigma) \\ 0 & 1 \end{pmatrix}.$$

Here, the map $\sigma \mapsto b(\sigma) \in \mathbf{Z}/n\mathbf{Z}$ is clearly a 1-cocycle: it verifies the identity

$$b(\sigma\tau) = b(\sigma) + \chi(\sigma)b(\tau)$$

for $\sigma, \tau \in \mathrm{Gal}(\overline{\mathbf{Q}}/\mathbf{Q})$.

4 Lenstra's input

The contents of this section were suggested to the author by H. W. Lenstra, Jr. The author thanks him heartily for his help.

Lemma 4.1. *For all $\sigma \in \mathrm{Gal}(\overline{\mathbf{Q}}/\mathbf{Q})$, we have $2b(\sigma) = 1 - \chi(\sigma)$.*

Proof. For each σ, $\rho(\sigma)$ fixes the vector $\begin{pmatrix} 1 \\ 2 \end{pmatrix} \in C$. The lemma follows immediately. ∎

Proposition 4.2. *The field $\mathbf{Q}(J[\mathscr{I}])$ is an abelian extension of $\mathbf{Q}$ which contains $\mathbf{Q}(\mu_n)$ and has degree 1 or 2 over $\mathbf{Q}(\mu_n)$.*

Proof. To say that $\mathbf{Q}(J[\mathscr{I}])$ is abelian over $\mathbf{Q}$ is to say that the image of ρ is abelian. This amounts to the identity $b(\sigma\tau) \stackrel{?}{=} b(\tau\sigma)$ for $\sigma, \tau \in \mathrm{Gal}(\overline{\mathbf{Q}}/\mathbf{Q})$. By the cocycle identity, the two sides of the equation are respectively $b(\sigma) + \chi(\sigma)b(\tau)$ and $b(\tau) + \chi(\tau)b(\sigma)$. These expressions are indeed equal, in view of the lemma above.

It is clear that the field $\mathbf{Q}(J[\mathscr{I}])$ contains $\mathbf{Q}(\mu_n)$ because the kernel of ρ is contained in the kernel of χ. Let H be this latter kernel; i.e., $H = \mathrm{Gal}(\overline{\mathbf{Q}}/\mathbf{Q}(\mu_n))$. On H, $\chi = 1$; hence we have $2b = 0$ in $\mathbf{Z}/n\mathbf{Z}$. In other words, the group $\rho(H)$ is a subgroup of the group of matrices $\begin{pmatrix} 1 & x \\ 0 & 1 \end{pmatrix}$ with $2x = 0$. Since this group has order $\gcd(2, n)$, the extension of $\mathbf{Q}$ cut out by ρ is an extension of $\mathbf{Q}(\mu_n)$ of degree 1 or 2. ∎

The proof of Proposition 4.2 (or, alternatively, the decomposition $J[\mathscr{I}] = \Sigma \oplus C$) shows that $\mathbf{Q}(J[\mathscr{I}]) = \mathbf{Q}(\mu_n)$ if n is odd. Suppose now that n is even; write $n = 2^k n_o$, where n_o is the "odd part" and $2^k \geq 2$ is the largest power of 2 dividing n. Then ρ is the direct sum of representations

$$\rho_2 \colon \mathrm{Gal}(\overline{\mathbf{Q}}/\mathbf{Q}) \to \mathbf{GL}(2, \mathbf{Z}/2^k\mathbf{Z}), \quad \rho_o \colon \mathrm{Gal}(\overline{\mathbf{Q}}/\mathbf{Q}) \to \mathbf{GL}(2, \mathbf{Z}/n_o\mathbf{Z}),$$

which are defined by the actions of $\mathrm{Gal}(\overline{\mathbf{Q}}/\mathbf{Q})$ on the 2-primary part and the odd part of $J[\mathscr{I}]$, respectively. It is evident that the latter representation cuts out $\mathbf{Q}(\mu_{n_o})$ and that the kernel of the former representation corresponds to an abelian extension K of $\mathbf{Q}$ which contains $\mathbf{Q}(\mu_{2^k})$ and has degree 1 or 2 over this cyclotomic field. Since ρ_2 is defined by the action of $\mathrm{Gal}(\overline{\mathbf{Q}}/\mathbf{Q})$ on a group of 2-power division points of J, this representation can be ramified only at 2 and at N. We have seen, however, that ρ is unramified at N (Proposition 3.1). Hence $K/\mathbf{Q}$ is an abelian extension of $\mathbf{Q}$ which is ramified only at 2; it follows (e.g., from the proof that the "local Kronecker-Weber theorem" implies the usual, global one [21, Ch. 14]) that K is contained in the cyclotomic field $\mathbf{Q}(\mu_{2^\infty})$. Hence we have either $K = \mathbf{Q}(\mu_{2^k})$ or $K = \mathbf{Q}(\mu_{2^{k+1}})$. Accordingly, we have

$$\mathbf{Q}(\mu_n) \subseteq \mathbf{Q}(J[\mathscr{I}]) \subseteq \mathbf{Q}(\mu_{2n}).$$

In summary, the displayed inclusions hold both in the case when n is odd and when n is even. In the former case, the two cyclotomic fields are equal, and they coincide with $\mathbf{Q}(J[\mathscr{I}])$. In the latter case, there remains an ambiguity which will be resolved by the proof of Theorem 1.7.

Before turning to this proof in the general case, we present a *simple proof of Theorem* 1.7 in the case where k is different from 2. To prove the Theorem is to show that ρ_2 cuts out the field $\mathbf{Q}(\mu_{2^{k+1}})$. This is perfectly clear if $k = 0$, in which case ρ_2 is the trivial representation: the field $K = \mathbf{Q}$ is indeed the field of second roots of 1. If $k = 1$, ρ_2 gives the action of $\mathrm{Gal}(\overline{\mathbf{Q}}/\mathbf{Q})$ on the group D which is described in [9, Ch. II, §12]; Lemma 12.4 of that section states that the field K is the field of fourth roots of unity.

Suppose now that k is at least 3, and choose $\sigma \in \mathrm{Gal}(\overline{\mathbf{Q}}/\mathbf{Q})$ so that $\chi(\sigma) \equiv 1 + 2^{k-1} \bmod 2^k$ and $\chi(\sigma) \equiv 1 \bmod n_o$. It is evident that $\chi(\sigma^2) = 1$; we will show, however, that $\rho_2(\sigma^2) \neq 1$. These two pieces of information imply that K is not contained in $\mathbf{Q}(\mu_n)$, which is precisely the information that we seek. To prove that $\rho_2(\sigma^2)$ is different from 1 is to show that $b(\sigma^2) \not\equiv 0 \bmod 2^k$. We have

$$b(\sigma^2) = (1+\chi(\sigma))b(\sigma) \equiv 2(1+2^{k-2})b(\sigma) \mod 2^k$$

by the cocycle identity and the choice of σ. Since k is at least 3, the factor $(1+2^{k-2})$ is odd. Now $2b(\sigma) = 1-\chi(\sigma) \equiv -2^{k-1} \bmod 2^k$ in view of Lemma 4.1. Thus $b(\sigma)$ is divisible by 2^{k-2} but not by 2^{k-1}. It follows that $b(\sigma^2)$ is divisible by 2^{k-1} but not by 2^k. ■

5 Proof of Theorem 1.7

We return to the discussion of the general case, removing the assumption $k \geq 3$. Recall that ρ is the representation of $\mathrm{Gal}(\overline{\mathbf{Q}}/\mathbf{Q})$ giving the action of $\mathrm{Gal}(\overline{\mathbf{Q}}/\mathbf{Q})$ on $J[\mathscr{I}]$ and that $\eta_\ell = 1+\ell-T_\ell$ for each $\ell \neq N$.

Lemma 5.1. *Let ℓ be a prime number prime to nN. Suppose that $\rho(\mathrm{Frob}_\ell) = 1$. Then η_ℓ belongs to $\mathscr{I}^2$.*

Proof. One has $\mathbf{T}/\mathscr{I}^2 = \bigoplus \mathbf{T}_\mathfrak{m}/\mathscr{I}^2\mathbf{T}_\mathfrak{m}$, where the sum is taken over the Eisenstein primes $\mathfrak{m}$ of $\mathbf{T}$. We must show that the image of η in $\mathbf{T}_\mathfrak{m}/\mathscr{I}^2\mathbf{T}_\mathfrak{m}$ is 0 for each such $\mathfrak{m}$. Fix $\mathfrak{m}$, and let p be the corresponding prime divisor of n. Consider the p-divisible group $J_\mathfrak{m} = \bigcup_\nu J[\mathfrak{m}^\nu]$ and its Tate module $\mathrm{Ta}_\mathfrak{m} := \mathrm{Hom}(\mathbf{Q}_p/\mathbf{Z}_p, J_\mathfrak{m})$. Let

$$\mathrm{Ta}_\mathfrak{m}^* := \mathrm{Hom}_{\mathbf{Z}_p}(\mathrm{Ta}_\mathfrak{m}, \mathbf{Z}_p) = \mathrm{Hom}(J_\mathfrak{m}, \mathbf{Q}_p/\mathbf{Z}_p);$$

the latter description of $\mathrm{Ta}_\mathfrak{m}^*$ presents this Tate module as the Pontryagin dual of $J_\mathfrak{m}$. Note that $\mathrm{Ta}_\mathfrak{m}$ and $\mathrm{Ta}_\mathfrak{m}^*$ have been shown to be free of rank 2 over $\mathbf{T}_\mathfrak{m}$ [9, Ch. II, Cor. 16.3]. The Tate pairing $\mathrm{Ta}_p(J) \times \mathrm{Ta}_p(J) \to \mathbf{Z}_p(1)$ may be viewed as an isomorphism $\mathrm{Ta}_\mathfrak{m} \approx \mathrm{Ta}_\mathfrak{m}^*(1)$ which is compatible with the natural actions of $\mathrm{Gal}(\overline{\mathbf{Q}}/\mathbf{Q})$ and $\mathbf{T}$ on the two modules.

Let $F = \mathrm{Frob}_\ell$. Since $1-F$ annihilates $J[\mathscr{I}]$, $1-F$ annihilates the Shimura subgroup $\Sigma \approx \mu_n$ of J, which is contained in $J[\mathscr{I}]$. Hence $\ell \equiv 1 \bmod n$. Accordingly, F acts as the identity on $\mathrm{Hom}(J[\mathscr{I}], \mu_n)$ and its p-primary subgroup $\mathrm{Hom}(J_\mathfrak{m}[\mathscr{I}], \mathbf{Q}_p/\mathbf{Z}_p)(1)$. We may view this dual as the quotient $\mathrm{Ta}_\mathfrak{m}^*(1)/\mathscr{I}\,\mathrm{Ta}_\mathfrak{m}^*(1) \approx \mathrm{Ta}_\mathfrak{m}/\mathscr{I}\,\mathrm{Ta}_\mathfrak{m}$. Hence we have

$$(1-F)(\mathrm{Ta}_\mathfrak{m}) \subseteq \mathscr{I}\cdot\mathrm{Ta}_\mathfrak{m}.$$

Since $\mathrm{Ta}_\mathfrak{m}$ is free of rank 2 over $\mathbf{T}_\mathfrak{m}$, we obtain

$$\det{}_{\mathbf{T}_\mathfrak{m}}(1-F \mid \mathrm{Ta}_\mathfrak{m}) \in \mathscr{I}^2\mathbf{T}_\mathfrak{m}.$$

This proves what is needed, since the determinant we have calculated is nothing but η_ℓ; indeed, the determinant and trace of F acting on $\mathrm{Ta}_\mathfrak{m}$ are ℓ and T_ℓ, respectively. ■

Theorem 5.2. *Assume that n is even. Let $\ell \neq N$ be a prime number which satisfies the congruence $\ell \equiv 1 \bmod n$ but not the congruence $\ell \equiv 1 \bmod 2n$. Assume further that the image of ℓ in $(\mathbf{Z}/N\mathbf{Z})^*$ is a generator of this cyclic group. Then $\rho(\mathrm{Frob}_\ell) \neq 1$.*

Proof. Let Δ be the unique quotient of $(\mathbf{Z}/N\mathbf{Z})^*$ of order n. To prove our result, we refer to §18 of [9, Ch. II]. In that section, one finds a homomorphism $\epsilon^+ : \mathscr{I}/\mathscr{I}^2 \to H^+/\mathscr{I}H^+$ and a map $\varphi: \Delta \to H^+/\mathscr{I}H^+$, both of which prove to be isomorphisms. The map $\kappa := \varphi^{-1} \circ \epsilon^+$ is an isomorphism $\mathscr{I}/\mathscr{I}^2 \xrightarrow{\sim} \Delta$. The *congruence formula for the winding homomorphism* yields

$$\kappa(\eta_\ell) = \tfrac{\ell-1}{2} \cdot \bar{\ell}.$$

Here, $\bar{\ell}$ is the image of $\ell \in (\mathbf{Z}/N\mathbf{Z})^*$ in Δ, and the operator $\frac{\ell-1}{2}$ is an exponent. (One is viewing the multiplicative abelian group Δ as a $\mathbf{Z}$-module.) Under our hypotheses, it is clear that $\frac{\ell-1}{2} \cdot \bar{\ell}$ has order 2 in Δ. Thus, by the congruence formula, η_ℓ is non-zero in $\mathscr{I}/\mathscr{I}^2$. Using Lemma 5.1, we deduce the required conclusion that $\rho(\mathrm{Frob}_\ell)$ is different from 1. ■

We now prove Theorem 1.7, i.e., the statement that $\mathbf{Q}(J[\mathscr{I}])$ coincides with the cyclotomic field $\mathbf{Q}(\mu_{2n})$.

As was explained above, the statement to be proved follows from the decomposition $J[\mathscr{I}] = \Sigma \oplus C$ when n is odd. Assume then that n is even. As we have discussed, the field $\mathbf{Q}(J[\mathscr{I}])$ is an extension of $\mathbf{Q}(\mu_n)$ of degree dividing 2. Moreover, if $\mathbf{Q}(J[\mathscr{I}])$ is indeed quadratic over $\mathbf{Q}(\mu_n)$, then $\mathbf{Q}(J[\mathscr{I}])$ has no choice but to be $\mathbf{Q}(\mu_{2n})$. To see that the extension $\mathbf{Q}(J[\mathscr{I}])/\mathbf{Q}(\mu_n)$ is non-trivial, we use the result above. Using the Chinese Remainder Theorem and Dirichlet's theorem on primes in an arithmetic progression, we may choose ℓ so as to satisfy the conditions of Theorem 5.2. A Frobenius element Frob_ℓ for ℓ in $\mathrm{Gal}(\overline{\mathbf{Q}}/\mathbf{Q})$ then acts trivially on μ_n, but non-trivially on $J[\mathscr{I}]$. ■

6 Adelic representations

Let ℓ be a prime. As usual, we consider the ℓ-divisible group $J_\ell = \bigcup_\nu J[\ell^\nu]$ and its Tate modules $\mathrm{Ta}_\ell := \mathrm{Hom}(\mathbf{Q}_\ell/\mathbf{Z}_\ell, J_\ell)$ and $\mathrm{Ta}_\ell \otimes_{\mathbf{Z}_\ell} \mathbf{Q}_\ell$. The ℓ-adic representation of $\mathrm{Gal}(\overline{\mathbf{Q}}/\mathbf{Q})$ attached to J is the continuous homomorphism

$$\rho_\ell : \mathrm{Gal}(\overline{\mathbf{Q}}/\mathbf{Q}) \to \mathrm{Aut}(\mathrm{Ta}_\ell) \hookrightarrow \mathrm{Aut}(\mathrm{Ta}_\ell \otimes \mathbf{Q}_\ell)$$

which arises from the action of $\mathrm{Gal}(\overline{\mathbf{Q}}/\mathbf{Q})$ on Ta_ℓ.

This action is $\mathbf{T}$-linear, where $\mathbf{T}$ is the Hecke ring introduced above. Thus ρ_ℓ takes values, for example, in the group $\mathrm{Aut}_{\mathbf{T}_\ell}(\mathrm{Ta}_\ell)$, where $\mathbf{T}_\ell = \mathbf{T} \otimes \mathbf{Z}_\ell$. Note that the $\mathbf{Z}_\ell$-algebra $\mathbf{T}_\ell$ is the product of the completions $\mathbf{T}_\mathfrak{m}$ of $\mathbf{T}$ at the maximal ideals $\mathfrak{m}$ of $\mathbf{T}$ which divide ℓ. The corresponding decomposition of Ta_ℓ into a product of modules over the individual factors $\mathbf{T}_\mathfrak{m}$ of $\mathbf{T}_\ell$ is the natural decomposition of $\mathrm{Ta}_\ell = \prod_\mathfrak{m} \mathrm{Ta}_\mathfrak{m}$, where the $\mathrm{Ta}_\mathfrak{m}$ are the $\mathfrak{m}$-adic Tate modules which were introduced earlier.

As we have noted, Mazur proves in [9, Ch. II, §15–§18] that $\mathrm{Ta}_\mathfrak{m}$ is free of rank 2 over $\mathbf{T}_\mathfrak{m}$ for each maximal ideal $\mathfrak{m}$ of $\mathbf{T}$ which is not simultaneously ordinary, non-Eisenstein and of residue characteristic 2. Thus, after a choice of basis, $\mathrm{Aut}_{\mathbf{T}_\ell}(\mathrm{Ta}_\ell)$ becomes $\mathbf{GL}(2, \mathbf{T}\otimes\mathbf{Z}_\ell)$ for each prime $\ell > 2$. Thus, if ℓ is odd, ρ_ℓ may be viewed as a homomorphism

$$\mathrm{Gal}(\overline{\mathbf{Q}}/\mathbf{Q}) \to \mathbf{GL}(2, \mathbf{T}\otimes\mathbf{Z}_\ell).$$

Similarly, we may view ρ_2 as taking values in $\mathbf{GL}(2, \mathbf{T}\otimes\mathbf{Q}_2)$. Accordingly, the image G_ℓ of ρ_ℓ is a subgroup of $\mathbf{GL}(2, \mathbf{T}\otimes\mathbf{Q}_\ell)$ in all cases and a subgroup of $\mathbf{GL}(2, \mathbf{T}\otimes\mathbf{Z}_\ell)$ when ℓ is odd. The determinant of ρ_ℓ is the ℓ-adic cyclotomic character.

The group G_ℓ is studied in [16], where the following two results are obtained as Proposition 7.1 and Theorem 6.4, respectively:

Theorem 6.1. *The group G_ℓ is open in the matrix group*

$$\{M \in \mathbf{GL}(2, \mathbf{T}\otimes\mathbf{Q}_\ell) \mid \det M \in \mathbf{Q}_\ell^*\}.$$

Theorem 6.2. *Suppose that ℓ is at least 5 and is prime to the discriminant of $\mathbf{T}$. Suppose further that no maximal ideal $\mathfrak{m}|\ell$ is an Eisenstein ideal of $\mathbf{T}$ (i.e., that ℓ is prime to n). Then*

$$G_\ell = \{M \in \mathbf{GL}(2, \mathbf{T}\otimes\mathbf{Z}_\ell) \mid \det M \in \mathbf{Z}_\ell^*\}.$$

Consider next the adelic representation $\rho_\mathfrak{f} := \prod_\ell \rho_\ell$, where the product is taken over the set of all prime numbers ℓ. The image $G_\mathfrak{f}$ of $\rho_\mathfrak{f}$ is a subgroup of the product $\prod_\ell G_\ell$, which in turn is contained in the group

$$\{M \in \mathbf{GL}(2, \mathbf{T}\otimes\mathbf{Q}_2) | \det M \in \mathbf{Q}_2^*\} \times \prod_{\ell\neq 2} \{M \in \mathbf{GL}(2, \mathbf{T}\otimes\mathbf{Z}_\ell) | \det M \in \mathbf{Z}_\ell^*\}.$$

According to [16, Th. 7.5], $G_\mathfrak{f}$ is open in the latter product.

For each prime ℓ, let H_ℓ be the intersection of $G_\mathfrak{f}$ with the group

$$1 \times \cdots \times 1 \times G_\ell \times 1 \times \cdots \times 1 \cdots,$$

where G_ℓ is placed in the ℓth factor. Thus H_ℓ is a subgroup of G_ℓ which may be viewed as the image of the restriction of ρ_ℓ to the kernel of the representation $\prod_{\ell'\neq\ell} \rho_{\ell'}$.

Theorem 6.3. *Assume that ℓ satisfies the conditions of Theorem 6.2, i.e., that ℓ is prime to $\operatorname{disc}\mathbf{T}$ and distinct from 2 and 3. Assume further that ℓ is different from N. Then $H_\ell = G_\ell = \{M \in \mathbf{GL}(2, \mathbf{T}\otimes\mathbf{Z}_\ell) \mid \det M \in \mathbf{Z}_\ell^*\}$.*

Proof. The proof of this result is explained in the course of the proof of Theorem 7.5 of [16]: Fix ℓ, and let X be the smallest closed subgroup of $\mathrm{Gal}(\overline{\mathbf{Q}}/\mathbf{Q})$ which contains all inertia groups of $\mathrm{Gal}(\overline{\mathbf{Q}}/\mathbf{Q})$ for the prime ℓ. Since $\rho_{\ell'}(X) = \{1\}$ for all primes $\ell' \neq \ell$, $\rho_\ell(X)$ is a subgroup of H_ℓ, which in turn is contained in G_ℓ. As the author observed at the end of §6 of [16], the desired equality $\rho_\ell(X) = G_\ell$ follows from Theorem 3.4 and Proposition 4.2 of [16]. ∎

We now present a variant of the result above for the prime $\ell = N$. For this, we let Γ be the subgroup $1 + N\mathbf{Z}_N$ of $\mathbf{Z}_N^*$, i.e., the N-Sylow subgroup of $\mathbf{Z}_N^*$.

Proposition 6.4. *Suppose that N is prime to the discriminant of* $\mathbf{T}$. *Then H_N contains the group* $\{M \in \mathbf{GL}(2, \mathbf{T} \otimes \mathbf{Z}_N) \mid \det M \in \Gamma\}$.

Proof. Let X now be the smallest closed subgroup of $\mathrm{Gal}(\overline{\mathbf{Q}}/\mathbf{Q})$ which contains the wild subgroups (i.e., N-Sylow subgroups) of all inertia groups for N in $\mathrm{Gal}(\overline{\mathbf{Q}}/\mathbf{Q})$. It follows from the exact sequence (2.4) that we have $\rho_\ell(X) = \{1\}$ for all $\ell \neq N$. (If $\ell \neq N$, inertia groups at N act unipotently in the ℓ-adic representations attached to J. Consequently, the image under ρ_ℓ of an inertia group at N is a pro-ℓ group.) Hence $\rho_N(X)$ is a subgroup of H_N, and it will suffice to show that

$$\rho_N(X) = \{M \in \mathbf{GL}(2, \mathbf{T} \otimes \mathbf{Z}_N) \mid \det M \in \Gamma\}.$$

We note that $\rho_N(X)$ is contained in this matrix group since the image of $\rho_N(X)$ under the determinant mapping $G_N \to \mathbf{Z}_N^*$ is a pro-N group. Since in fact the group $\det \rho_N(X)$ is all of Γ, the equality $\rho_N(X) = \{M \in \mathbf{GL}(2, \mathbf{T} \otimes \mathbf{Z}_N) \mid \det M \in \Gamma\}$ means that $\rho_N(X)$ contains $\mathbf{SL}(2, \mathbf{T} \otimes \mathbf{Z}_N)$.

Because $\mathbf{T}$ is unramified at N, [16, Prop. 4.2] implies that the inclusion

$$\rho_N(X) \supseteq \mathbf{SL}(2, \mathbf{T} \otimes \mathbf{Z}_N)$$

holds if and only if it holds "mod N" in the sense that the image of X in $\mathbf{GL}(2, \mathbf{T}/N\mathbf{T})$ contains $\mathbf{SL}(2, \mathbf{T}/N\mathbf{T})$. To say that this image contains $\mathbf{SL}(2, \mathbf{T}/N\mathbf{T})$ is in fact to say that the image coincides with $\mathbf{SL}(2, \mathbf{T}/N\mathbf{T})$; indeed, Γ maps to the trivial subgroup of $(\mathbf{Z}/N\mathbf{Z})^*$. The image in question is certainly a normal subgroup of $\mathbf{SL}(2, \mathbf{T}/N\mathbf{T})$ since X is normal in $\mathrm{Gal}(\overline{\mathbf{Q}}/\mathbf{Q})$ and G_N contains $\mathbf{SL}(2, \mathbf{T} \otimes \mathbf{Z}_N)$. The ring $\mathbf{T}/N\mathbf{T}$ is a product of finite fields of characteristic N because $\mathbf{T}$ is unramified at N; intrinsically, $\mathbf{T}/N\mathbf{T} = \prod_{\mathfrak{m}} \mathbf{T}/\mathfrak{m}$, where $\mathfrak{m}$ runs over the maximal ideals of $\mathbf{T}$ which divide N.

Fix $\mathfrak{m}$ for the moment and let $\rho_{\mathfrak{m}} : \mathrm{Gal}(\overline{\mathbf{Q}}/\mathbf{Q}) \to \mathbf{GL}(2, \mathbf{T}/\mathfrak{m})$ be the mod $\mathfrak{m}$ reduction of the N-adic Galois representation ρ_N. This reduction is an irreducible two-dimensional representation because $\mathfrak{m}$ cannot be an Eisenstein prime; indeed, $\mathfrak{m}$ does not divide $N - 1$. As we have seen in Proposition 2.2, $\rho_{\mathfrak{m}}$ cannot be "finite" (or *peu ramifiée*) in the sense of [18]; recall that the

Main Theorem of [14] implies that $\rho_{\mathfrak{m}}$ would be modular of level 1 if it were finite. Thus $\rho_{\mathfrak{m}}$ is wildly ramified at N, so that the group $\rho_{\mathfrak{m}}(X)$ has order divisible by N. But $\rho_{\mathfrak{m}}(X)$ is a normal subgroup of $\mathbf{SL}(2, \mathbf{T}/\mathfrak{m})$; we conclude that $\rho_{\mathfrak{m}}(X) = \mathbf{SL}(2, \mathbf{T}/\mathfrak{m})$.

Thus the image of $\rho_N(X)$ in $\mathbf{SL}(2, \mathbf{T}/N\mathbf{T}) = \prod_{\mathfrak{m}} \mathbf{SL}(2, \mathbf{T}/\mathfrak{m})$ is a normal subgroup of $\mathbf{SL}(2, \mathbf{T}/N\mathbf{T})$ which maps surjectively to each factor $\mathbf{SL}(2, \mathbf{T}/\mathfrak{m})$. By taking commutators with elements of the form

$$1 \times \cdots \times 1 \times g \times 1 \times \cdots \times 1,$$

we find that $\rho_N(X)$ maps surjectively to $\mathbf{SL}(2, \mathbf{T}/N\mathbf{T})$. Therefore, as was explained above, $\rho_N(X)$ contains $\mathbf{SL}(2, \mathbf{T} \otimes \mathbf{Z}_N)$. ■

Returning briefly to the group $G_{\mathfrak{f}}$, we note that we have

$$(\prod_{\ell} H_\ell) \subseteq G_{\mathfrak{f}} \subseteq (\prod_{\ell} G_\ell),$$

where the products are taken over all prime numbers ℓ. A theorem of B. Kaskel [16, Th. 7.3] implies that the image of $G_{\mathfrak{f}}$ in the group $G^N := \prod_{\ell \neq N} G_\ell$ is all of G^N. This suggests viewing the full product $\prod_\ell G_\ell$ as the binary product $G^N \times G_N$. Then $G_{\mathfrak{f}}$ is a subgroup of this product which maps surjectively to each of the two factors. The group H_N may be viewed as the kernel of the projection map $G_{\mathfrak{f}} \to G^N$; symmetrically, we let $H^N \subset G^N$ be the kernel of the second projection map. As is well known (see "Goursat's Lemma," an exercise in Bourbaki's Algèbre, Ch. I, §4), the projections from $G_{\mathfrak{f}}$ onto G^N and G_N induce natural isomorphisms $G^N/H^N \approx G_{\mathfrak{f}}/(H^N \times H_N)$ and $G_{\mathfrak{f}}/(H^N \times H_N) \approx G_N/H_N$. We obtain as a consequence an isomorphism

$$\alpha : G^N/H^N \xrightarrow{\sim} G_N/H_N.$$

The group $G_{\mathfrak{f}}$ contains $H^N \times H_N$ as a normal subgroup, and the image of $G_{\mathfrak{f}}$ in

$$(G^N \times G_N)/(H^N \times H_N) = (G^N/H^N) \times (G_N/H_N)$$

is the graph of the isomorphism α.

It is worth remarking that $G_{\mathfrak{f}}$ is open in $G^N \times G_N$ by [16, Th. 7.5]. Hence the groups H^N and H_N are open in G^N and G_N respectively. Thus the groups $G_f/(H^N \times H_N)$, G^N/H^N and G_N/H_N are finite groups which have the same order. The order of $(G^N \times G_N)/(H^N \times H_N)$ is the square of the orders of the three other groups. If N is prime to disc $\mathbf{T}$, then the order of G_N/H_N is a divisor of $N-1$ by Prop. 6.4. Moreover as we will see below, the order of G_N/H_N is always divisible by Mazur's constant $n = \text{num}(\frac{N-1}{12})$.

Adopting a Galois-theoretic point of view, we let K be the subfield of $\overline{\mathbf{Q}}$ corresponding to the finite quotient $G_{\mathfrak{f}}/(H^N \times H_N)$ of $\text{Gal}(\overline{\mathbf{Q}}/\mathbf{Q})$. Let K_N be the extension of $\mathbf{Q}$ generated by the coordinates of the N-power torsion points on J and let K^N be the extension of $\mathbf{Q}$ which is defined similarly, using prime-to-N torsion points in place of N-power torsion points. Then the compositum $K_\infty = K^N K_N$ is the subfield of $\overline{\mathbf{Q}}$ corresponding to the quotient $G_\mathfrak{f}$ of $\mathrm{Gal}(\overline{\mathbf{Q}}/\mathbf{Q})$, and it is clear that we have $\mathrm{Gal}(K_\infty/K_N) = H^N$ and $\mathrm{Gal}(K_\infty/K^N) = H_N$. Thus

$$G_\mathfrak{f}/(H^N \times H_N) = G_N/H_N = G^N/H^N = \mathrm{Gal}(K/\mathbf{Q}).$$

What information do we have about K? We may restate Proposition 6.4 as follows: If $\mathbf{T}$ is unramified at N, then K is contained in the field of Nth roots of unity. Indeed, in that case, G_N/H_N is a quotient of

$$\{M \in \mathbf{GL}(2, \mathbf{T} \otimes \mathbf{Z}_N) \mid \det M \in \mathbf{Z}_N^*\}/\{M \in \mathbf{GL}(2, \mathbf{T} \otimes \mathbf{Z}_N) \mid \det M \in \Gamma\},$$

which corresponds (via the determinant) to the Galois group $\mathrm{Gal}(\mathbf{Q}(\mu_N)/\mathbf{Q})$. Without the assumption on $\operatorname{disc} \mathbf{T}$, we can remark, at least, that K is ramified only at N; it is a subfield of K_N, which is ramified only at N.

We now exhibit the lower bound for $[K : \mathbf{Q}]$ which was alluded to above, proving that K contains the unique subfield of $\mathbf{Q}(\mu_N)$ with degree n over $\mathbf{Q}$. (Since n is 1 only when $X_0(N)$ has genus 0, it follows that K is a non-trivial extension of $\mathbf{Q}$ whenever $J_0(N)$ is non-zero.) For this, we note first that K_N contains the field $\mathbf{Q}(\mu_N)$ of Nth roots of 1; indeed, K_N contains the field generated by the N-power roots of 1 in $\overline{\mathbf{Q}}$, since the determinant of ρ_N is the N-adic cyclotomic character. The Galois group $\mathrm{Gal}(\mathbf{Q}(\mu_N)/\mathbf{Q}) = (\mathbf{Z}/N\mathbf{Z})^*$ has a unique quotient of order n. As in the proof of Theorem 5.2, we refer to this quotient as Δ; field-theoretically, Δ corresponds to a Galois extension K_Δ of $\mathbf{Q}$ with

$$K_\Delta \subseteq \mathbf{Q}(\mu_N) \subset K_N.$$

Since $\mathrm{Gal}(K_\Delta/\mathbf{Q}) = \Delta$, $[K_\Delta : \mathbf{Q}] = n$.

Theorem 6.5. *The field K contains K_Δ.*

Proof. Let $\mathfrak{m}$ be an Eisenstein prime (i.e., maximal ideal) of $\mathbf{T}$; let ℓ be the corresponding divisor of n. The Tate module $\mathrm{Ta}_\mathfrak{m}$ which was introduced in the proof of Lemma 5.1 is free of rank two over $\mathbf{T}_\mathfrak{m}$, the completion of $\mathbf{T}$ at $\mathfrak{m}$. The action of $\mathrm{Gal}(\overline{\mathbf{Q}}/\mathbf{Q})$ on $\mathrm{Ta}_\mathfrak{m}$ is given by a representation

$$\rho_\mathfrak{m} : \mathrm{Gal}(\overline{\mathbf{Q}}/\mathbf{Q}) \to \mathbf{GL}(2, \mathbf{T}_\mathfrak{m})$$

whose determinant is the ℓ-adic cyclotomic character; if p is prime to ℓN, then the trace of $\rho_\mathfrak{m}(\mathrm{Frob}_p)$ is $T_p \in \mathbf{T}_\mathfrak{m}$, T_p being the pth Hecke operator. Taking the sum of the $\rho_\mathfrak{m}$ and then reducing mod $\mathscr{I}^2$, we obtain a representation

$$\rho : \mathrm{Gal}(\overline{\mathbf{Q}}/\mathbf{Q}) \to \mathbf{GL}(2, \mathbf{T}/\mathscr{I}^2)$$

with analogous properties. In particular, for each prime p prime to nN, the trace and determinant of $\rho(\mathrm{Frob}_p)$ are the images of T_p and p, respectively, in $\mathbf{T}/\mathscr{I}^2$.

Let η: Gal$(\overline{\mathbf{Q}}/\mathbf{Q}) \to \mathbf{T}/\mathscr{I}^2$ be the function $1 + \det\rho - \operatorname{tr}\rho$. For p prime to nN, $\eta(\mathrm{Frob}_p)$ is the image in $\mathscr{I}/\mathscr{I}^2$ of the element $\eta_p = 1 + p - t_p$ of $\mathscr{I}$. In particular, the Cebotarev density theorem implies that η is a function Gal$(\overline{\mathbf{Q}}/\mathbf{Q}) \to \mathscr{I}/\mathscr{I}^2$.

As we recalled in the proof of Theorem 5.2, there is an isomorphism $\kappa: \mathscr{I}/\mathscr{I}^2 \xrightarrow{\sim} \Delta$ which satisfies the congruence formula

$$\kappa(\eta_p) = \tfrac{p-1}{2} \cdot \overline{p}$$

for all primes p not dividing nN. In this formula, $\overline{p}$ represents the image in Δ of the congruence class of p mod N. Let α: Gal$(\overline{\mathbf{Q}}/\mathbf{Q}) \to \mathscr{I}/\mathscr{I}^2$ be the composite of: (1) the mod N cyclotomic character χ_N : Gal$(\overline{\mathbf{Q}}/\mathbf{Q}) \to (\mathbf{Z}/N\mathbf{Z})^*$; (2) the quotient map $(\mathbf{Z}/N\mathbf{Z})^* \to \Delta$; (3) the inverse of κ. Then we may write alternatively

$$\eta_p = \tfrac{p-1}{2} \cdot \alpha(p),$$

where the left-hand side is interpreted in $\mathscr{I}/\mathscr{I}^2$. If now $\chi = \chi_{2n}$ is the mod $2n$ cyclotomic character, then the formula for η_p and the Cebotarev density theorem imply the identity

$$\eta = \tfrac{\chi-1}{2} \cdot \alpha$$

of $\mathscr{I}/\mathscr{I}^2$-valued functions on Gal$(\overline{\mathbf{Q}}/\mathbf{Q})$.

Let H be the kernel of $\rho \times \chi$. Then $\eta(hg) = \eta(g)$ for all $g \in$ Gal$(\overline{\mathbf{Q}}/\mathbf{Q})$, since $\rho(hg) = \rho(g)$ in that case. Let h be an element of H and take g to be a complex conjugation in Gal$(\overline{\mathbf{Q}}/\mathbf{Q})$. Since $\chi(g) = -1$ and $\chi(h) = 1$, the equation $\eta(hg) = \eta(g)$ amounts to the identity $\alpha(hg) = \alpha(g)$. Since α is a homomorphism, we deduce that $\alpha(h) = 1$.

In other words, if $h \in$ Gal$(\overline{\mathbf{Q}}/\mathbf{Q})$ is trivial under $\rho \times \chi$, then h is trivial in Gal$(K_\Delta/\mathbf{Q})$. In particular, if $\rho^N(h) = 1$, then h fixes K_Δ. Accordingly, K_Δ is contained in the fixed field K^N of the kernel of ρ^N. Since, by construction, K_Δ is a subfield of K_N, K_Δ is contained in K. ■

Theorem 6.5, which will not be used in the proof of Theorem 1.6, suggests the problem of pinpointing K completely. According to Proposition 6.4 and Theorem 6.5, we have $K_\Delta \subseteq K \subseteq \mathbf{Q}(\mu_N)$ under the apparently mild assumption that N does not divide disc$(\mathbf{T})$. Since Gal $(\mathbf{Q}(\mu_N)/K_\Delta)$ is cyclic of order $(N-1)/n = \gcd(N-1, 12)$, to identify K under these circumstances is to calculate a divisor of $\gcd(N-1, 12)$, namely $[K : K_\Delta]$. In the cases where $X_0(N)$ has genus 0 (i.e., $N < 11$ and $N = 13$), we clearly have $K = \mathbf{Q} = K_\Delta$. In the case $N = 11$, K is constrained by our results to be either $\mathbf{Q}(\mu_{11})$ or the maximal real subfield of $\mathbf{Q}(\mu_{11})$. As was noted by Lang and Trotter [8] (see also [17, §5.3]), $K = \mathbf{Q}(\mu_{11})$ because the field generated by the 2-division points of $J_0(11)$ contains $\mathbf{Q}(\sqrt{-11})$. In the case $N = 37$, we have $\gcd(N-1, 12) = 12$, so that there are six a priori possibilities for K. In fact, Kaskel [7] shows that K is the maximal real subfield of $\mathbf{Q}(\mu_{37})$; the divisor in question is 6.

7 Proof of Theorem 1.6

We recall the statement to be proved: *Let P be a point of finite order on J whose ℓ-primary component P_ℓ is not rational point. Assume that at least one of the following statements is true:* (1) *N is prime to the order of P;* (2) *ℓ is prime to $N-1$;* (3) *N is prime to the discriminant of* $\mathbf{T}$ (*i.e., Hypothesis* 1.4 *holds*). *Then there is a $\sigma \in \mathrm{Gal}(\overline{\mathbf{Q}}/\mathbf{Q})$ such that $\sigma P - P$ has order ℓ.*

In the proof that follows, we write P^ℓ for the sum of the p-primary components of P for primes different from ℓ. Thus $P = P_\ell + P^\ell$. Similarly, we put $P^N = P - P_N$.

Consider the extension $\mathbf{Q}(P_\ell)/\mathbf{Q}$, which is non-trivial by hypothesis. To orient the reader, we note that this extension can be ramified only at ℓ and at N, the latter prime being the unique prime of bad reduction of J. According to [2, Th. 2.2], $\mathbf{Q}(P_\ell)/\mathbf{Q}$ is automatically ramified at ℓ except perhaps when $\ell = 2$.

On the other hand, it is plausible that $\mathbf{Q}(P_\ell)/\mathbf{Q}$ is unramified at N. Let us first deal with this possibility, which turns out to be especially simple; here the hypotheses (1)–(3) are irrelevant. According to Proposition 3.3, P_ℓ lies in $J[\mathscr{I}]$. This latter group contains the Shimura subgroup Σ and the cuspidal group C of J. The source and target of the resulting natural map $\Sigma \oplus C \to J[\mathscr{I}]$ have order n; the kernel and cokernel of this map have order 1 if n is odd and order 2 if n is even.

To fix ideas, we assume for the moment that ℓ is an odd prime. Then P_ℓ lies in the ℓ-primary part of $J[\mathscr{I}]$, which is the direct sum of the ℓ-primary parts of Σ and C. Hence P_ℓ is the sum of a rational point of J and an element of ℓ-power order of $\Sigma \approx \mu_n$. Since P_ℓ is not rational, this latter element is non-trivial; its order may be written ℓ^a with $a \geq 1$. Let σ be an element of $\mathrm{Gal}(\overline{\mathbf{Q}}/\mathbf{Q}(\mu_{\ell^{a-1}}))$ which has non-trivial image in $\mathrm{Gal}(\mathbf{Q}(\mu_{\ell^a})/\mathbf{Q}(\mu_{\ell^{a-1}}))$. Then it is evident that $\sigma P_\ell - P_\ell$ has order ℓ on J. Indeed, this element is non-trivial since σ does not fix P_ℓ, but it is of order dividing ℓ since σ does fix ℓP_ℓ. Now the extension $\mathbf{Q}(\mu_{\ell^a})/\mathbf{Q}(\mu_{\ell^{a-1}})$ is ramified at ℓ; thus we may take σ to be in an inertia group for a prime of $\mathbf{Q}(\mu_{\ell^{a-1}})$ which lies over ℓ. This choice ensures that P^ℓ is fixed by σ. Then $\sigma P - P = \sigma P_\ell - P_\ell$ is a point of order ℓ, as desired.

Next, we suppose that $\ell = 2$; we continue to suppose that P_ℓ is unramified at N. Then $J[\mathscr{I}]$ has even order; i.e., n is even. If $P_\ell = P_2$ lies in $\Sigma + C$, then things proceed as in the case $\ell > 2$. However, as we recalled above, the sum $\Sigma + C$, which is not direct, represents a proper subgroup of $J[\mathscr{I}]$ (namely, one of index 2.) Hence we must discuss the case where P_2, which is a point in $J[\mathscr{I}]$, does not lie in the sum $\Sigma + C$.

In this case, the group $J[\mathscr{I}]$ is generated by its subgroup $\Sigma + C$ of index 2 together with the point P_2. Using Theorem 1.7, we find that

$$\mathbf{Q}(\mu_{2n}) = \mathbf{Q}(J[\mathscr{I}]) = K(P_2),$$

where $K = \mathbf{Q}(\Sigma + C) = \mathbf{Q}(\mu_n)$. The extension $\mathbf{Q}(\mu_{2n})/\mathbf{Q}(\mu_n)$ is a quadratic extension which is ramified at 2. We take σ in an inertia group for 2 which fixes K but not P_2. Since $2P_2$ lies in $\Sigma + C$, the difference $\sigma P_2 - P_2$ is of order 2. We have $\sigma P - P = \sigma P_2 - P_2$ in analogy with the situation already considered.

Having treated the relatively simple case where $\mathbf{Q}(P_\ell)/\mathbf{Q}$ is unramified at N, we assume from now on that P_ℓ is ramified at N. This assumption means that there is an inertia subgroup $I \subset \mathrm{Gal}(\overline{\mathbf{Q}}/\mathbf{Q})$ for the prime N which acts non-trivially on P_ℓ. Hence there is a $\tau \in I$ such that the order of $\tau P - P$ is divisible by ℓ. We seek to construct a $\sigma \in I$ for which $\sigma P - P$ has order precisely ℓ.

Assume first that (1) holds, i.e., that the order of P is prime to N. Let m be this order, and let ℓd be the order of $\tau P - P$; thus, ℓd divides m. Recall the exact sequence of I-modules

$$0 \to \mathrm{Hom}(\mathscr{X}/m\mathscr{X}, \mu_m) \to J[m] \to \mathscr{X}/m\mathscr{X} \to 0. \tag{2.4}$$

Since m is prime to N, the two flanking groups are unramified. It follows, as is well known, that $A := \tau - 1$ acts on $J[m]$ as an endomorphism with square 0. By the binomial theorem, we find the equation $\tau^d = 1 + dA$ in $\mathrm{End}\, J[m]$. Therefore

$$\tau^d P - P = dAP = d(\tau - 1)P = d(\tau P - P)$$

is a point of order ℓ. We take $\sigma = \tau^d$.

Next, assume that (2) holds. Arguing as above, we may find an $s \in I$ such that $sP^N - P^N$ has order ℓ. Moreover, for each $i \geq 1$, we have $s^i P^N - P^N = i(sP^N - P^N)$. Consider again (2.4), with m replaced by m', the order of P_N. Let $j = \phi(m')$ (Euler ϕ-function). Then s^j acts trivially on the groups $\mathrm{Hom}(\mathscr{X}/m'\mathscr{X}, \mu_{m'})$ and $\mathscr{X}/m'\mathscr{X}$ in (2.4), so that $s^{jm'}$ fixes P_N. By (2), j is prime to ℓ, and thus $i := jm'$ is prime to ℓ as well. Taking $\sigma = s^i$, we find that $\sigma P - P$ has order ℓ, as required.

We now turn to the most complicated case, that where (3) holds, but where (1) and (2) are no longer assumed. We change notation slightly, writing m (rather than m') for the order of P_N. Thus m is a power of N. Let s again be an element of I such that $sP^N - P^N$ has order ℓ.

We fix our attention once again on (2.4), which we view as a sequence of I-modules. Concerning the Hecke action, we note that the two groups

$$M := \mathrm{Hom}(\mathscr{X}/m\mathscr{X}, \mu_m), \quad M' := \mathscr{X}/m\mathscr{X}$$

are each free of rank 1 over $\mathbf{T}/m\mathbf{T}$ in view of Theorem 2.3 and the fact that $\mathbf{T}$ is Gorenstein away from the prime 2. The central group $J[m]$ is free of rank 2 over $\mathbf{T}/m\mathbf{T}$ because of [9, Ch. II, Cor. 15.2]. The inertia group I acts trivially on $\mathscr{X}$ and as the mod m cyclotomic character χ on μ_m. Thus M' is unramified, and M is ramified if m is different from 1.

We will be interested in the value of $\chi(s) \in (\mathbf{Z}/m\mathbf{Z})^*$. Let i be the prime-to-ℓ part of the order of $\chi(s)$, and replace s by s^i. After this replacement, the order of $\chi(s)$ is a power of ℓ. Also, as we have discussed, this replacement multiplies $sP^N - P^N$ by i. Since i is prime to ℓ, $sP^N - P^N$ remains of order ℓ.

If $\chi(s)$ is now 1, then the situation is similar to that which we just discussed. Namely, s^m is the identity on $J[m]$, and we may take $\sigma = s^m$.

Assume now that $\chi(s)$ is different from 1; thus $\chi(s)$ is a non-trivial ℓ-power root of 1. In this case, the $\mathbf{T}$-module $J[m]$ is the *direct sum* of two subspaces: the space where s acts as 1 and the space where s acts as $\chi(s)$ (which is not congruent to 1 mod N). Indeed, the endomorphism $\dfrac{s-\chi(s)}{1-\chi(s)}$ of $J[m]$ is zero on $M = \mathrm{Hom}(\mathscr{X}/m\mathscr{X}, \mu_m)$ and the identity on $M' = \mathscr{X}/m\mathscr{X}$. It splits the exact sequence which is displayed above, giving us an isomorphism of $\mathbf{T}$-modules:

$$J[m] \approx M \oplus M'.$$

The module M', viewed as a submodule of $J[m]$, is the fixed part of $J[m]$ relative to the action of s.

We claim that there is an $h \in \mathrm{Gal}(\overline{\mathbf{Q}}/\mathbf{Q})$ such that $hP^N = P^N$ and such that $hP_N \in M'$. This claim will prove what is wanted, since the choice $\sigma = h^{-1}sh$ will guarantee that the difference $\sigma P - P$ is the ℓ-division point

$$h^{-1}(shP^N - hP^N) + h^{-1}(shP_N - hP_N) = h^{-1}(sP^N - P^N).$$

To find the desired h it suffices to produce an element of $\mathbf{SL}_{\mathbf{T}/m\mathbf{T}}J[m] \approx \mathbf{SL}(2, \mathbf{T}/m\mathbf{T})$ which maps P_N into M'. Indeed, Proposition 6.4 implies that all such elements arise from H_N, i.e., from elements of $\mathrm{Gal}(\overline{\mathbf{Q}}/\mathbf{Q})$ which fix torsion points of J with order prime to N.

To produce the required element of $\mathbf{SL}(2, \mathbf{T}/m\mathbf{T})$, we work explicitly. Choose $\mathbf{T}/m\mathbf{T}$-bases e' and e of the free rank 1 modules M' and M, and use $\{e', e\}$ as a basis of $J[m]$. Then M' is the span of the vector $(1,0)$ and M is the span of $(0,1)$. Let u and v be the coordinates of P_N relative to the chosen basis. We must exhibit a matrix in $\mathbf{SL}(2, \mathbf{T}/m\mathbf{T})$ which maps (u,v) to a vector with second component 0.

Because of the hypothesis that N is prime to $\operatorname{disc}\mathbf{T}$, $\mathbf{T} \otimes \mathbf{Z}_N$ is a finite product of rings of integers of finite unramified extensions of $\mathbf{Q}_N$. Thus $\mathbf{T}/m\mathbf{T}$ is a product of rings of the form $R = \mathscr{O}/m\mathscr{O}$, where $\mathscr{O}$ is the ring of integers of a finite unramified extension of $\mathbf{Q}_N$. It suffices to solve our problem factor by factor: given $(u,v) \in R^2$, we must find an element of $\mathbf{SL}(2,R)$ which maps (u,v) into the line generated by $(1,0)$. It is clear that we may write (u,v) in the form $N^t(u',v')$, where t is a non-negative integer and at least one of u', v' is a unit in R. Solving the problem for (u',v') solves it for (u,v), so we may, and do, assume that either u or v is a unit.

If u is a unit, then

$$\begin{pmatrix} 1 & 0 \\ -vu^{-1} & 1 \end{pmatrix} \begin{pmatrix} u \\ v \end{pmatrix} = \begin{pmatrix} u \\ 0 \end{pmatrix}.$$

If v is a unit, then

$$\begin{pmatrix} 0 & 1 \\ 1 & -uv^{-1} \end{pmatrix} \begin{pmatrix} u \\ v \end{pmatrix} = \begin{pmatrix} v \\ 0 \end{pmatrix}. \qquad \blacksquare$$

References

[1] N. Boston, H. W. Lenstra, Jr., and K. A. Ribet, *Quotients of group rings arising from two-dimensional representations*, C. R. Acad. Sci. Paris Sér. I Math. **312** (1991), 323–328.

[2] R. Coleman, B. Kaskel, and K. Ribet, *Torsion points on* $X_0(N)$, Contemporary Math. (to appear).

[3] J. A. Csirik, *The Galois structure of* $J_0(N)[I]$ (to appear).

[4] P. Deligne and M. Rapoport, *Les schémas de modules de courbes elliptiques*, Lecture Notes in Math., vol. 349, Springer-Verlag, Berlin and New York, 1973, pp. 143–316.

[5] E. De Shalit, *A note on the Shimura subgroup of* $J_0(p)$, Journal of Number Theory **46** (1994), 100–107.

[6] A. Grothendieck, *SGA* 7 I, *Exposé* IX, Lecture Notes in Math., vol. 288, Springer-Verlag, Berlin and New York, 1972, pp. 313–523.

[7] B. Kaskel, *The adelic representation associated to* $X_0(37)$, Ph. D. thesis, UC Berkeley, May, 1996.

[8] S. Lang and H. Trotter, *Frobenius distributions in* GL_2*-extensions*, Lecture Notes in Math., vol. 504, Springer-Verlag, Berlin and New York, 1976.

[9] B. Mazur, *Modular curves and the Eisenstein ideal*, Publ. Math. IHES **47** (1977), 33–186.

[10] B. Mazur, *Rational isogenies of prime degree*, Invent. Math. **44** (1978), 129–162.

[11] A. Ogg, *Hyperelliptic modular curves*, Bull. Soc. Math. France **102** (1974), 449–462.

[12] A. Ogg, *Automorphismes de courbes modulaires*, Sém. Delange-Pisot-Poitou (1974/1975, exposé 7).

[13] M. Raynaud, *Courbes sur une variété abélienne et points de torsion*, Invent. Math. **71** (1983), 207–233.

[14] K. A. Ribet, *On modular representations of* $\mathrm{Gal}(\overline{\mathbf{Q}}/\mathbf{Q})$ *arising from modular forms*, Invent. Math. **100** (1990), 431–476.

[15] K. A. Ribet, *Report on mod* ℓ *representations of* $\mathrm{Gal}(\overline{\mathbf{Q}}/\mathbf{Q})$, Proceedings of Symposia in Pure Mathematics **55 (2)** (1994), 639–676.

[16] K. A. Ribet, *Images of semistable Galois representations*, Pacific Journal of Math. **81** (1997), 277–297.

[17] J.-P. Serre, *Propriétés galoisiennes des points d'ordre fini des courbes elliptiques*, Invent. Math. **15** (1972), 259–331.

[18] J.-P. Serre, *Sur les représentations modulaires de degré* 2 *de* $\mathrm{Gal}(\overline{\mathbf{Q}}/\mathbf{Q})$, Duke Math. J. **54** (1987), 179–230.

[19] G. Shimura, *A reciprocity law in non-solvable extensions*, Journal für die reine und angewandte Mathematik **221** (1966), 209–220.

[20] J. Tate, *The non-existence of certain Galois extensions of* $\mathbf{Q}$ *unramified outside* 2, Contemporary Mathematics **174** (1994), 153–156.

[21] L. C. Washington, *Introduction to cyclotomic fields*, Graduate Texts in Math., vol. 83 (second edition), Springer-Verlag, Berlin-Heidelberg-New York, 1997.

Elliptic Curves with Complex Multiplication and the Conjecture of Birch and Swinnerton-Dyer

Karl Rubin * **

Department of Mathematics, Ohio State University, 231 W. 18 Avenue, Columbus, Ohio 43210 USA, rubin@math.ohio-state.edu

The purpose of these notes is to present a reasonably self-contained exposition of recent results concerning the Birch and Swinnerton-Dyer conjecture for elliptic curves with complex multiplication. The goal is the following theorem.

Theorem. *Suppose E is an elliptic curve defined over an imaginary quadratic field K, with complex multiplication by K, and $L(E,s)$ is the L-function of E. If $L(E,1) \neq 0$ then*

(i) *$E(K)$ is finite,*
(ii) *for every prime $p > 7$ such that E has good reduction above p, the p-part of the Tate-Shafarevich group of E has the order predicted by the Birch and Swinnerton-Dyer conjecture.*

The first assertion of this theorem was proved by Coates and Wiles in [CW1]. We will prove this in §10 (Theorem 10.1). A stronger version of (ii) (with no assumption that E have good reduction above p) was proved in [Ru2]. The program to prove (ii) was also begun by Coates and Wiles; it can now be completed thanks to the recent Euler system machinery of Kolyvagin [Ko]. This proof will be given in §12, Corollary 12.13 and Theorem 12.19.

The material through §4 is background which was not in the Cetraro lectures but is included here for completeness. In those sections we summarize, generally with references to [Si] instead of proofs, the basic properties of elliptic curves that will be needed later. For more details, including proofs, see Silverman's book [Si], Chapter 4 of Shimura's book [Sh], Lang's book [La], or Cassels' survey article [Ca].

The content of the lectures was essentially §§5-12.

---

* Partially supported by the National Science Foundation. The author also gratefully acknowledges the CIME for its hospitality.

** current address: Department of Mathematics, Stanford University, Stanford, CA 94305 USA, rubin@math.stanford.edu

1 Quick Review of Elliptic Curves

1.1 Notation

Suppose F is a field. An elliptic curve E defined over F is a nonsingular curve defined by a generalized Weierstrass equation

$$y^2 + a_1xy + a_3y = x^3 + a_2x^2 + a_4x + a_6 \tag{1}$$

with $a_1, a_2, a_3, a_4, a_6 \in F$. The points $E(F)$ have a natural, geometrically-defined group structure, with the point at infinity O as the identity element. The discriminant $\Delta(E)$ is a polynomial in the a_i and the j-invariant $j(E)$ is a rational function in the a_i. (See §III.1 of [Si] for explicit formulas.) The j-invariant of an elliptic curve depends only on the isomorphism class of that curve, but the discriminant Δ depends on the particular Weierstrass model.

Example 1.1. Suppose that E is defined by a Weierstrass equation

$$y^2 = x^3 + a_2x^2 + a_4x + a_6$$

and $d \in F^\times$. The *twist of E by $\sqrt{d}$* is the elliptic curve E_d defined by

$$y^2 = x^3 + a_2dx^2 + a_4d^2x + a_6d^3.$$

Then (exercise:) E_d is isomorphic to E over the field $F(\sqrt{d})$, $\Delta(E_d) = d^6\Delta(E)$, and $j(E_d) = j(E)$. See also the proof of Corollary 5.22.

1.2 Differentials

See [Si] §II.4 for the definition and basic background on differentials on curves.

Proposition 1.2. *Suppose E is an elliptic curve defined by a Weierstrass equation* (1). *Then the space of holomorphic differentials on E defined over F is a one-dimensional vector space over F with basis*

$$\omega_E = \frac{dx}{2y + a_1x + a_3}.$$

Further, ω_E is invariant under translation by points of $E(\bar{F})$.

Proof. See [Si] Propositions III.1.5 and III.5.1. That ω_E is holomorphic is an exercise, using that ω_E is also equal to $dy/(3x^2 + 2a_2x + a_4 - a_1y)$. □

1.3 Endomorphisms

Definition 1.3. Suppose E is an elliptic curve. An *endomorphism* of E is a morphism from E to itself which maps O to O.

An endomorphism of E is also a homomorphism of the abelian group structure on E (see [Si] Theorem III.4.8).

Example 1.4. For every integer m, multiplication by m on E is an endomorphism of E, which we will denote by $[m]$. If $m \neq 0$ then the endomorphism $[m]$ is nonzero; in fact, it has degree m^2 and, if m is prime to the characteristic of F then the kernel of $[m]$ is isomorphic to $(\mathbf{Z}/m\mathbf{Z})^2$. (See [Si] Proposition III.4.2 and Corollary III.6.4.)

Example 1.5. Suppose F is finite, $\#(F) = q$. Then the map $\varphi_q : (x, y, z) \mapsto (x^q, y^q, z^q)$ is a (purely inseparable) endomorphism of E, called the q-th power Frobenius morphism.

Definition 1.6. If E is an elliptic curve defined over F, we write $\mathrm{End}_F(E)$ for the ring (under addition and composition) of endomorphisms of E defined over F. Then $\mathrm{End}_F(E)$ has no zero divisors, and by Example 1.4 there is an injection $\mathbf{Z} \hookrightarrow \mathrm{End}_F(E)$.

Definition 1.7. Write $\mathcal{D}(E/F)$ for one-dimensional vector space (see Proposition 1.2) of holomorphic differentials on E defined over F. The map $\phi \mapsto \phi^*$ defines a homomorphism of abelian groups

$$\iota = \iota_F : \mathrm{End}_F(E) \to \mathrm{End}_F(\mathcal{D}(E/F)) \cong F.$$

The kernel of ι is the ideal of inseparable endomorphisms. In particular if F has characteristic zero, then ι_F is injective.

Lemma 1.8. *Suppose* $\mathrm{char}(F) = 0$, *L is a field containing F, and* $\phi \in \mathrm{End}_L(E)$. *If* $\iota_L(\phi) \in F$ *then* $\phi \in \mathrm{End}_F(E)$.

Proof. If $\sigma \in \mathrm{Aut}(\bar{L}/F)$ then

$$\iota_L(\phi^\sigma) = \sigma(\iota_L(\phi)) = \iota_L(\phi).$$

Since L has characteristic zero, ι_L is injective so we conclude that $\phi^\sigma = \phi$. □

Definition 1.9. If $\phi \in \mathrm{End}_F(E)$ we will write $E[\phi] \subset E(\bar{F})$ for the kernel of ϕ and $F(E[\phi])$ for the extension of F generated by the coordinates of the points in $E[\phi]$. Note that $F(E[\phi])$ is independent of the choice of a Weierstrass model of E over F. By [Si] Theorem III.4.10, $\#(E[\phi])$ divides $\deg(\phi)$, with equality if and only if ϕ is separable.

Definition 1.10. If ℓ is a rational prime define the ℓ-adic *Tate module* of E

$$T_\ell(E) = \varprojlim_n E[\ell^n],$$

inverse limit with respect to the maps $\ell : E[\ell^{n+1}] \to E[\ell^n]$. If $\ell \neq \mathrm{char}(F)$ then Example 1.4 shows that

$$T_\ell(E) \cong \mathbf{Z}_\ell^2.$$

The Galois group G_F acts $\mathbf{Z}_\ell$-linearly on $T_\ell(E)$, giving a representation

$$\rho_\ell : G_F \to \mathrm{Aut}(T_\ell(E)) \cong \mathrm{GL}_2(\mathbf{Z}_\ell)$$

when $\ell \neq \mathrm{char}(F)$.

Theorem 1.11. *If E is an elliptic curve then* $\mathrm{End}_{\bar{F}}(E)$ *is one of the following types of rings.*

(i) $\mathbf{Z}$,
(ii) *an order in an imaginary quadratic field,*
(iii) *an order in a division quaternion algebra over* $\mathbf{Q}$.

Proof. See [Si] §III.9. □

Example 1.12. Suppose $\mathrm{char}(F) \neq 2$ and E is the curve $y^2 = x^3 - dx$ where $d \in F^\times$. Let ϕ be defined by $\phi(x, y) = (-x, iy)$ where $i = \sqrt{-1} \in \bar{F}$. Then $\phi \in \mathrm{End}_{\bar{F}}(E)$, and $\iota(\phi) = i$ so $\phi \in \mathrm{End}_F(E) \Leftrightarrow i \in F$. Also, ϕ has order 4 in $\mathrm{End}_{\bar{F}}(E)^\times$ so we see that $\mathbf{Z}[\phi] \cong \mathbf{Z}[i] \subset \mathrm{End}_{\bar{F}}(E)$. (In fact, $\mathbf{Z}[\phi] = \mathrm{End}_{\bar{F}}(E)$ if $\mathrm{char}(F) = 0$ or if $\mathrm{char}(F) \equiv 1 \pmod 4$, and $\mathrm{End}_{\bar{F}}(E)$ is an order in a quaternion algebra if $\mathrm{char}(F) \equiv 3 \pmod 4$.) The next lemma gives a converse to this example.

Lemma 1.13. *Suppose E is given by a Weierstrass equation $y^2 = x^3 + ax + b$. If* $\mathrm{Aut}(E)$ *contains an element of order* 4 *(resp.* 3*) then $b = 0$ (resp. $a = 0$).*

Proof. The only automorphisms of such a Weierstrass elliptic curve are of the form $(x, y) \mapsto (u^2 x, u^3 y)$ (see [Si] Remark III.1.3). The order of such an automorphism is the order of u in $F^\times$, and when u has order 3 or 4 this change of variables preserves the equation if and only if $a = 0$ (resp. $b = 0$). □

2 Elliptic Curves over C

Remark 2.1. Note that an elliptic curve defined over a field of characteristic zero can be defined over $\mathbf{Q}[a_1, a_2, a_3, a_4, a_6]$, and this field can be embedded in $\mathbf{C}$. In this way many of the results of this section apply to all elliptic curves in characteristic zero.

2.1 Lattices

Definition 2.2. Suppose L is a lattice in $\mathbf{C}$. Define the Weierstrass $\wp$-function, the Weierstrass σ-function, and the Eisenstein series attached to L

$$\wp(z;L) = \frac{1}{z^2} + \sum_{0\neq\omega\in L} \frac{1}{(z+\omega)^2} - \frac{1}{\omega^2}$$

$$\sigma(z;L) = z \prod_{0\neq\omega\in L} \left(1-\frac{z}{\omega}\right) e^{(z/\omega)+(z/\omega)^2/2}$$

$$G_k(L) = \sum_{0\neq\omega\in L} \frac{1}{\omega^k} \quad \text{for } k \text{ even, } k \geq 4.$$

We will suppress the L from the notation in these functions when there is no danger of confusion, See [Si] Theorem VI.3.1, Lemma VI.3.3, and Theorem VI.3.5 for the convergence and periodicity properties of these functions.

Theorem 2.3. (i) *If L is a lattice in $\mathbf{C}$ then the map*

$$z \mapsto (\wp(z;L), \wp'(z;L)/2)$$

is an analytic isomorphism (and a group homomorphism) from $\mathbf{C}/L$ to $E(\mathbf{C})$ where E is the elliptic curve $y^2 = x^3 - 15G_4(L)x - 35G_6(L)$.

(ii) *Conversely, if E is an elliptic curve defined over $\mathbf{C}$ given by an equation $y^2 = x^3+ax+b$ then there is a unique lattice $L \subset \mathbf{C}$ such that $15G_4(L) = -a$ and $35G_6(L) = -b$, so* (i) *gives an isomorphism from $\mathbf{C}/L$ to $E(\mathbf{C})$.*

(iii) *The correspondence above identifies the holomorphic differential ω_E with dz.*

Proof. The first statement is Proposition VI.3.6 of [Si] and the second is proved in [Sh] §4.2. For (iii), we have that

$$dx/2y = d(\wp(z))/\wp'(z) = dz.$$

□

Remark 2.4. If E is the elliptic curve defined over $\mathbf{C}$ with a Weierstrass model $y^2 = x^3+ax+b$ and ω_E is the differential $dx/2y$ of Proposition 1.2, then the lattice L associated to E by Theorem 2.3(ii) is

$$\left\{\int_\gamma \omega_E : \gamma \in H_1(E,\mathbf{Z})\right\}$$

and the map

$$P \mapsto \int_O^P \omega_E$$

is the isomorphism from $E(\mathbf{C})$ to $\mathbf{C}/L$ which is the inverse of the map of Theorem 2.3(i).

Definition 2.5. If $L \subset \mathbf{C}$ is a lattice define

$$\Delta(L) = (60G_4(L))^3 - 27(140G_6(L))^2$$
$$j(L) = -1728(60G_4(L))^3/\Delta(L).$$

Then $\Delta(L)$ is the discriminant and $j(L)$ the j-invariant of the elliptic curve corresponding to L by Theorem 2.3.

Proposition 2.6. *Suppose E is an elliptic curve defined over $\mathbf{C}$, corresponding to a lattice L under the bijection of Theorem 2.3. Then the map ι of Definition 1.7 is an isomorphism*

$$\mathrm{End}_{\mathbf{C}}(E) \xrightarrow{\sim} \{\alpha \in \mathbf{C} : \alpha L \subset L\}.$$

Proof. See [Si] Theorem VI.4.1. □

Corollary 2.7. *If E is an elliptic curve defined over a field F of characteristic zero, then $\mathrm{End}_F(E)$ is either $\mathbf{Z}$ or an order in an imaginary quadratic field.*

Proof. If E is defined over a subfield of $\mathbf{C}$ then Proposition 2.6 identifies $\mathrm{End}_{\mathbf{C}}(E)$ with $\{\alpha \in \mathbf{C} : \alpha L \subset L\}$. The latter object is a discrete subring of $\mathbf{C}$, and hence is either $\mathbf{Z}$ or an order in an imaginary quadratic field.

Using the principle of Remark 2.1 at the beginning of this section, the same holds for all fields F of characteristic zero. □

The following table gives a dictionary between elliptic curves over an arbitrary field and elliptic curves over $\mathbf{C}$.

over abitrary field	over $\mathbf{C}$
(E, ω_E)	$(\mathbf{C}/L, dz)$
x, y	$\wp(z; L), \wp'(z; L)/2$
isomorphism class of E	$\{\alpha L : \alpha \in \mathbf{C}^\times\}$
$\mathrm{End}_{\mathbf{C}}(E)$	$\{\alpha \in \mathbf{C} : \alpha L \subset L\}$
$\mathrm{Aut}_{\mathbf{C}}(E)$	$\{\alpha \in \mathbf{C}^\times : \alpha L = L\}$
$E[m]$	$m^{-1}L/L$

3 Elliptic Curves over Local Fields

For this section suppose

- p is a rational prime,
- F is a finite extension of $\mathbf{Q}_p$,
- $\mathcal{O}$ is the ring of integers of F,
- $\mathfrak{p}$ is the maximal ideal of F,
- π is a generator of $\mathfrak{p}$
- $\mathbf{k} = \mathcal{O}/\mathfrak{p}$ is the residue field of $\mathcal{O}$
- $v : F \to \mathbf{Z} \cup \{\infty\}$ is the valuation on F, $v(\pi) = 1$.

We fix an elliptic curve E defined over F.

3.1 Reduction

Definition 3.1. A Weierstrass equation (1) for E is *minimal* if

- $a_1, a_2, a_3, a_4, a_6 \in \mathcal{O}$,
- the valuation of the discriminant of this equation is minimal in the set of valuations of all Weierstrass equations for E with coefficients in $\mathcal{O}$.

Every elliptic curve E has a minimal Weierstrass equation, or minimal model, and the *minimal discriminant* of E is the ideal of $\mathcal{O}$ generated by the discriminant of a minimal Weierstrass model of E.

The *reduction* $\tilde{E}$ of E is the curve defined over the residue field $\mathbf{k}$ by the Weierstrass equation

$$y^2 + \tilde{a}_1 xy + \tilde{a}_3 y = x^3 + \tilde{a}_2 x^2 + \tilde{a}_4 x + \tilde{a}_6 \tag{2}$$

where the a_i are the coefficients of a minimal Weierstrass equation for E and $\tilde{a}_i$ denotes the image of a_i in $\mathbf{k}$. The reduction $\tilde{E}$ is independent (up to isomorphism) of the particular minimal equation chosen for E (see [Si] Proposition VII.1.3(b)).

The curve $\tilde{E}$ may be singular, but it has at most one singular point ([Si] Proposition III.1.4(a)). In that case the quasi-projective curve

$$\tilde{E}_{\mathrm{ns}} = \tilde{E} - \{\text{singular point on } \tilde{E}\}$$

has a geometrically-defined group law just as an elliptic curve does (see [Si] Proposition III.2.5).

If Δ is the minimal discriminant of E, then one of the following three possibilities holds (see for example [Si] Proposition III.2.5):

(i) $\Delta \in \mathcal{O}^\times$ and $\tilde{E}$ is nonsingular, i.e., $\tilde{E} = \tilde{E}_{\mathrm{ns}}$ is an elliptic curve,
(ii) $\Delta \notin \mathcal{O}^\times$, $\tilde{E}$ is singular, and $\tilde{E}_{\mathrm{ns}}(\mathbf{k}) \cong \mathbf{k}^\times$, or
(iii) $\Delta \notin \mathcal{O}^\times$, $\tilde{E}$ is singular, and $\tilde{E}_{\mathrm{ns}}(\mathbf{k}) \cong \mathbf{k}$.

We say that E has *good* (resp. *multiplicative*, resp. *additive*) *reduction* if (i) (resp. (ii), resp. (iii)) is satisfied.

We say that E has *potentially good reduction* if there is a finite extension F' of F such that E has good reduction over F'.

Lemma 3.2. (i) *E has potentially good reduction if and only if $j(E) \in \mathcal{O}$.*
(ii) *If E has potentially good reduction then E has either good or additive reduction.*

Proof. See [Si] Propositions VII.5.4 and IV.5.5. □

Definition 3.3. There is a natural *reduction map*

$$\mathbf{P}^2(F) \to \mathbf{P}^2(\mathbf{k}).$$

By restriction this defines a reduction map from $E(F)$ to $\tilde{E}(\mathbf{k})$. We define $E_0(F) \subset E(F)$ to be the inverse image of $\tilde{E}_{\mathrm{ns}}(\mathbf{k})$ and $E_1(F) \subset E(F)$ to be the inverse image of $\tilde{O} \in \tilde{E}_{\mathrm{ns}}(\mathbf{k})$.

Proposition 3.4. *There is an exact sequence of abelian groups*

$$0 \to E_1(F) \to E_0(F) \to \tilde{E}_{\mathrm{ns}}(\mathbf{k}) \to 0$$

where the map on the right is the reduction map. If E has good reduction then the reduction map induces an injective homomorphism

$$\mathrm{End}_F(E) \to \mathrm{End}_{\mathbf{k}}(\tilde{E}).$$

Proof. See [Si] Proposition VII.2.1. □

If E has good reduction and $\phi \in \mathrm{End}_F(E)$, we will write $\tilde{\phi}$ for the endomorphism of $\tilde{E}$ which is the reduction of ϕ.

Lemma 3.5. *If E is defined by a minimal Weierstrass equation then*

$$E_1(F) = \{(x,y) \in E(F) : v(x) < 0\} = \{(x,y) \in E(F) : v(y) < 0\}.$$

If $(x,y) \in E_1(F)$ then $3v(x) = 2v(y) < 0$.

Proof. It is clear from the definition of the reduction map that $(x, y, 1)$ reduces to $(0,1,0)$ if and only if $v(y) < 0$ and $v(y) < v(x)$. If $(x,y) \in E(F)$ then, since x and y satisfy a Weierstrass equation with coefficients in $\mathcal{O}$, it is clear that

$$v(x) < 0 \Leftrightarrow v(y) < 0$$

and in that case $v(y) = (3/2)v(x) < v(x)$. □

Lemma 3.6. *Suppose E has good reduction, $\phi \in \mathrm{End}_F(E)$, and $\tilde{\phi}$ is purely inseparable. Then*

(i) *$\tilde{\phi}$ is injective on $\tilde{E}(\mathbf{k})$.*
(ii) $\ker(\phi) \subset E_1(F)$

Proof. Clear. □

3.2 The Formal Group

Theorem 3.7. *Fix a minimal Weierstrass model* (1) *of E. There is a formal group $\hat{E}$ defined by a power series $\mathcal{F}_E \in \mathcal{O}[[Z, Z']]$, and a power series*

$$w(Z) = Z^3 + a_1 Z^4 + (a_1^2 + a_2)Z^5 + \cdots \in \mathcal{O}[[Z]],$$

such that if we define

$$x(Z) = Z/w(Z) \in Z^{-2}\mathcal{O}[[Z]], \qquad y(Z) = -1/w(Z) \in Z^{-3}\mathcal{O}[[Z]]$$

then

(i) $(x(Z), y(Z)) \in E(\mathcal{O}((Z)))$,

(ii) $$(x(Z), y(Z)) + (x(Z'), y(Z')) = (x(\mathcal{F}_E(Z, Z')), y(\mathcal{F}_E(Z, Z')))$$

as points on E with coordinates in the fraction field of $F((Z, Z'))$,

(iii) *there is a map* $\mathrm{End}_F(E) \to \mathrm{End}(\hat{E})$ *(which we will denote by $\phi \mapsto \phi(Z) \in \mathcal{O}[[Z]]$) such that for every $\phi \in \mathrm{End}_F(E)$,*

$$\phi((x(Z), y(Z))) = (x(\phi(Z)), y(\phi(Z)))$$

in $E(F((Z)))$.

Proof. See [Ta] or [Si], §IV.1 for an explicit construction of the power series $w(Z)$ and $\mathcal{F}_E(Z, Z)$. The idea is that $Z = -x/y$ is a uniformizing parameter at the origin of E, and everything (x, y, the group law, endomorphisms) can be expanded as power series in Z. □

For every $n \geq 1$ write $\hat{E}(\mathfrak{p}^n)$ for the commutative group whose underlying set is $\mathfrak{p}^n$, with the operation $(z, z') \mapsto \mathcal{F}_E(z, z')$.

Corollary 3.8. *With notation as in Theorem 3.7,*

$$Z \mapsto (Z/w(Z), -1/w(Z))$$

is an isomorphism from $\hat{E}(\mathfrak{p})$ to $E_1(F)$ with inverse given by

$$(x, y) \mapsto -x/y.$$

Proof. See [Si] Proposition VII.2.2. The first map is a map into $E_1(F)$ by Lemma 3.5 and Theorem 3.7(i), and is a homomorphism by Theorem 3.7(ii). It is injective because the only zero of $w(Z)$ in $\mathfrak{p}$ is $Z = 0$. The second map is clearly a left-inverse of the first, and it maps into $\mathfrak{p}$ by Lemma 3.5. We only need show that the second map is also one-to-one.

If we rewrite our Weierstrass equation for E with variables $w = -1/y$ and $z = -x/y$ we get a new equation

$$a_6 w^3 + (a_4 z + a_3) w^2 + (a_2 z^2 + a_1 z - 1) w + z^3 = 0.$$

Fix a value of $z \in \mathfrak{p}$ and consider the set S of roots w of this equation. If (z, w) corresponds to a point in $E_1(F)$ then by Lemma 3.5, $v(w) = v(z^3) > 0$. It follows easily that S contains at most one root w corresponding to a point of $E_1(F)$, and hence the map $(x, y) \mapsto -x/y$ is one-to-one on $E_1(F)$. □

Corollary 3.9. *Suppose $\#(\mathbf{k}) = q$, E has good reduction, and $\phi \in \mathrm{End}_K(E)$ reduces to the Frobenius endomorphism $\varphi_q \in \mathrm{End}_{\mathbf{k}}(\tilde{E})$. Then*

$$\phi(Z) \equiv Z^q \pmod{\mathfrak{p}\mathcal{O}[[Z]]}.$$

Proof. If the reduction of ϕ is φ_q then by Theorem 3.7(iii)

$$\begin{aligned}(x(\phi(Z)), y(\phi(Z))) = \phi((x(Z), y(Z))) &\equiv (x(Z)^q, y(Z)^q) \\ &\equiv (x(Z^q), y(Z^q)) \pmod{\mathfrak{p}\mathcal{O}((Z))}.\end{aligned}$$

Since $y(Z)$ is invertible in $\mathcal{O}((Z))$, we conclude that

$$\phi(Z) = -x(\phi(Z))/y(\phi(Z)) \equiv -x(Z^q)/y(Z^q) = Z^q \pmod{\mathfrak{p}\mathcal{O}((Z))}.$$

□

Definition 3.10. Recall that

$$\omega_E = \frac{dx}{2y + a_1x + a_3} = \frac{dy}{3x^2 + 2a_2x + a_4 - a_1y}$$

is the holomorphic, translation-invariant differential on E from Proposition 1.2. Define

$$\hat{\omega}(Z) = \frac{\frac{d}{dZ}x(Z)}{2y(Z) + a_1x(Z) + a_3} \in 1 + Z\mathcal{O}[[Z]].$$

Let $\lambda_{\hat{E}}(Z)$ be the unique element of $Z + Z^2F[[Z]]$ such that $\frac{d}{dZ}\lambda_{\hat{E}}(Z) = \hat{\omega}(Z)$.

Lemma 3.11. (i) *The power series $\lambda_{\hat{E}}$ is the logarithm map of $\hat{E}$, the isomorphism from $\hat{E}$ to the additive formal group $\mathbf{G}_a$ such that $\lambda'_E(0) = 1$.*
(ii) *The power series $\lambda_{\hat{E}}$ converges on $\mathfrak{p}$. If $\mathrm{ord}_{\mathfrak{p}}(p) < p - 1$ then $\lambda_{\hat{E}}$ is an isomorphism from $\hat{E}(\mathfrak{p})$ to the additive group $\mathfrak{p}$.*

Proof. Let $\mathcal{F}_E \in \mathcal{O}[[Z, Z']]$ be the addition law for $\hat{E}$. We need to show that

$$\lambda_E(\mathcal{F}_E(Z, Z')) = \lambda_E(Z) + \lambda_E(Z').$$

Since ω_E is translation invariant (Proposition 1.2),

$$\hat{\omega}(\mathcal{F}_E(Z, Z'))d(\mathcal{F}_E(Z, Z')) = \hat{\omega}(Z)dZ$$

and therefore

$$\tfrac{d}{dZ}\lambda_E(\mathcal{F}_E(Z, Z')) = \tfrac{d}{dZ}\lambda_E(Z).$$

Therefore $\lambda_E(\mathcal{F}_E(Z, Z')) = \lambda_E(Z) + c(Z')$ with $c(Z') \in F[[Z']]$. Evaluating at $Z = 0$ shows $c(Z') = \lambda_E(Z')$ as desired.

The uniqueness of the logarithm map and (ii) are standard elementary results in the theory of formal groups. □

Definition 3.12. Define $\lambda_E : E_1(F) \to F$ to be the composition of the inverse of the isomorphism of Corollary 3.8 with $\lambda_{\hat{E}}$.

Corollary 3.13. *If $\mathrm{ord}_{\mathfrak{p}}(p) < p-1$ then $\lambda_E : E_1(F) \to \mathfrak{p}$ is an isomorphism.*

Proof. This is immediate from Lemma 3.11. □

Recall the map $\iota : \mathrm{End}_F(E) \to F$ of Definition 1.7 defined by the action of an endomorphism on holomorphic differentials.

Proposition 3.14. *For every $\phi \in \mathrm{End}_F(E)$, $\phi(Z) = \iota(\phi)Z + O(Z^2)$.*

Proof. By definition of ι, $\hat{\omega}(\phi(Z)) = \iota(\phi)\hat{\omega}(Z)$, i.e.,

$$\frac{d(x(\phi(Z)))}{2y(\phi(Z)) + a_1 x(\phi(Z)) + a_3} = \iota(\phi)\frac{d(x(Z))}{2y(Z) + a_1 x(Z) + a_3}.$$

Using the definitions of $x(Z)$ and $y(Z)$, the right-hand side is $(\iota(\phi)+O(Z))dZ$, and the left-hand side is $(\phi'(0) + O(Z))dZ$. This completes the proof. □

3.3 Applications to Torsion Subgroups

Theorem 3.15. *Suppose $\phi \in \mathrm{End}_F(E)$ and $\iota(\phi) \in \mathcal{O}^\times$.*

(i) *ϕ is an automorphism of $E_1(F)$.*
(ii) *If E has good reduction then the reduction map $E[\phi] \cap E(F) \to \tilde{E}(\mathbf{k})$ is injective.*

Proof. By definition of a formal group, $\mathcal{F}_E(X,Y) = X+Y+O(X^2,XY,Y^2)$. Using Proposition 3.14, for every $n \geq 1$ we have a commutative diagram

$$\begin{array}{ccccc}
\hat{E}(\mathfrak{p}^n)/\hat{E}(\mathfrak{p}^{n+1}) & \xrightarrow{\sim} & \mathfrak{p}^n/\mathfrak{p}^{n+1} & \xrightarrow{\sim} & \mathbf{k} \\
\phi\big\downarrow & & \iota(\phi)\big\downarrow & & \iota(\phi)\big\downarrow \\
\hat{E}(\mathfrak{p}^n)/\hat{E}(\mathfrak{p}^{n+1}) & \xrightarrow{\sim} & \mathfrak{p}^n/\mathfrak{p}^{n+1} & \xrightarrow{\sim} & \mathbf{k}
\end{array}$$

Since $\iota(\phi) \in \mathcal{O}^\times$ we see that ϕ is an automorphism of $\hat{E}(\mathfrak{p}^n)/\hat{E}(\mathfrak{p}^{n+1})$ for every $n \geq 1$, and from this it is not difficult to show that ϕ is an automorphism of $\hat{E}(\mathfrak{p})$. Therefore by Corollary 3.8, ϕ is an automorphism of $E_1(F)$. This proves (i), and (ii) as well since $E_1(F)$ is the kernel of the reduction map and (i) shows that $E_1(F) \cap E[\phi] = 0$. □

Remark 3.16. Theorem 3.15 shows in particular that if E has good reduction and m is prime to p, then the reduction map $E[m] \to \tilde{E}[m]$ is injective.

Corollary 3.17. *Suppose E has good reduction, $\phi \in \mathrm{End}_F(E)$, and $\iota(\phi) \in \mathcal{O}^\times$. If $P \in E(\bar{F})$ and $\phi(P) \in E(F)$, then $F(E[\phi], P)/F$ is unramified.*

Proof. Let $F' = F(E[\phi], P)$ and let $\mathbf{k}'$ be its residue field. Then F'/F is Galois and we let $I \subset \mathrm{Gal}(F'/F)$ denote the inertia group.

Suppose $\sigma \in I$. Then the reduction $\tilde{\sigma}$ of σ is the identity on $\mathbf{k}'$, so if $R \in E(\bar{F})$ and $\phi(R) \in E(F)$ then $\sigma R - R \in E[\phi]$ and

$$\widetilde{\sigma R - R} = \tilde{\sigma}\tilde{R} - \tilde{R} = 0.$$

By Theorem 3.15(ii), since $\iota(\phi) \in \mathcal{O}^\times$ we conclude that $\sigma R = R$. In other words σ fixes $E[\phi]$ and P, so σ fixes F', i.e., $\sigma = 1$. Hence I is trivial and F'/F is unramified. □

Corollary 3.18. *Suppose $\ell \neq p$, and let I denote the inertia subgroup of G_F.*

(i) *If E has good reduction then I acts trivially on $T_\ell(E)$.*
(ii) *If E has potentially good reduction then I acts on $T_\ell(E)$ through a finite quotient.*

Proof. This is clear by Corollary 3.17. □

The converse of Corollary 3.18 is the following.

Theorem 3.19 (Criterion of Néron-Ogg-Shafarevich). *Let $I \subset G_F$ denote the inertia group.*

(i) *If $\ell \neq p$ and I acts trivially on $T_\ell(E)$, then E has good reduction.*
(ii) *If $\ell \neq p$ and $T_\ell(E)^I \neq 0$, then E has good or multiplicative reduction.*

Proof. See [Si] Theorem VII.7.1 for (i). The proof of (ii) is the same except that we use the fact that if E has additive reduction, then over any unramified extension F' of F with residue field $\mathbf{k}'$, $\tilde{E}_{\mathrm{ns}}(\mathbf{k}')$ is killed by p and hence has no points of order ℓ. □

4 Elliptic Curves over Number Fields

For this section suppose F is a number field and E is an elliptic curve defined over F. Our main interest is in studying the Mordell-Weil group $E(F)$.

If $\mathfrak{q}$ is a prime of F we say that E has *good* (resp. *potentially good, bad, additive, multiplicative*) *reduction* at $\mathfrak{q}$ if E, viewed as an elliptic curve over the local field $F_\mathfrak{q}$ (F completed at $\mathfrak{q}$) does. We will write $\Delta(E)$ for the minimal discriminant of E, the ideal of F which is the product over all primes $\mathfrak{q}$ of the minimal discriminant of E over $F_\mathfrak{q}$. This is well-defined because (every Weierstrass model of) E has good reduction outside of a finite set of primes.

Since F has characteristic zero, the map $\iota : \mathrm{End}_F(E) \to F$ of Definition 1.7 (giving the action of $\mathrm{End}_F(E)$ on differentials) is injective, and from now on we will *identify* $\mathrm{End}_F(E)$ with its image $\mathcal{O} \subset F$. By Corollary 2.7, $\mathcal{O}$ is either $\mathbf{Z}$ or an order in an imaginary quadratic field.

If $\alpha \in \mathcal{O}$ we will also write α for the corresponding endomorphism of E, so $E[\alpha] \subset E(\bar{F})$ is the kernel of α and $F(E[\alpha])$ is the extension of F generated by the coordinates of the points in $E[\alpha]$.

Definition 4.1. Suppose $\alpha \in \mathcal{O}$, $\alpha \neq 0$. Multiplication by α is surjective on $E(\bar{F})$, so there is an exact sequence

$$0 \to E[\alpha] \to E(\bar{F}) \xrightarrow{\alpha} E(\bar{F}) \to 0.$$

Taking G_F-cohomology yields a long exact sequence

$$E(F) \xrightarrow{\alpha} E(F) \to H^1(F, E[\alpha]) \to H^1(F, E) \xrightarrow{\alpha} H^1(F, E)$$

where $H^1(F, E) = H^1(F, E(\bar{F}))$. We can rewrite this as

$$0 \to E(F)/\alpha E(F) \to H^1(F, E[\alpha]) \to H^1(F, E)_\alpha \to 0 \tag{3}$$

where $H^1(F, E)_\alpha$ denotes the kernel of α on $H^1(F, E)$. Concretely, the connecting map $E(F)/\alpha E(F) \hookrightarrow H^1(F, E[\alpha])$ is the "Kummer theory" map defined by

$$P \mapsto (\sigma \mapsto \sigma Q - Q) \tag{4}$$

where $Q \in E(\bar{F})$ satisfies $\alpha Q = P$.

In exactly the same way, if $\mathfrak{q}$ is a prime (finite or infinite) of F we can replace F by the completion $F_\mathfrak{q}$ in (3), and this leads to the diagram

$$\begin{array}{ccccc} 0 \to E(F)/\alpha E(F) & \longrightarrow & H^1(F, E[\alpha]) & \longrightarrow & H^1(F, E)_\alpha \to 0 \\ \downarrow & & \downarrow \scriptstyle{\mathrm{res}_\mathfrak{q}} & & \downarrow \scriptstyle{\mathrm{res}_\mathfrak{q}} \\ 0 \to E(F_\mathfrak{q})/\alpha E(F_\mathfrak{q}) & \longrightarrow & H^1(F_\mathfrak{q}, E[\alpha]) & \longrightarrow & H^1(F_\mathfrak{q}, E)_\alpha \to 0. \end{array} \tag{5}$$

We define the *Selmer group* (relative to α)

$$\mathcal{S}_\alpha(E) = \mathcal{S}_\alpha(E_{/F}) \subset H^1(F, E[\alpha])$$

by

$$\begin{aligned} \mathcal{S}_\alpha(E) &= \{c \in H^1(F, E[\alpha]) : \mathrm{res}_\mathfrak{q}(c) \in \mathrm{image}(E(F_\mathfrak{q})/\alpha E(F_\mathfrak{q})) \text{ for every } \mathfrak{q}\} \\ &= \{c \in H^1(F, E[\alpha]) : \mathrm{res}_\mathfrak{q}(c) = 0 \text{ in } H^1(F_\mathfrak{q}, E) \text{ for every } \mathfrak{q}\}. \end{aligned}$$

Proposition 4.2. *Suppose $\alpha \in \mathcal{O}$, $\alpha \neq 0$. Under the Kummer map* (3), *$\mathcal{S}_\alpha(E)$ contains the image of $E(F)/\alpha E(F)$.*

Proof. Clear. □

Remark 4.3. One should think of the Selmer group $\mathcal{S}_\alpha(E)$ as the smallest subgroup of $H^1(F, E[\alpha])$ defined by natural local conditions which contains the image of $E(F)/\alpha E(F)$.

Proposition 4.4. *Suppose $\alpha \in \mathcal{O}$, $\alpha \neq 0$. Then the Selmer group $\mathcal{S}_\alpha(E)$ is finite.*

Proof. Suppose first that $E[\alpha] \subset E(F)$, so $H^1(F, E[\alpha]) = \mathrm{Hom}(G_F, E[\alpha])$. Let L be the maximal abelian extension of F of exponent $\deg(\alpha)$ which is unramified outside of the (finite) set of primes

$$\Sigma_{E,\alpha} = \{\mathfrak{p} \text{ of } F : \mathfrak{p} \text{ divides } \alpha\Delta(E) \text{ or } \mathfrak{p} \text{ is infinite}\}.$$

If $c \in \mathcal{S}_\alpha(E) \subset \mathrm{Hom}(G_F, E[\alpha])$ then c is trivial on

- commutators,
- $\deg(\alpha)$-th powers,
- inertia groups of primes outside of $\Sigma_{E,\alpha}$,

the first two because $E[\alpha]$ is abelian and annihilated by $\deg(\alpha)$, and the last because of (4) and Corollary 3.17. Therefore c factors through $\mathrm{Gal}(L/F)$, so

$$\mathcal{S}_\alpha(E) \subset \mathrm{Hom}(\mathrm{Gal}(L/F), E[\alpha]).$$

Class field theory shows that L/F is finite, so this proves the proposition in this case.

In general, the restriction map

$$0 \to H^1(F(E[\alpha])/F, E[\alpha]) \to H^1(F, E[\alpha]) \xrightarrow{\mathrm{res}} H^1(F(E[\alpha]), E[\alpha])$$

sends $\mathcal{S}_\alpha(E_{/F})$ into $\mathcal{S}_\alpha(E_{/F(E[\alpha])})$. The case above shows that $\mathcal{S}_\alpha(E_{/F(E[\alpha])})$ is finite, and $H^1(F(E[\alpha])/F, E[\alpha])$ is finite, so $\mathcal{S}_\alpha(E_{/F})$ is finite. □

Corollary 4.5 (Weak Mordell-Weil Theorem). *For every nonzero $\alpha \in \mathcal{O}$, $E(F)/\alpha E(F)$ is finite.*

Proof. This is clear from Propositions 4.2 and 4.4. □

Theorem 4.6 (Mordell-Weil). *$E(F)$ is finitely generated.*

Proof. See [Si] §VIII.6. □

Definition 4.7. The *Tate-Shafarevich group* $\text{Ш}(E)$ of E over F is the subgroup of $H^1(F, E(\bar{F}))$ defined by

$$\text{Ш}(E) = \ker\left(H^1(F, E(\bar{F})) \to \prod_{v \text{ of } F} H^1(F_v, E)\right).$$

Proposition 4.8. *If $\alpha \in \mathcal{O}$, $\alpha \neq 0$, then the exact sequence* (3) *restricts to an exact sequence*

$$0 \to E(F)/\alpha E(F) \to \mathcal{S}_\alpha(E) \to \text{Ш}(E)_\alpha \to 0$$

where $\text{Ш}(E)_\alpha$ is the subgroup of elements of $\text{Ш}(E)$ killed by α.

Proof. This is clear from the definitions and the diagram (5). □

5 Elliptic Curves with Complex Multiplication

Fix a subfield F of $\mathbf{C}$ and an elliptic curve E defined over F.

Definition 5.1. We say E has *complex multiplication* over F if $\mathrm{End}_F(E)$ is an order in an imaginary quadratic field, i.e., if $\mathrm{End}_F(E) \neq \mathbf{Z}$.

Assume from now on that E has complex multiplication, and let

$$\mathcal{O} = \iota(\mathrm{End}_F(E)) \subset F.$$

As in §4 we will use ι to *identify* $\mathrm{End}_F(E)$ with $\mathcal{O}$. Let $K = \mathbf{Q}\mathcal{O} \subset F$ be the imaginary quadratic field containing $\mathcal{O}$, and denote the full ring of integers of K by $\mathcal{O}_K$. If $\mathfrak{a}$ is an ideal of $\mathcal{O}$ we will write $E[\mathfrak{a}] = \cap_{\alpha\in\mathfrak{a}} E[\alpha]$.

Fix an embedding of F into $\mathbf{C}$. Viewing E as an elliptic curve over $\mathbf{C}$ and using Proposition 2.6 we can write

$$E(\mathbf{C}) \cong \mathbf{C}/L \text{ where } L \subset K \subset \mathbf{C} \text{ and } \mathcal{O}L = L. \tag{6}$$

(A priori L is just a lattice in $\mathbf{C}$, but replacing L by λL where $\lambda^{-1} \in L$ we may assume that $L \subset K$.) Thus if $\mathcal{O} = \mathcal{O}_K$, then L is a fractional ideal of K.

5.1 Preliminaries

In this section we record the basic consequences of complex multiplication. Put most simply, if E has complex multiplication over F then all torsion points in $E(\bar{F})$ are defined over abelian extensions of F.

Remark 5.2. It will simplify the exposition to assume that $\mathcal{O} = \mathcal{O}_K$. The following proposition shows that this restriction is not too severe. Two elliptic curves are ***isogenous*** if there is an isogeny (a nonzero morphism sending one origin to the other) from one to the other.

Proposition 5.3. *There is an elliptic curve E', defined over F and isogenous over F to E, such that* $\mathrm{End}_F(E) \cong \mathcal{O}_K$.

Proof. Suppose the conductor of $\mathcal{O}$ is c, i.e., $\mathcal{O} = \mathbf{Z}+c\mathcal{O}_K$, and let $\mathfrak{c} = c\mathcal{O}_K \subset \mathcal{O}$. The subgroup $E[\mathfrak{c}]$ is stable under G_F, so by [Si] Proposition III.4.12 and Exercise III.3.13 there is an elliptic curve E' over F and an isogeny from E to E' with kernel $E[\mathfrak{c}]$. We only need to check that $\mathrm{End}_F(E') = \mathcal{O}_K$.

With the identification (6), $E'(\mathbf{C}) \cong \mathbf{C}/L'$ where

$$L' = \{z \in \mathbf{C} : z\mathfrak{c} \subset L\}.$$

Suppose $\alpha \in \mathcal{O}_K$. For every $z \in L'$,

$$(\alpha z)\mathfrak{c} = z(\alpha\mathfrak{c}) \in z\mathfrak{c} \subset L$$

so $\alpha z \in L'$. Therefore by Proposition 2.6, $\alpha \in \mathrm{End}_{\mathbf{C}}(E')$. By Lemma 1.8, since $\alpha \in K \subset F$ we conclude that $\alpha \in \mathrm{End}_F(E')$. □

From now on we will assume that $\mathcal{O}$ is the maximal order $\mathcal{O}_K$.

Proposition 5.4. *If $\mathfrak{a}$ is a nonzero ideal of $\mathcal{O}$ then $E[\mathfrak{a}] \cong \mathcal{O}/\mathfrak{a}$ as $\mathcal{O}$-modules.*

Proof. Using the identification (6) we see that $E[\mathfrak{a}] \cong \mathfrak{a}^{-1}L/L$ where L is a fractional ideal of K, and then $\mathfrak{a}^{-1}L/L \cong \mathcal{O}/\mathfrak{a}$. □

Corollary 5.5. *If $\mathfrak{a}$ is a nonzero ideal of $\mathcal{O}$ then the action of G_F on $E[\mathfrak{a}]$ induces an injection*

$$\mathrm{Gal}(F(E[\mathfrak{a}])/F) \hookrightarrow (\mathcal{O}/\mathfrak{a})^\times.$$

In particular $F(E[\mathfrak{a}])/F$ is abelian.

Proof. If $\beta \in \mathcal{O}$, $\sigma \in G_F$, and $P \in E(\bar{F})$ then, since the endomorphism β is defined over F, $\sigma(\beta P) = \beta(\sigma P)$. Thus there is a map

$$\mathrm{Gal}(F(E[\mathfrak{a}])/F) \hookrightarrow \mathrm{Aut}_{\mathcal{O}}(E[\mathfrak{a}]).$$

By Proposition 5.4,

$$\mathrm{Aut}_{\mathcal{O}}(E[\mathfrak{a}]) \cong \mathrm{Aut}_{\mathcal{O}}(\mathcal{O}/\mathfrak{a}) = (\mathcal{O}/\mathfrak{a})^\times.$$

□

If $\mathfrak{a}$ is a nonzero ideal of $\mathcal{O}$ let $E[\mathfrak{a}^\infty] = \cup_n E[\mathfrak{a}^n]$.

Corollary 5.6. *The action of G_F on $E[\mathfrak{a}^\infty]$ induces an injection*

$$\mathrm{Gal}(F(E[\mathfrak{a}^\infty])/F) \hookrightarrow (\varprojlim_n \mathcal{O}/\mathfrak{a}^n)^\times.$$

In particular for every prime p,

$$\mathrm{Gal}(F(E[p^\infty])/F) \hookrightarrow (\mathcal{O} \otimes \mathbf{Z}_p)^\times.$$

Proof. Immediate from Corollary 5.5. □

Theorem 5.7. *Suppose F is a finite extension of $\mathbf{Q}_\ell$ for some ℓ.*

(i) *E has potentially good reduction.*
(ii) *Suppose $\mathfrak{p}$ is a prime of $\mathcal{O}$ and $n \in \mathbf{Z}^+$ is such that the multiplicative group $1 + \mathfrak{p}^n\mathcal{O}_\mathfrak{p}$ is torsion-free (where $\mathcal{O}_\mathfrak{p}$ is the completion of $\mathcal{O}$ at $\mathfrak{p}$). If $\mathfrak{p} \nmid \ell$ then E has good reduction over $F(E[\mathfrak{p}^n])$ at all primes not dividing $\mathfrak{p}$.*

Proof. Suppose p is a rational prime. By Corollary 5.6, the Galois group $\mathrm{Gal}(F(E[p^\infty])/F(E[p]))$ is isomorphic to a subgroup of the multiplicative group $1 + p\mathcal{O} \otimes \mathbf{Z}_p$. If $p > 3$ then the p-adic logarithm map shows that $1 + p\mathcal{O} \otimes \mathbf{Z}_p \cong p\mathcal{O}_\mathfrak{p} \cong \mathbf{Z}_p^2$. Thus

$$\mathrm{Gal}(F(E[p^\infty])/F(E[p])) \cong \mathbf{Z}_p^d$$

with $d \leq 2$. If $p \neq \ell$, class field theory shows that such an extension is unramified. Thus by the criterion of Néron-Ogg-Shafarevich (Theorem 3.19(i)) E has good reduction over $F(E[p])$. This proves (i).

The proof of (ii) is similar. Write $F_\infty = F(E[\mathfrak{p}^\infty])$ and $F_n = F(E[\mathfrak{p}^n])$, and suppose $\mathfrak{q}$ is a prime of F_n not dividing $\mathfrak{p}$. By (i) and Corollary 3.17, the inertia group $I_\mathfrak{q}$ of $\mathfrak{q}$ in $\mathrm{Gal}(F_\infty/F_n)$ is finite. But Corollary 5.6 shows that

$$\mathrm{Gal}(F_\infty/F_n) \subset 1 + \mathfrak{p}^n\mathcal{O}_\mathfrak{p},$$

which has no finite subgroups, so $I_\mathfrak{q}$ acts trivially on $E[\mathfrak{p}^\infty]$. Therefore by Theorem 3.19(ii), E has good or multiplicative reduction at $\mathfrak{q}$. Since we already know that the reduction is potentially good, Lemma 3.2(ii) allows us to conclude that E has good reduction at $\mathfrak{q}$. □

Remark 5.8. The hypothesis of Theorem 5.7(ii) is satisfied with $n = 1$ if the residue characteristic of $\mathfrak{p}$ is greater than 3.

Proposition 5.9. *Suppose $\mathfrak{q}$ is a prime of F where E has good reduction and $q = \mathbf{N}_{F/\mathbf{Q}}\mathfrak{q}$. There is an endomorphism $\alpha \in \mathcal{O}$ whose reduction modulo $\mathfrak{q}$ is the Frobenius endomorphism φ_q of $\tilde{E}$.*

Proof. If $\varphi_q = [m]$ for some $m \in \mathbf{Z}$ then the proposition is clear. So suppose now that $\varphi_q \notin \mathbf{Z}$, and write $\mathbf{k}$ for the residue field of F at $\mathfrak{q}$. Since φ_q commutes with every endomorphism of $\tilde{E}$, we see from Theorem 1.11 that the only possibility is that $\mathrm{End}_\mathbf{k}(\tilde{E})$ is an order in an imaginary quadratic field. But the reduction map $\mathrm{End}_F(E) \to \mathrm{End}_\mathbf{k}(\tilde{E})$ is injective (Proposition 3.4) so its image, the maximal order of K, must be all of $\mathrm{End}_\mathbf{k}(\tilde{E})$. This proves the proposition. □

5.2 The Main Theorem of Complex Multiplication

In this section we study further the action of G_F on torsion points of E. We will see that not only are torsion points abelian over F, in fact they are "almost" abelian over K, so that (using class field theory) we can describe the action of G_F on torsion points in terms of an action of the ideles of K.

The reference for this section is [Sh] Chapter 5; see also [ST]. We continue to suppose that E has complex multiplication by the *maximal* order of K.

Definition 5.10. Let $\mathbf{A}_K^\times$ denote the group of ideles of K. There is a natural map from $\mathbf{A}_K^\times$ to the group of fractional ideals of K, and if $x \in \mathbf{A}_K^\times$ and $\mathfrak{a}$ is a fractional ideal of K we will write $x\mathfrak{a}$ for the product of $\mathfrak{a}$ and the fractional ideal corresponding to x.

If $\mathfrak{p}$ is a prime of K let $\mathcal{O}_\mathfrak{p} \subset K_\mathfrak{p}$ denote the completions of $\mathcal{O}$ and K at $\mathfrak{p}$. If $\mathfrak{a}$ is a fractional ideal of K, write $\mathfrak{a}_\mathfrak{p} = \mathfrak{a}\mathcal{O}_\mathfrak{p}$ and then

$$K/\mathfrak{a} = \mathfrak{a} \otimes (K/\mathcal{O}) = \mathfrak{a} \otimes (\oplus_\mathfrak{p}(K_\mathfrak{p}/\mathcal{O}_\mathfrak{p})) = \oplus_\mathfrak{p} K_\mathfrak{p}/\mathfrak{a}_\mathfrak{p}. \tag{7}$$

If $x = (x_\mathfrak{p}) \in \mathbf{A}_K^\times$ then multiplication by $x_\mathfrak{p}$ gives an isomorphism from $K_\mathfrak{p}/\mathfrak{a}_\mathfrak{p}$ to $K_\mathfrak{p}/x_\mathfrak{p}\mathfrak{a}_\mathfrak{p} = K_\mathfrak{p}/(x\mathfrak{a})_\mathfrak{p}$, so putting these maps together in (7) we get an isomorphism

$$x : K/\mathfrak{a} \xrightarrow{\sim} K/x\mathfrak{a}.$$

The following theorem is Theorem 5.4 in Shimura's book [Sh]. Let K^{ab} denote the maximal abelian extension of K and $[\,\cdot\,, K^{\mathrm{ab}}/K]$ the Artin map of global class field theory. If σ is an automorphism of $\mathbf{C}$ let E^σ denote the elliptic curve obtained by applying σ to the coefficients of an equation for E.

Theorem 5.11 (Main theorem of complex multiplication). *Fix a fractional ideal $\mathfrak{a}$ of K and an analytic isomorphism*

$$\xi : \mathbf{C}/\mathfrak{a} \to E(\mathbf{C})$$

as in (6). Suppose $\sigma \in \mathrm{Aut}(\mathbf{C}/K)$ and $x \in \mathbf{A}_K^\times$ satisfies $[x, K^{\mathrm{ab}}/K] = \sigma|_{K^{\mathrm{ab}}}$. Then there is a unique isomorphism $\xi' : \mathbf{C}/x^{-1}\mathfrak{a} \to E^\sigma(\mathbf{C})$ such that the following diagram commutes

$$\begin{array}{ccc} K/\mathfrak{a} & \xrightarrow{\ \xi\ } & E_{\mathrm{tors}} \\ {\scriptstyle x^{-1}}\big\downarrow & & \big\downarrow{\scriptstyle \sigma} \\ K/x^{-1}\mathfrak{a} & \xrightarrow{\ \xi'\ } & E^\sigma_{\mathrm{tors}} \end{array}$$

where E_{tors} denotes the torsion in $E(\mathbf{C})$ and similarly for E^σ_{tors}.

Proof. See [Sh] Theorem 5.4. □

Let H denote the Hilbert class field H of K.

Corollary 5.12. (i) $K(j(E)) = H \subset F$,
(ii) $j(E)$ *is an integer of* H.

Proof. Suppose $\sigma \in \mathrm{Aut}(\mathbf{C}/K)$. With the notation of Theorem 5.11, as in Proposition 2.6 we see that

$$\begin{aligned} j(E) = j(E)^\sigma &\Leftrightarrow E \cong E^\sigma \Leftrightarrow \mathbf{C}/\mathfrak{a} \cong \mathbf{C}/x\mathfrak{a} \Leftrightarrow x\mathfrak{a} = \lambda\mathfrak{a} \text{ for some } \lambda \in \mathbf{C} \\ &\Leftrightarrow x \in K^\times \prod_{\mathfrak{p}\nmid\infty} \mathcal{O}_\mathfrak{p}^\times \prod_{\mathfrak{p}\mid\infty} K_\mathfrak{p}^\times \Leftrightarrow [x, H/K] = 1 \Leftrightarrow \sigma \text{ is the identity on } H. \end{aligned}$$

This proves (i), and (ii) follows from Theorem 5.7(i) and Lemma 3.2(i). □

Corollary 5.13. *There is an elliptic curve defined over H with endomorphism ring $\mathcal{O} = \mathcal{O}_K$.*

Proof. By Theorem 2.3(i) there is an elliptic curve E' defined over $\mathbf{C}$ with $E'(\mathbf{C}) \cong \mathbf{C}/\mathcal{O}$, and by Proposition 2.6, $\mathrm{End}_{\mathbf{C}}(E') \cong \mathcal{O}$. Corollary 5.12 shows that $j(E') \in H$, so (see Proposition III.1.4 of [Si]) there is an elliptic curve E defined over H with $j(E) = j(E')$. Hence E is isomorphic over $\mathbf{C}$ to E', so $\mathrm{End}_{\mathbf{C}}(E) \cong \mathcal{O}$.

The map $\iota : \mathrm{End}_{\mathbf{C}}(E) \to \mathbf{C}$ of Definition 1.7 is injective, so the image is $\mathcal{O} \subset H$. By Lemma 1.8 we conclude that $\mathrm{End}_{\mathbf{C}}(E) = \mathrm{End}_H(E)$. Thus E has the desired properties. □

Exercise 5.14. Let A be the ideal class group of K. If $E \cong \mathbf{C}/\mathfrak{a}$, $\mathfrak{b}$ is an ideal of K, $\sigma_{\mathfrak{b}}$ is its image under the isomorphism $A_K \xrightarrow{\sim} \mathrm{Gal}(H/K)$, and $\sigma \in G_K$ restricts to $\sigma_{\mathfrak{b}}$ on H, then

$$E^{\sigma}(\mathbf{C}) \cong \mathbf{C}/\mathfrak{b}^{-1}\mathfrak{a}.$$

For the rest of this section we suppose that F is a number field.

Theorem 5.15. *There is a Hecke character*

$$\psi = \psi_E : \mathbf{A}_F^{\times}/F^{\times} \to \mathbf{C}^{\times}$$

with the following properties.

(i) *If $x \in \mathbf{A}_F^{\times}$ and $y = \mathbf{N}_{F/K}x \in \mathbf{A}_K^{\times}$, then*

$$\psi(x)\mathcal{O} = y_{\infty}^{-1}(y\mathcal{O}) \subset \mathbf{C}.$$

(ii) *If $x \in \mathbf{A}_F^{\times}$ is a finite idele (i.e., the archimedean component is 1) and $\mathfrak{p}$ is a prime of K, then $\psi(x)(\mathbf{N}_{F/K}x)_{\mathfrak{p}}^{-1} \in \mathcal{O}_{\mathfrak{p}}^{\times}$ and for every $P \in E[\mathfrak{p}^{\infty}]$*

$$[x, F^{\mathrm{ab}}/F]P = \psi(x)(\mathbf{N}_{F/K}x)_{\mathfrak{p}}^{-1}P.$$

(iii) *If $\mathfrak{q}$ is a prime of F and $U_{\mathfrak{q}}$ denotes the local units in the completion of F at $\mathfrak{q}$, then*

$$\psi(U_{\mathfrak{q}}) = 1 \Leftrightarrow E \textit{ has good reduction at } \mathfrak{q}.$$

Proof. Suppose $x \in \mathbf{A}_F^{\times}$, and let $y = \mathbf{N}_{F/K}x$, $\sigma = [x, F^{\mathrm{ab}}/F]$. Then σ restricted to K^{ab} is $[y, K^{\mathrm{ab}}/K]$ so we can apply Theorem 5.11 with σ and y. Since σ fixes F, $E^{\sigma} = E$ so Theorem 5.11 gives a diagram with isomorphisms $\xi : \mathbf{C}/\mathfrak{a} \to E(\mathbf{C})$ and $\xi' : \mathbf{C}/y^{-1}\mathfrak{a} \to E(\mathbf{C})$. Then $\xi^{-1} \circ \xi' : \mathbf{C}/y^{-1}\mathfrak{a} \xrightarrow{\sim} \mathbf{C}/\mathfrak{a}$ is an isomorphism, so it must be multiplication by an element $\psi_{\mathrm{fin}}(x) \in K^{\times}$ satisfying $\psi_{\mathrm{fin}}(x)\mathcal{O} = y\mathcal{O}$. Define

$$\psi(x) = y_{\infty}^{-1}\psi_{\mathrm{fin}}(x).$$

It is clear that $\psi : \mathbf{A}_F^\times/F^\times \to \mathbf{C}^\times$ is a homomorphism and that (i) is satisfied. If $\mathfrak{p}$ is a prime of K and $k > 0$ then Theorem 5.11 gives a diagram

$$\begin{array}{ccccc} \mathfrak{p}^{-k}\mathfrak{a}_\mathfrak{p}/\mathfrak{a}_\mathfrak{p} & \xrightarrow{\sim} & \mathfrak{p}^{-k}\mathfrak{a}/\mathfrak{a} & \xrightarrow{\xi} & E[\mathfrak{p}^k] \\ {\scriptstyle y_\mathfrak{p}^{-1}}\downarrow & & {\scriptstyle y^{-1}}\downarrow & & \downarrow{\scriptstyle \sigma} \\ \mathfrak{p}^{-k}y_\mathfrak{p}^{-1}\mathfrak{a}_\mathfrak{p}/y_\mathfrak{p}^{-1}\mathfrak{a}_\mathfrak{p} & \xrightarrow{\sim} & \mathfrak{p}^{-k}y^{-1}\mathfrak{a}/y^{-1}\mathfrak{a} & \xrightarrow{\psi_{\mathrm{fin}}(x)\xi} & E[\mathfrak{p}^k] \end{array}$$

(where the left-hand square comes from the definition of the action of y on $K/\mathfrak{a}$) which proves (ii).

Suppose $\mathfrak{q}$ is a prime of F and p is a rational prime not lying below $\mathfrak{q}$. By (ii), if $u \in U_\mathfrak{q}$ then $[u, F^{\mathrm{ab}}/F]$ acts on $T_p(E)$ as multiplication by $\psi(u)$. Since $[U_\mathfrak{q}, F^{\mathrm{ab}}/F]$ is the inertia group at $\mathfrak{q}$, (iii) follows from Theorem 3.19 and Corollary 3.18(i).

Thus for almost all $\mathfrak{q}$, $\psi(U_\mathfrak{q}) = 1$. Even for primes $\mathfrak{q}$ of bad reduction, since the reduction is potentially good (Theorem 5.7(i)) the action of $[U_\mathfrak{q}, F^{\mathrm{ab}}/F]$ on $T_p(E)$ factors through a finite quotient (Corollary 3.18(ii)) so the argument above shows that ψ vanishes on an open subgroup of $U_\mathfrak{q}$. Therefore ψ is continuous, and the proof of the theorem is complete. □

Let $\mathfrak{f} = \mathfrak{f}_E$ denote the conductor of the Hecke character ψ of Theorem 5.15. We can view ψ as a character of fractional ideals of F prime to $\mathfrak{f}$ in the usual way.

Corollary 5.16. *As a character on ideals, ψ satisfies*

(i) *if $\mathfrak{b}$ is an ideal of F prime to $\mathfrak{f}$ then $\psi(\mathfrak{b})\mathcal{O} = \mathbf{N}_{F/K}\mathfrak{b}$,*
(ii) *if $\mathfrak{q}$ is a prime of F not dividing $\mathfrak{f}$ and $\mathfrak{b}$ is an ideal of $\mathcal{O}$ prime to $\mathfrak{q}$, then $[\mathfrak{q}, F(E[\mathfrak{b}])/F]$ acts on $E[\mathfrak{b}]$ by multiplication by $\psi(\mathfrak{q})$.*
(iii) *if $\mathfrak{q}$ is a prime of F where E has good reduction and $q = \mathbf{N}_{F/\mathbf{Q}}\mathfrak{q}$ then $\psi(\mathfrak{q}) \in \mathcal{O}$ reduces modulo $\mathfrak{q}$ to the Frobenius endomorphism φ_q of $\tilde{E}$.*

Proof. The first two assertions are just translations of Theorem 5.15(i) and (ii). If $P \in E_{\mathrm{tors}}$ has order prime to $\mathfrak{q}$, $\tilde{P}$ denotes its reduction modulo a prime of $\bar{F}$ above $\mathfrak{q}$, and $\sigma_\mathfrak{q} = [\mathfrak{q}, F(E[\mathfrak{b}])/F]$, then

$$\widetilde{\psi(\mathfrak{q})}\tilde{P} = \widetilde{\sigma_\mathfrak{q}P} = \varphi_q\tilde{P}$$

where the first equality is from (ii) and the second is the definition of the Artin symbol $[\mathfrak{q}, F(E[\mathfrak{b}])/F]$. Since the reduction map is injective on prime-to-$\mathfrak{q}$ torsion (Theorem 3.15) this proves (iii). □

Remark 5.17. Note that Corollary 5.16(iii) gives an explicit version of Proposition 5.9. Proposition 5.9 is one of the key points in the proof of the Main Theorem of Complex Multiplication, of which Corollary 5.16 is a direct consequence.

Corollary 5.18. *Suppose $F = K$ and $\mathfrak{p}$ is a prime of K such that the map $\mathcal{O}^\times \to (\mathcal{O}/\mathfrak{p})^\times$ is not surjective. Then $E[\mathfrak{p}] \not\subset E(K)$.*

Proof. By Theorem 5.15(ii), $[\mathcal{O}_\mathfrak{p}^\times, K^{\mathrm{ab}}/K]$ acts on $E[\mathfrak{p}]$ via the character $\psi(x)x^{-1}$ of $\mathcal{O}_\mathfrak{p}^\times$, and by Theorem 5.15(i), $\psi(\mathcal{O}_\mathfrak{p}^\times) \subset \mathcal{O}^\times$. The corollary follows. □

Corollary 5.19. *Suppose $F = K$. Then the map $\mathcal{O}^\times \to (\mathcal{O}/\mathfrak{f})^\times$ is injective. In particular E cannot have good reduction at all primes of K.*

Proof. Let $u \in \mathcal{O}^\times$, $u \neq 1$ and let x be the idele defined by $x_\infty = 1$ and $x_\mathfrak{p} = u$ for all finite $\mathfrak{p}$. Then $\psi(x) = \psi(u^{-1}x) = u \neq 1$, so by definition of $\mathfrak{f}$, $u \not\equiv 1 \pmod{\mathfrak{f}}$. The second assertion now follows from Theorem 5.15(iii). □

If $\mathfrak{a}$ is an ideal of K let $K(\mathfrak{a})$ denote the ray class field of K modulo $\mathfrak{a}$.

Corollary 5.20. *Suppose E is defined over K, $\mathfrak{a}$ is an ideal of K prime to $6\mathfrak{f}$, and $\mathfrak{p}$ is a prime of K not dividing $6\mathfrak{f}$.*

(i) $E[\mathfrak{af}] \subset E(K(\mathfrak{af}))$.
(ii) *The map* $\mathrm{Gal}(K(E[\mathfrak{a}])/K) \to (\mathcal{O}/\mathfrak{a})^\times$ *of Corollary 5.5 is an isomorphism.*
(iii) *If* $\mathfrak{b} \mid \mathfrak{a}$ *then the natural map* $\mathrm{Gal}(K(\mathfrak{af})/K(\mathfrak{bf})) \to \mathrm{Gal}(K(E[\mathfrak{a}])/K(E[\mathfrak{b}]))$ *is an isomorphism.*
(iv) $K(E[\mathfrak{ap}^n])/K(E[\mathfrak{a}])$ *is totally ramified above* $\mathfrak{p}$.
(v) *If the map* $\mathcal{O}^\times \to (\mathcal{O}/\mathfrak{a})^\times$ *is injective then* $K(E[\mathfrak{ap}^n])/K(E[\mathfrak{a}])$ *is unramified outside of* $\mathfrak{p}$.

Proof. Suppose $x \in \mathbf{A}_K^\times$, $x_\mathfrak{p} \in \mathcal{O}_\mathfrak{p}^\times$ for all finite $\mathfrak{p}$ and $x_\infty = 1$. If $x \equiv 1 \pmod{^\times \mathfrak{f}}$ then Theorem 5.15(ii) shows that $[x, K^{\mathrm{ab}}/K]$ acts on E_{tors} as multiplication by x^{-1}. If $x \equiv 1 \pmod{^\times \mathfrak{a}}$ Theorem 5.15 shows that $[x, K^{\mathrm{ab}}/K]$ acts on $E[\mathfrak{a}]$ as multiplication by $\psi(x)$. Thus

- if $\mathfrak{p} \mid \mathfrak{f}$ then the kernel of $\mathcal{O}_\mathfrak{p}^\times \to [\mathcal{O}_\mathfrak{p}^\times, K(E[\mathfrak{a}])/K]$ is the kernel of the composition $\mathcal{O}_\mathfrak{p}^\times \xrightarrow{\psi} \mathcal{O}^\times \to (\mathcal{O}/\mathfrak{a})^\times$;
- if $\mathfrak{p}^n \mid \mathfrak{a}$ and $\mathfrak{p}^{n+1} \nmid \mathfrak{a}$ then $\mathcal{O}_\mathfrak{p}^\times/(1 + \mathfrak{p}^n\mathcal{O}_\mathfrak{p}) \hookrightarrow [\mathcal{O}_\mathfrak{p}^\times, K(E[\mathfrak{a}])/K] \hookrightarrow (\mathcal{O}/\mathfrak{p}^n)^\times$ is an isomorphism.

All assertions of the corollary follow without difficulty from this. □

Remark 5.21. In fact, without much more difficulty one can strengthen Corollary 5.20(i) (see [CW1] Lemma 4) to show that $E[\mathfrak{af}] = E(K(\mathfrak{af}))$, but we will not need this.

Corollary 5.22. *Suppose $\mathfrak{q}$ is a prime of F. There is an elliptic curve E' defined over F, such that*

- *E' is isomorphic to E over $\bar{F}$,*
- *E' has good reduction at $\mathfrak{q}$.*

Proof. Let ψ_E be the Hecke character attached to E and $U_\mathfrak{q}$ the group of local units at $\mathfrak{q}$, viewed as a subgroup of $\mathbf{A}_F^\times$. By Theorem 5.15(i), $\psi_E(U_\mathfrak{q}) \subset \mathcal{O}^\times$. Therefore we can find a continuous map

$$\chi : \mathbf{A}_F^\times / F^\times \to \mathcal{O}^\times$$

such that $\chi = \psi_E$ on $U_\mathfrak{q}$. We will take E' to be the twist of E by χ^{-1} (see [Si] §X.5).

Explicitly, suppose E is given by a Weierstrass equation

$$y^2 = x^3 + ax + b.$$

and let $w = \#(\mathcal{O}^\times)$. By class field theory we can view χ as an element of

$$\mathrm{Hom}(G_F, \mathcal{O}^\times) = H^1(F, \boldsymbol{\mu}_w) \cong F^\times/(F^\times)^w.$$

In other words, there is a $d \in F^\times$ such that

$$(d^{1/w})^\sigma = \chi(\sigma) d^{1/w} \quad \text{for every } \sigma \in G_F.$$

Define

$$E' = \begin{cases} y^2 = x^3 + d^2ax + d^3b & \text{if } w = 2 \\ y^2 = x^3 + dax & \text{if } w = 4 \\ y^2 = x^3 + db & \text{if } w = 6 \end{cases}$$

(see Example 1.1). The map

$$(x, y) \mapsto \begin{cases} (dx, d^{3/2}y) & \text{if } w = 2 \\ (d^{1/2}x, d^{3/4}y) & \text{if } w = 4 \\ (d^{1/3}x, d^{1/2}y) & \text{if } w = 6 \end{cases}$$

defines an isomorphism $\phi : E \xrightarrow{\sim} E'$ over $F(d^{1/w})$ (where we are using Lemma 1.13). If $P \in E(\bar{F})$ and $\sigma \in G_F$, then

$$\sigma(\phi(P)) = \phi^\sigma(\sigma P) = \chi(\sigma)^{-1}\phi(\sigma P).$$

From the definition of the Hecke character $\psi_{E'}$ of E' we see that $\psi_{E'} = \chi^{-1}\psi_E$. By construction this is trivial on $U_\mathfrak{q}$, so by Theorem 5.15(iii) E' has good reduction at $\mathfrak{q}$. □

6 Descent

In this section we use the results of §5 to compute the Selmer group of an elliptic curve with complex multiplication. After some cohomological lemmas in §6.1, we define an enlarged Selmer group $\mathcal{S}'(E)$ in §6.2 which is easier to compute (Lemma 6.4 and Theorem 6.5) than the true Selmer group $\mathcal{S}(E)$.

The main result describing the Selmer group $\mathcal{S}(E)$ is Theorem 6.9. The methods of this section closely follow the original work of Coates and Wiles [CW1] (see for example [Co]).

We continue to assume that E is an elliptic curve defined over a field F of characteristic 0, with complex multiplication by the maximal order $\mathcal{O}$ of an imaginary quadratic field K.

6.1 Preliminaries

Lemma 6.1. *Suppose $\mathfrak{p}$ is a prime of K lying above a rational prime $p > 3$, and $n \geq 0$. Let C be a subgroup of $(\mathcal{O}/\mathfrak{p}^n)^\times$, acting on $\mathcal{O}/\mathfrak{p}^n$ via multiplication. If either C is not a p-group or C is cyclic, then for every $i > 0$*

$$H^i(C, \mathcal{O}/\mathfrak{p}^n) = 0.$$

Proof. If C is cyclic this is a simple exercise. If C', the prime-to-p-part of C, is nontrivial, then $(\mathcal{O}/\mathfrak{p}^n)^{C'} = 0$ and $H^i(C', \mathcal{O}/\mathfrak{p}^n) = 0$ for every i, so the inflation-restriction exact sequence

$$0 \to H^i(C/C', (\mathcal{O}/\mathfrak{p}^n)^{C'}) \to H^i(C, \mathcal{O}/\mathfrak{p}^n) \to H^i(C', \mathcal{O}/\mathfrak{p}^n)$$

shows that $H^i(C, \mathcal{O}/\mathfrak{p}^n) = 0$. □

Lemma 6.2. *Suppose $\mathfrak{p}$ is a prime of K lying above a rational prime $p > 3$, and $n \geq 0$.*

(i) *If $\mathcal{O}_\mathfrak{p} = \mathbf{Z}_p$ or if $E[\mathfrak{p}] \not\subset E(F)$ then the restriction map gives an isomorphism*

$$H^1(F, E[\mathfrak{p}^n]) \cong H^1(F(E[\mathfrak{p}^n]), E[\mathfrak{p}^n])^{\mathrm{Gal}(F(E[\mathfrak{p}^n])/F)}.$$

(ii) *Suppose F is a finite extension of $\mathbf{Q}_\ell$ for some $\ell \neq p$. Then the restriction map gives an injection*

$$H^1(F, E)_{\mathfrak{p}^n} \hookrightarrow H^1(F(E[\mathfrak{p}^n]), E)_{\mathfrak{p}^n}.$$

Proof. Use Proposition 5.4 and Corollary 5.5 to identify $E[\mathfrak{p}^n]$ with $\mathcal{O}/\mathfrak{p}^n$ and $\mathrm{Gal}(F(E[\mathfrak{p}^n])/F)$ with a subgroup C of $(\mathcal{O}/\mathfrak{p}^n)^\times$. Then C is cyclic if $\mathcal{O}_\mathfrak{p} = \mathbf{Z}_p$, and C is a p-group if and only if $E[\mathfrak{p}] \subset E(F)$ (since $\mathrm{Gal}(F(E[\mathfrak{p}])/F) \subset (\mathcal{O}/\mathfrak{p})^\times$ has order prime to p). Thus (i) follows from Lemma 6.1 and the inflation-restriction exact sequence.

The kernel of the restriction map in (ii) is $H^1(F_n/F, E(F_n))_{\mathfrak{p}^n}$, where $F_n = F(E[\mathfrak{p}^n])$. We may as well assume that $n \geq 1$, or there is nothing to prove. By Theorem 5.7(ii), E has good reduction over F_n, so by Proposition 3.4 there is a reduction exact sequence

$$0 \to E_1(F_n) \to E(F_n) \to \tilde{E}(\mathbf{k}_n) \to 0$$

where $\mathbf{k}_n$ is the residue field of F_n. Thus $E_1(F_n)$ is a profinite $\mathcal{O}$-module, of finite index in $E(F_n)$, on which (by Theorem 3.15(i)) every α prime to ℓ acts invertibly. It follows that the pro-$\mathfrak{p}$ part of $E(F_n)$ is finite, say $E[\mathfrak{p}^m]$ for some $m \geq n$, and hence

$$H^1(F_n/F, E(F_n))_{\mathfrak{p}^n} \subset H^1(F_n/F, E[\mathfrak{p}^m]) = H^1(F(E[\mathfrak{p}^m])/F, E[\mathfrak{p}^m]).$$

If $E[\mathfrak{p}] \subset E(F)$ then E has good reduction by Theorem 5.7(ii) (and Remark 5.8) so F_n/F is unramified and hence cyclic. Hence exactly as in (i), Lemma 6.1 shows that $H^1(F(E[\mathfrak{p}^m])/F, E[\mathfrak{p}^m]) = 0$, and (ii) follows. □

6.2 The Enlarged Selmer Group

Suppose for the rest of this section that F is a number field.

Definition 6.3. If $\alpha \in \mathcal{O}$ define $\mathcal{S}'_\alpha(E) = \mathcal{S}'_\alpha(E_{/F}) \subset H^1(F, E[\alpha])$ by

$$\begin{aligned}\mathcal{S}'_\alpha(E) &= \{c \in H^1(F, E[\alpha]) : \mathrm{res}_\mathfrak{q}(c) \in \mathrm{image}(E(F_\mathfrak{q})/\alpha E(F_\mathfrak{q})) \text{ for every } \mathfrak{q} \nmid \alpha\} \\ &= \{c \in H^1(F, E[\alpha]) : \mathrm{res}_\mathfrak{q}(c) = 0 \text{ in } H^1(F_\mathfrak{q}, E(\bar{F}_\mathfrak{q})) \text{ for every } \mathfrak{q} \nmid \alpha\}\end{aligned}$$

in the diagram (5). Clearly $\mathcal{S}_\alpha(E) \subset \mathcal{S}'_\alpha(E)$.

Lemma 6.4. *Suppose $\mathfrak{p}$ is a prime of K not dividing* 6, $n \geq 1$, $E[\mathfrak{p}^n] \subset E(F)$ *and $\mathfrak{p}^n = \alpha\mathcal{O}$. Then*

$$\mathcal{S}'_\alpha(E_{/F}) = \mathrm{Hom}(\mathrm{Gal}(M/F), E[\mathfrak{p}^n])$$

where M is the maximal abelian p-extension of F unramified outside of primes above $\mathfrak{p}$.

Proof. Since $E[\mathfrak{p}^n] \subset E(F)$,

$$H^1(F, E[\mathfrak{p}^n]) = \mathrm{Hom}(G_F, E[\mathfrak{p}^n]), \quad H^1(F_\mathfrak{q}, E[\mathfrak{p}^n]) = \mathrm{Hom}(G_{F_\mathfrak{q}}, E[\mathfrak{p}^n]).$$

Suppose $\mathfrak{q}$ is a prime of F not dividing $\mathfrak{p}$. By Theorem 5.7(ii), E has good reduction at $\mathfrak{p}$ so by (4) and Corollary 3.17, the image of $E(F_\mathfrak{q})/\alpha E(F_\mathfrak{q})$ under (5) is contained in $\mathrm{Hom}(G_{F_\mathfrak{q}}/I_\mathfrak{q}, E[\mathfrak{p}^n])$, where $I_\mathfrak{q}$ is the inertia group in $G_{F_\mathfrak{q}}$, and we have $\mathcal{O}$-module isomorphisms

$$\mathrm{Hom}(G_{F_\mathfrak{q}}/I_\mathfrak{q}, E[\mathfrak{p}^n]) \cong E[\mathfrak{p}^n] \cong \mathcal{O}/\mathfrak{p}^n\mathcal{O}.$$

On the other hand, using Theorem 3.15 and writing $\mathbf{k}$ for the residue field of $F_\mathfrak{q}$,

$$E(F_\mathfrak{q})/\alpha E(F_\mathfrak{q}) \cong \tilde{E}(\mathbf{k})/\alpha\tilde{E}(\mathbf{k}) \cong \mathcal{O}/\mathfrak{p}^n\mathcal{O}.$$

Thus the image of $E(F_\mathfrak{q})/\alpha E(F_\mathfrak{q}) \hookrightarrow H^1(F_\mathfrak{q}, E[\mathfrak{p}^n])$ under (5) must be equal to $\mathrm{Hom}(G_{F_\mathfrak{q}}/I_\mathfrak{q}, E[\mathfrak{p}^n])$, and the lemma follows from the definition of $\mathcal{S}'_\alpha$. □

Theorem 6.5. *Suppose E is defined over K, $\mathfrak{p}$ is a prime of K not dividing 6, $n \geq 1$, and $\mathfrak{p}^n = \alpha\mathcal{O}$. Let $K_n = K(E[\mathfrak{p}^n])$. Then*

$$\mathcal{S}'_\alpha(E_{/K}) = \mathrm{Hom}(\mathrm{Gal}(M_n/K_n), E[\mathfrak{p}^n])^{\mathrm{Gal}(K_n/K)}$$

where M_n is the maximal abelian p-extension of K_n unramified outside of primes above $\mathfrak{p}$.

Proof. Let $G = \mathrm{Gal}(K_n/K)$. By Lemma 6.2(ii) and Corollary 5.18, the restriction map gives an isomorphism

$$H^1(K, E[\mathfrak{p}^n]) \cong H^1(K_n, E[\mathfrak{p}^n])^G.$$

Clearly the image of $\mathcal{S}'_\alpha(E_{/K})$ under this restriction isomorphism is contained in $\mathcal{S}'_\alpha(E_{/K_n})$. Conversely, every class in $H^1(K, E[\mathfrak{p}^n])$ whose restriction lies in $\mathcal{S}'_\alpha(E_{/K_n})$ already lies in $\mathcal{S}'_\alpha(E_{/K})$, because by Lemma 6.2(iii) the restriction map

$$H^1(K_\mathfrak{q}, E(\bar{K}_\mathfrak{q})) \to H^1(K_\mathfrak{q}(E[\mathfrak{p}^n]), E(\bar{K}_\mathfrak{q}))$$

is injective for every prime $\mathfrak{q}$ not dividing $\mathfrak{p}$. This proves that

$$\mathcal{S}'_\alpha(E_{/K}) = \mathcal{S}'_\alpha(E_{/K_n})^G,$$

and so the theorem follows from Lemma 6.4. □

6.3 The True Selmer Group

For the rest of this section we will suppose that E is defined over K, i.e., $F = K$. Recall that by Corollary 5.12 this implies that K has class number one. Fix a prime $\mathfrak{p}$ of K not dividing $6\mathfrak{f}$ and a generator π of $\mathfrak{p}$. Let $\lambda_E : E_1(K_\mathfrak{p}) \to \mathfrak{p}\mathcal{O}_\mathfrak{p}$ be the logarithm map of Definition 3.12.

Lemma 6.6. *The map λ_E extends uniquely to a surjective map $E(K_\mathfrak{p}) \twoheadrightarrow \mathfrak{p}\mathcal{O}_\mathfrak{p}$ whose kernel is finite and has no $\mathfrak{p}$-torsion.*

Proof. By Corollary 3.13, $\lambda_E : E_1(K_\mathfrak{p}) \to \mathfrak{p}\mathcal{O}_\mathfrak{p}$ is an isomorphism, and by Lemma 3.6(i) and Corollary 5.16(iii), $E(K_\mathfrak{p})/E_1(K_\mathfrak{p})$ is finite and has no $\mathfrak{p}$-torsion. □

Definition 6.7. For every $n \geq 1$ let $K_{n,\mathfrak{p}} = K_\mathfrak{p}(E[\mathfrak{p}^n])$ and define a Kummer pairing

$$\begin{aligned} \langle \cdot\, , \cdot \rangle_{\pi^n} : E(K_\mathfrak{p}) \times K_{n,\mathfrak{p}}^\times &\to E[\mathfrak{p}^n] \\ P\ ,\ x\quad &\mapsto [x, K_{n,\mathfrak{p}}^{\mathrm{ab}}/K_{n,\mathfrak{p}}]Q - Q \end{aligned}$$

where $[\,\cdot\, , K_{n,\mathfrak{p}}^{\mathrm{ab}}/K_{n,\mathfrak{p}}]$ is the local Artin map and $Q \in E(\bar{K}_\mathfrak{p})$ satisfies $\pi^n Q = P$.

Lemma 6.8. *For every n there is a unique Galois-equivariant homomorphism $\delta_n : K_{n,\mathfrak{p}}^\times \to E[\mathfrak{p}^n]$ such that if $P \in E(K_\mathfrak{p})$ and $x \in K_{n,\mathfrak{p}}^\times$,*

$$\langle P, x \rangle_{\pi^n} = (\pi^{-1}\lambda_E(P))\delta_n(x).$$

Further, if $\mathcal{O}_{n,\mathfrak{p}}$ denotes the ring of integers of $K_{n,\mathfrak{p}}$ then $\delta_n(\mathcal{O}_{n,\mathfrak{p}}^\times) = E[\mathfrak{p}^n]$.

Proof. Define $\delta_n(x) = \langle R, x \rangle_{\pi^n}$ where $\lambda_E(R) = \pi$, and then everything except the surjectivity assertion is clear.

First note that by Theorem 5.15(ii), if $x \in \mathcal{O}_\mathfrak{p}^\times$ then $[x, K_{n,\mathfrak{p}}/K_\mathfrak{p}]$ acts on $E[\mathfrak{p}^n]$ as multiplication by x^{-1}. Therefore $E(K_\mathfrak{p})$ has no $\mathfrak{p}$-torsion and $E[\mathfrak{p}]$ has no proper $G_{K_\mathfrak{p}}$-stable subgroups.

By Lemma 6.6, $E(K_\mathfrak{p})/\mathfrak{p}^n E(K_\mathfrak{p}) \xrightarrow{\sim} \mathcal{O}/\mathfrak{p}^n$. Since

$$E(K_\mathfrak{p})/\mathfrak{p}^n E(K_\mathfrak{p}) \hookrightarrow H^1(K_\mathfrak{p}, E[\mathfrak{p}^n]) \hookrightarrow \mathrm{Hom}(K_{n,\mathfrak{p}}^\times, E[\mathfrak{p}^n])$$

is injective (the first map by (5) and the second by Lemmas 6.2(ii) and 6.6), the image of δ_n is not contained in $E[\mathfrak{p}^{n-1}]$. Since the image of δ_n is stable under $G_{K_\mathfrak{p}}$, it must be all of $E[\mathfrak{p}^n]$. But $\delta_n(K_{n,\mathfrak{p}}^\times)/\delta_n(\mathcal{O}_{\mathfrak{p},n}^\times)$ is a quotient of $E[\mathfrak{p}^n]$ on which $G_{K_\mathfrak{p}}$ acts trivially, and (as above) such a quotient must be trivial, so $\delta_n(\mathcal{O}_{\mathfrak{p},n}^\times) = E[\mathfrak{p}^n]$ as well. □

Theorem 6.9. *With notation as above, let $K_n = K(E[\mathfrak{p}^n])$ and $\mathcal{O}_n$ its ring of integers, and define*

$$W_n = K_n^\times \prod_{v \mid \infty} K_{n,v}^\times \prod_{v \nmid \mathfrak{p}\infty} \mathcal{O}_{n,v}^\times \cdot \ker(\delta_n) \subset \mathbf{A}_{K_n}^\times.$$

Then

$$\mathcal{S}_{\pi^n}(E_{/K}) = \mathrm{Hom}(\mathbf{A}_{K_n}^\times / W_n, E[\mathfrak{p}^n])^{\mathrm{Gal}(K_n/K)}.$$

Proof. By definition we have an injective map

$$E(K_\mathfrak{p})/\pi^n E(K_\mathfrak{p}) \hookrightarrow \mathrm{Hom}(K_{n,\mathfrak{p}}^\times/\ker(\delta_n), E[\mathfrak{p}^n])^{\mathrm{Gal}(K_{n,\mathfrak{p}}/K_\mathfrak{p})}.$$

By Lemma 6.6, $E(K_\mathfrak{p})/\pi^n E(K_\mathfrak{p}) \cong \mathcal{O}/\mathfrak{p}^n$. By Lemma 6.8 $K_{n,\mathfrak{p}}^\times/\ker(\delta_n) \cong E[\mathfrak{p}^n]$, and by Theorem 5.15(ii),

$$\mathrm{Hom}(E[\mathfrak{p}^n], E[\mathfrak{p}^n])^{\mathrm{Gal}(K_{n,\mathfrak{p}}/K_\mathfrak{p})} = \mathrm{Hom}_\mathcal{O}(E[\mathfrak{p}^n], E[\mathfrak{p}^n]) \cong \mathcal{O}/\mathfrak{p}^n.$$

Therefore the injection above is an isomorphism, and the theorem follows from Proposition 6.5 and class field theory. □

Let A denote the ideal class group of $K(E[\mathfrak{p}])$, and $\mathcal{E}$ the group of global units of $K(E[\mathfrak{p}])$.

Corollary 6.10. *With notation as above,*

$$\mathcal{S}_\pi(E) = 0 \Leftrightarrow \left(\mathrm{Hom}(A, E[\mathfrak{p}])^{\mathrm{Gal}(K(E[\mathfrak{p}])/K)} = 0 \quad \textit{and} \quad \delta_1(\mathcal{E}) \neq 0\right).$$

Proof. By Corollary 5.20, $K(E[\mathfrak{p}])/K$ is totally ramified at $\mathfrak{p}$, of degree $\mathbf{N}\mathfrak{p}-1$. We identify $K_{1,\mathfrak{p}}$ with the completion of $K(E[\mathfrak{p}])$ at the unique prime above $\mathfrak{p}$, and let $\mathcal{O}_{1,\mathfrak{p}}$ denote its ring of integers and $\bar{\mathcal{E}}$ the closure of $\mathcal{E}$ in $\mathcal{O}_{1,\mathfrak{p}}$. Let $V = \ker(\delta_1) \cap \mathcal{O}_{1,\mathfrak{p}}^{\times}$ and $\Delta = \mathrm{Gal}(K(E[\mathfrak{p}])/K)$. We have an exact sequence

$$0 \to \mathcal{O}_{1,\mathfrak{p}}^{\times}/\bar{\mathcal{E}}V \to \mathbf{A}_{K_1}^{\times}/W_1 \to A' \to 0$$

where W_1 is as in Theorem 6.9 and A' is a quotient of A by some power of the class of the prime $\mathcal{P}$ above $\mathfrak{p}$. Since $\mathcal{P}^{\mathbf{N}\mathfrak{p}-1} = \mathfrak{p}$ is principal, $\mathrm{Hom}(A', E[\mathfrak{p}]) = \mathrm{Hom}(A, E[\mathfrak{p}])$. Using Theorem 6.9 we conclude that

$$\mathcal{S}_{\pi}(E) = 0 \Leftrightarrow \left(\mathrm{Hom}(A, E[\mathfrak{p}])^{\Delta} = 0 \quad \text{and} \quad \mathrm{Hom}(\mathcal{O}_{1,\mathfrak{p}}^{\times}/\bar{\mathcal{E}}V, E[\mathfrak{p}])^{\Delta} = 0\right).$$

By Lemma 6.8, $\delta_1 : \mathcal{O}_{1,\mathfrak{p}}^{\times}/V \to E[\mathfrak{p}]$ is an isomorphism. Since $E[\mathfrak{p}]$ has no proper Galois-stable submodules, it follows that

$$\mathrm{Hom}(\mathcal{O}_{1,\mathfrak{p}}^{\times}/\bar{\mathcal{E}}V, E[\mathfrak{p}])^{\Delta} = 0 \quad \Leftrightarrow \quad \bar{\mathcal{E}} \not\subset V \quad \Leftrightarrow \quad \delta_1(\mathcal{E}) \neq 0.$$

This completes the proof of the corollary. □

7 Elliptic Units

In this section we define elliptic units and relate them to special values of L-functions. Elliptic units will be defined as certain rational functions of x-coordinates of torsion points on a CM elliptic curve. The results of §5 will allow us determine the action of the Galois group on these numbers, and hence their fields of definition. We follow closely [CW1] §5; see also [dS] Chapter II and Robert's original memoir [Ro].

Throughout this section we fix an imaginary quadratic field K with ring of integers $\mathcal{O}$, an elliptic curve E over $\mathbf{C}$ with complex multiplication by $\mathcal{O}$, and a nontrivial ideal $\mathfrak{a}$ of $\mathcal{O}$ prime to 6. For simplicity we will assume that the class number of K is one; see [dS] for the general case.

7.1 Definition and Basic Properties

Definition 7.1. Choose a Weierstrass equation (1) for E with coordinate functions x, y on E. Define a rational function on E

$$\Theta_{E,\mathfrak{a}} = \alpha^{-12}\Delta(E)^{\mathbf{N}\mathfrak{a}-1} \prod_{P \in E[\mathfrak{a}]-O} (x - x(P))^{-6}$$

where α is a generator of $\mathfrak{a}$ and $\Delta(E)$ is the discriminant of the chosen model of E. Clearly this is independent of the choice of α.

Lemma 7.2. (i) *$\Theta_{E,\mathfrak{a}}$ is independent of the choice of Weierstrass model.*
(ii) *If $\phi : E' \xrightarrow{\sim} E$ is an isomorphism of elliptic curves then $\Theta_{E',\mathfrak{a}} = \Theta_{E,\mathfrak{a}} \circ \phi$.*

(iii) *If E is defined over F then the rational function $\Theta_{E,\mathfrak{a}}$ is defined over F.*

Proof. Any other Weierstrass model has coordinate functions x', y' given by

$$x' = u^2 x + r, \quad y' = u^3 y + sx + t$$

where $u \in \mathbf{C}^\times$ ([Si] Remark III.1.3), and then $a_i' = u^i a_i$ and

$$\Delta(E') = u^{12}\Delta(E).$$

Since $\#(E[\mathfrak{a}]) = \mathbf{N}\mathfrak{a}$, this proves (i), and (ii) is just a different formulation of (i). For (iii) we need only observe that $\alpha \in F$, $\Delta(E) \in F$, and G_F permutes the set $\{x(P) : P \in E[\mathfrak{a}] - O\}$, so G_F fixes $\Theta_{E,\mathfrak{a}}$. □

Lemma 7.3. *Suppose E is defined over K and $\mathfrak{p}$ is a prime of K where E has good reduction. Fix a Weierstrass model for E which is minimal at $\mathfrak{p}$. Let $\mathfrak{b}$ and $\mathfrak{c}$ be nontrivial relatively prime ideals of $\mathcal{O}$ and $P \in E[\mathfrak{b}]$, $Q \in E[\mathfrak{c}]$ points in $E(\bar{K})$ of exact orders $\mathfrak{b}$ and $\mathfrak{c}$, respectively. Fix an extension of the $\mathfrak{p}$-adic order $\mathrm{ord}_\mathfrak{p}$ to $\bar{K}$, normalized so $\mathrm{ord}_\mathfrak{p}(\mathfrak{p}) = 1$.*

(i) *If $n > 0$ and $\mathfrak{b} = \mathfrak{p}^n$ then* $\mathrm{ord}_\mathfrak{p}(x(P)) = -2/(\mathbf{N}\mathfrak{p}^{n-1}(\mathbf{N}\mathfrak{p} - 1))$.
(ii) *If $\mathfrak{b}$ is not a power of $\mathfrak{p}$ then* $\mathrm{ord}_\mathfrak{p}(x(P)) \geq 0$.
(iii) *If $\mathfrak{p} \nmid \mathfrak{b}\mathfrak{c}$ then* $\mathrm{ord}_\mathfrak{p}(x(P) - x(Q)) = 0$.

Proof. Suppose that $\mathfrak{b} = \mathfrak{p}^n$ with $n \geq 1$. Let $\hat{E}$ be the formal group over $\mathcal{O}_\mathfrak{p}$ associated to E in Theorem 3.7. Let $\pi = \psi_E(\mathfrak{p})$, let $[\pi^m](X) \in \mathcal{O}[[X]]$ be the endomorphism of $\hat{E}$ corresponding to π^m for every m, and define

$$f(X) = [\pi^n](X)/[\pi^{n-1}](X) \in \mathcal{O}[[X]].$$

Since π reduces to the Frobenius endomorphism of the reduction $\tilde{E}$ of E modulo $\mathfrak{p}$ (Corollary 5.16(iii)), it follows from Corollary 3.9 and Proposition 3.14 that

- $f(X) \equiv X^{\mathbf{N}\mathfrak{p}^n - \mathbf{N}\mathfrak{p}^{n-1}} \pmod{\mathfrak{p}}$
- $f(X) \equiv \pi \pmod{X}$.

Thus by the Weierstrass preparation theorem,

$$f(X) = e(X)u(X)$$

where $e(X)$ is an Eisenstein polynomial of degree $\mathbf{N}\mathfrak{p}^{n-1}(\mathbf{N}\mathfrak{p} - 1)$ and $u(X) \in \mathcal{O}[[X]]^\times$.

Since the reduction of π is a purely inseparable endomorphism of $\hat{E}$, Lemma 3.6 shows that $E[\mathfrak{p}^n] \subset E_1(\bar{K}_\mathfrak{p})$. Thus $z = -x(P)/y(P)$ is a root of $f(X)$, and hence of $e(X)$, so $\mathrm{ord}_\mathfrak{p}(x(P)/y(P)) = 1/(\mathbf{N}\mathfrak{p}^{n-1}(\mathbf{N}\mathfrak{p} - 1))$. Now (i) follows from Lemma 3.5.

If $\mathfrak{b}$ is not a power of $\mathfrak{p}$ then by Theorem 3.15(i), $P \notin E_1(\bar{K}_\mathfrak{p})$. Hence by Lemma 3.5, $\mathrm{ord}_\mathfrak{p}(x(P)) \geq 0$, which is (ii). Further, writing $\tilde{P}$ and $\tilde{Q}$ for the reductions of P and Q, we have

$$\mathrm{ord}_\mathfrak{p}(x(P) - x(Q)) > 0 \Leftrightarrow x(\tilde{P}) = x(\tilde{Q}) \Leftrightarrow \tilde{P} = \pm\tilde{Q} \Leftrightarrow$$
$$\Leftrightarrow \widetilde{P \mp Q} = \tilde{O} \Leftrightarrow P \mp Q \in E_1(\bar{K}_\mathfrak{p}).$$

Since $\mathfrak{b}$ and $\mathfrak{c}$ are relatively prime, the order of $P \pm Q$ is not a power of $\mathfrak{p}$. So again by Theorem 3.15(i), $P \pm Q \notin E_1(\bar{K}_\mathfrak{p})$, and (iii) follows. □

For every ideal $\mathfrak{b}$ of $\mathcal{O}$ write $K(\mathfrak{b})$ for the ray class field of K modulo $\mathfrak{b}$.

Theorem 7.4. *Suppose $\mathfrak{b}$ is a nontrivial ideal of $\mathcal{O}$ relatively prime to $\mathfrak{a}$, and $Q \in E[\mathfrak{b}]$ is an $\mathcal{O}$-generator of $E[\mathfrak{b}]$.*

(i) $\Theta_{E,\mathfrak{a}}(Q) \in K(\mathfrak{b})$.

(ii) *If $\mathfrak{c}$ is an ideal of $\mathcal{O}$ prime to $\mathfrak{b}$, c is a generator of $\mathfrak{c}$, and $\sigma_\mathfrak{c} = [\mathfrak{c}, K(\mathfrak{b})/K]$, then*
$$\Theta_{E,\mathfrak{a}}(Q)^{\sigma_\mathfrak{c}} = \Theta_{E,\mathfrak{a}}(cQ).$$

(iii) *If $\mathfrak{b}$ is not a prime power then $\Theta_{E,\mathfrak{a}}(Q)$ is a global unit. If $\mathfrak{b}$ is a power of a prime $\mathfrak{p}$ then $\Theta_{E,\mathfrak{a}}(Q)$ is a unit at primes not dividing $\mathfrak{p}$.*

Proof. Since we assumed that K has class number one, by Corollary 5.13 and Lemma 7.2(i) we may assume that E is defined over K by a Weierstrass model (1). Then by Lemma 7.2(iii) $\Theta_{E,\mathfrak{a}}$ belongs to the function field $K(E)$.

Let ψ be the Hecke character associated to E by Theorem 5.15. Suppose $x \in \prod_\mathfrak{p} \mathcal{O}_\mathfrak{p}^\times \subset \mathbf{A}_K^\times$ and $x \equiv 1 \bmod^\times \mathfrak{b}$, and let $\sigma_x = [x, K^{\mathrm{ab}}/K]$. By Theorem 5.15, $\psi(x) \in \mathcal{O}^\times = \mathrm{Aut}(E)$ and $\sigma_x Q = \psi(x)Q$. Therefore

$$\Theta_{E,\mathfrak{a}}(Q)^{\sigma_x} = \Theta_{E,\mathfrak{a}}(Q^{\sigma_x}) = \Theta_{E,\mathfrak{a}}(\psi(x)Q) = \Theta_{E,\mathfrak{a}}(Q),$$

the last equality by Lemma 7.2(ii). Since these σ_x generate $\mathrm{Gal}(\bar{K}/K(\mathfrak{b}))$, this proves (i).

For (ii), let $x \in \mathbf{A}_K^\times$ be an idele with $x\mathcal{O} = \mathfrak{c}$ and $x_\mathfrak{p} = 1$ for $\mathfrak{p}$ dividing $\mathfrak{b}$. Then Theorem 5.15 shows that $\psi(x) \in c\mathcal{O}^\times$ and $\sigma_\mathfrak{c} Q = \psi(x)Q$. So again using Lemma 7.2(ii),

$$\Theta_{E,\mathfrak{a}}(Q)^{\sigma_\mathfrak{c}} = \Theta_{E,\mathfrak{a}}(\psi(x)Q) = \Theta_{E,\mathfrak{a}}(cQ).$$

This is (ii).

For (iii), let $\mathfrak{p}$ be a prime of K such that $\mathfrak{b}$ is not a power of $\mathfrak{p}$. By Corollary 5.22 and Lemma 7.2, we may assume that our Weierstrass equation for E has good reduction at $\mathfrak{p}$, so that $\Delta(E)$ is prime to $\mathfrak{p}$. Let $n = \mathrm{ord}_\mathfrak{p}(\mathfrak{a})$. Then

$$\mathrm{ord}_\mathfrak{p}(\Theta_{E,\mathfrak{a}}(Q))/6 = -2n - \sum_{P \in E[\mathfrak{p}^n] - O} \mathrm{ord}_\mathfrak{p}(x(Q) - x(P))$$
$$- \sum_{P \in E[\mathfrak{a}] - E[\mathfrak{p}^n]} \mathrm{ord}_\mathfrak{p}(x(Q) - x(P)).$$

By Lemma 7.3, since $\mathfrak{b}$ is not a power of $\mathfrak{p}$,

$$\operatorname{ord}_{\mathfrak{p}}(x(Q)-x(P)) = \begin{cases} -2/(\mathbf{N}\mathfrak{p}^m - \mathbf{N}\mathfrak{p}^{m-1}) & \text{if } P \text{ has order exactly } \mathfrak{p}^m,\ m>0 \\ 0 & \text{if the order of } P \text{ is not a power of } \mathfrak{p}. \end{cases}$$

From this one verifies easily that $\operatorname{ord}_{\mathfrak{p}}(\Theta_{E,\mathfrak{a}}(Q)) = 0$. □

7.2 The Distribution Relation

Lemma 7.5. *$\Theta_{E,\mathfrak{a}}$ is a rational function on E with divisor*

$$12\mathbf{N}\mathfrak{a}[O] - 12 \sum_{P\in E[\mathfrak{a}]} [P].$$

Proof. The coordinate function x is an even rational function with a double pole at O and no other poles. Thus for every point P, the divisor of $x - x(P)$ is $[P]+[-P]-2[O]$ and the lemma follows easily. □

Theorem 7.6. *Suppose $\mathfrak{b}$ is and ideal of $\mathcal{O}$ relatively prime to $\mathfrak{a}$, and β is a generator of $\mathfrak{b}$. Then for every $P \in E(\bar{K})$,*

$$\prod_{R\in E[\mathfrak{b}]} \Theta_{E,\mathfrak{a}}(P+R) = \Theta_{E,\mathfrak{a}}(\beta P).$$

Proof. Lemmas 7.2(iii) and 7.5 show that both sides of the equation in the theorem are rational functions on E, defined over K, with divisor

$$12 \sum_{Q\in E[\mathfrak{a}\mathfrak{b}]} [Q] - 12\mathbf{N}\mathfrak{a} \sum_{R\in E[\mathfrak{b}]} [R].$$

Thus their ratio is a constant $\lambda \in K^\times$, and we need to show that $\lambda = 1$.

Let $w_K = \#(\mathcal{O}^\times)$ and fix a generator α of $\mathfrak{a}$. Evaluating this ratio at $P = O$ one sees that

$$\lambda = \frac{\Delta(E)^{(\mathbf{N}\mathfrak{a}-1)(\mathbf{N}\mathfrak{b}-1)}}{\alpha^{12(\mathbf{N}\mathfrak{b}-1)}\beta^{12(\mathbf{N}\mathfrak{a}-1)}} \prod_{\substack{R\in E[\mathfrak{b}]\\ R\neq O}} \prod_{\substack{P\in E[\mathfrak{a}]\\ P\neq O}} (x(R)-x(P))^{-6} = \mu^{w_K}$$

with

$$\mu = \frac{\Delta(E)^{(\mathbf{N}\mathfrak{a}-1)(\mathbf{N}\mathfrak{b}-1)/w_K}}{\alpha^{12(\mathbf{N}\mathfrak{b}-1)/w_K}\beta^{12(\mathbf{N}\mathfrak{a}-1)/w_K}} \prod (x(R)-x(P))^{-12/w_K},$$

where the final product is over $R \in E[\mathfrak{b}] - O$ and $P \in (E[\mathfrak{a}]-O)/\pm 1$ (recall $\mathfrak{a}$ is prime to 6). Since w_K divides 12, all of the exponents in the definition of μ are integers.

Exactly as in the proof of Theorem 7.4(iii), one can show that $\mu \in \mathcal{O}^\times$, and therefore $\lambda = 1$. □

Corollary 7.7. *Suppose $\mathfrak{b}$ is an ideal of $\mathcal{O}$ prime to $\mathfrak{a}$, $Q \in E[\mathfrak{b}]$ has order exactly $\mathfrak{b}$, $\mathfrak{p}$ is a prime dividing $\mathfrak{b}$, π is a generator of $\mathfrak{p}$, and $\mathfrak{b}' = \mathfrak{b}/\mathfrak{p}$. If the reduction map $\mathcal{O}^\times \to (\mathcal{O}/\mathfrak{b}')^\times$ is injective then*

$$\mathbf{N}_{K(\mathfrak{b})/K(\mathfrak{b}')}\Theta_{E,\mathfrak{a}}(Q) = \begin{cases} \Theta_{E,\mathfrak{a}}(\pi Q) & \text{if } \mathfrak{p} \mid \mathfrak{b}' \\ \Theta_{E,\mathfrak{a}}(\pi Q)^{1-\mathrm{Frob}_\mathfrak{p}^{-1}} & \text{if } \mathfrak{p} \nmid \mathfrak{b}' \end{cases}$$

where in the latter case $\mathrm{Frob}_\mathfrak{p}$ *is the Frobenius of* $\mathfrak{p}$ *in* $\mathrm{Gal}(K(\mathfrak{b}')/K)$.

Proof. Let C denote the multiplicative group $1 + \mathfrak{b}'(\mathcal{O}/\mathfrak{b})$. Because of our hypotheses that $\mathcal{O}^\times$ injects into $(\mathcal{O}/\mathfrak{b}')^\times$, C is isomorphic to the kernel of the map

$$(\mathcal{O}/\mathfrak{b})^\times/\mathcal{O}^\times \to (\mathcal{O}/\mathfrak{b}')^\times/\mathcal{O}^\times.$$

Thus class field theory gives an isomorphism

$$C \xrightarrow{\sim} \mathrm{Gal}(K(\mathfrak{b})/K(\mathfrak{b}'))$$

which we will denote by $c \mapsto \sigma_c$. Therefore

$$\mathbf{N}_{K(\mathfrak{b})/K(\mathfrak{b}')}\Theta_{E,\mathfrak{a}}(Q) = \prod_{c\in C} \Theta_{E,\mathfrak{a}}(Q)^{\sigma_c} = \prod_{c \in C} \Theta_{E,\mathfrak{a}}(cQ)$$

by Theorem 7.4(ii).

One sees easily that

$$\begin{aligned}\{cQ : c \in C\} &= \{P \in E[\mathfrak{b}] : \pi P = \pi Q \text{ and } P \notin E[\mathfrak{b}']\} \\ &= \begin{cases} \{Q + R : R \in E[\mathfrak{p}]\} & \text{if } \mathfrak{p} \mid \mathfrak{b}' \\ \{Q + R : R \in E[\mathfrak{p}], R \not\equiv -Q \pmod{E[\mathfrak{b}']}\} & \text{if } \mathfrak{p} \nmid \mathfrak{b}' \end{cases}\end{aligned}$$

Thus if $\mathfrak{p} \mid \mathfrak{b}'$

$$\mathbf{N}_{K(\mathfrak{b})/K(\mathfrak{b}')}\Theta_{E,\mathfrak{a}}(Q) = \prod_{R\in E[\mathfrak{p}]} \Theta_{E,\mathfrak{a}}(Q+R) = \Theta_{E,\mathfrak{a}}(\pi Q)$$

by Theorem 7.6. Similarly, if $\mathfrak{p} \nmid \mathfrak{b}'$

$$\Theta_{E,\mathfrak{a}}(Q + R_0)\mathbf{N}_{K(\mathfrak{b})/K(\mathfrak{b}')}\Theta_{E,\mathfrak{a}}(Q) = \Theta_{E,\mathfrak{a}}(\pi Q)$$

where $R_0 \in E[\mathfrak{p}]$ satisfies $Q + R_0 \in E[\mathfrak{b}']$. But then by Theorem 7.4(ii) (note that our assumption on $\mathfrak{b}'$ implies that $\mathfrak{b}' \neq \mathcal{O}$)

$$\Theta_{E,\mathfrak{a}}(Q + R_0)^{\mathrm{Frob}_\mathfrak{p}} = \Theta_{E,\mathfrak{a}}(\pi Q + \pi R_0) = \Theta_{E,\mathfrak{a}}(\pi Q)$$

so this completes the proof. □

7.3 Elliptic Curves over K

Since the function $\Theta_{E,\mathfrak{a}}$ depends only on the isomorphism class of E over $\mathbf{C}$, we need to provide it with information that depends on E itself to make it sensitive enough to "see" the value of the L-function of E at 1. Following Coates and Wiles [CW1] we will write down a product of translates of $\Theta_{E,\mathfrak{a}}$ and then show that it has the connections we need with L-values.

From now on suppose that our elliptic curve E is defined over K, ψ is the Hecke character attached to E by Theorem 5.15, $\mathfrak{f}$ is the conductor of ψ, and $\mathfrak{a}$ is prime to $\mathfrak{f}$ as well as to 6. For $P \in E(\bar{K})$ let τ_P denote translation P, so τ_P is a rational function defined over $K(P)$.

Fix an $\mathcal{O}$-generator S of $E[\mathfrak{f}]$. By Corollary 5.20(i) $S \in E(K(\mathfrak{f}))$, and we define

$$\Lambda_{E,\mathfrak{a}} = \Lambda_{E,\mathfrak{a},S} = \prod_{\sigma \in \mathrm{Gal}(K(\mathfrak{f})/K)} \Theta_{E,\mathfrak{a}} \circ \tau_{S^\sigma}.$$

Proposition 7.8. (i) *$\Lambda_{E,\mathfrak{a}}$ is a rational function defined over K.*

(ii) *If B is a set of ideals of $\mathcal{O}$, prime to $\mathfrak{af}$, such that the Artin map $\mathfrak{b} \mapsto [\mathfrak{b}, K(\mathfrak{f})/K]$ is a bijection from B to $\mathrm{Gal}(K(\mathfrak{f})/K)$, then*

$$\Lambda_{E,\mathfrak{a}}(P) = \prod_{\mathfrak{b} \in B} \Theta_{E,\mathfrak{a}}(\psi(\mathfrak{b})S + P).$$

(iii) *If $\mathfrak{r}$ is an ideal of $\mathcal{O}$ and $Q \in E[\mathfrak{r}]$, $Q \notin E[\mathfrak{f}]$, then $\Lambda_{E,\mathfrak{a}}(Q)$ is a global unit in $K(E[\mathfrak{r}])$.*

Proof. The first assertion is clear, (ii) is immediate from Corollary 5.16(ii), and (iii) follows from Theorem 7.4(iii). □

7.4 Expansions over C

We continue to suppose that E is defined over K. Fix a Weierstrass model of E (over K) and let $L \subset \mathbf{C}$ be the corresponding lattice given by Theorem 2.3(ii); then $\mathcal{O}L = L$ (Proposition 2.6) so we can choose $\Omega \in \mathbf{C}^\times$ such that $L = \Omega\mathcal{O}$. The map $\xi(z) = (\wp(z;L), \wp'(z;L)/2)$ is an isomorphism $\mathbf{C}/L \xrightarrow{\sim} E(\mathbf{C})$, and we define $\Theta_{L,\mathfrak{a}} = \Theta_{E,\mathfrak{a}} \circ \xi$, i.e.,

$$\Theta_{L,\mathfrak{a}}(z) = \alpha^{-12}\Delta(L)^{\mathrm{N}\mathfrak{a}-1} \prod_{u \in \mathfrak{a}^{-1}L/L-0} (\wp(z;L) - \wp(u;L))^{-6}.$$

Definition 7.9. Define

$$\begin{aligned}
A(L) &= \pi^{-1}\mathrm{area}(\mathbf{C}/L),\\
s_2(L) &= \lim_{s\to 0^+} \sum_{0\neq\omega\in L} \omega^{-2}|\omega|^{-2s},\\
\eta(z;L) &= A(L)^{-1}\bar{z} + s_2(L)z,\\
\theta(z;L) &= \Delta(L)e^{-6\eta(z;L)z}\sigma(z;L)^{12}.
\end{aligned}$$

Lemma 7.10. $\Theta_{L,\mathfrak{a}}(z) = \theta(z;L)^{\mathbf{N}\mathfrak{a}}/\theta(z;\mathfrak{a}^{-1}L)$.

Proof. Write $f(z) = \theta(z;L)^{\mathbf{N}\mathfrak{a}}/\theta(z;\mathfrak{a}^{-1}L)$. Note that although $\theta(z;L)$ is not holomorphic (because of the $\bar{z}$ in the definition of $\eta(z;L)$), $f(z)$ is holomorphic. One can check explicitly, using well-known properties of $\sigma(z;L)$ (see [dS] §II.2.1), that $f(z)$ is periodic with respect to L and its divisor on $\mathbf{C}/L$ is $12\mathbf{N}\mathfrak{a}[0] - 12\sum_{v\in\mathfrak{a}^{-1}L/L}[v]$.

Thus by Lemma 7.5, $\Theta_{L,\mathfrak{a}} = \lambda f$ for some $\lambda \in \mathbf{C}^\times$. At $z = 0$, both functions have Laurent series beginning $\alpha^{-12}\Delta(L)^{\mathbf{N}\mathfrak{a}-1}z^{12(\mathbf{N}\mathfrak{a}-1)}$, so $\lambda = 1$. □

Definition 7.11. For $k \geq 1$ define the Eisenstein series

$$E_k(z;L) = \lim_{s\to k}\sum_{\omega\in L}\frac{(\bar{z}+\bar{\omega})^k}{|z+\omega|^{2s}}$$
$$= \sum_{\omega\in L}\frac{1}{(z+\omega)^k} \quad \text{if } k \geq 3$$

where the limit means evaluation of the analytic continuation at $s = k$.

Proposition 7.12.

$$E_1(z;L) = \log(\sigma(z;L))' - s_2(L)z - A(L)^{-1}\bar{z},$$
$$E_2(z;L) = \wp(z;L) + s_2(L),$$
$$E_k(z;L) = \frac{(-1)^k}{(k-1)!}\left(\frac{d}{dz}\right)^{k-2}\wp(z;L) \quad \textit{if } k \geq 3.$$

Proof. The third equality is immediate from the definition of $\wp(z;L)$. For the first two, see [CW1] pp. 242–243 or [GS] Proposition 1.5. □

Theorem 7.13. *For every* $k \geq 1$,

$$\left(\frac{d}{dz}\right)^k \log\Theta_{L,\mathfrak{a}}(z) = 12(-1)^{k-1}(k-1)!(\mathbf{N}\mathfrak{a}E_k(z;L) - E_k(z;\mathfrak{a}^{-1}L)).$$

Proof. By Lemma 7.10

$$\left(\frac{d}{dz}\right)^k \log\Theta_{L,\mathfrak{a}}(z) = \left(\frac{d}{dz}\right)^{k-1}\left(\mathbf{N}\mathfrak{a}\frac{d}{dz}\log(\theta(z;L)) - \frac{d}{dz}\log(\theta(z;\mathfrak{a}^{-1}L))\right).$$

The definition of θ shows that

$$\log(\theta(z;L)) = \log(\Delta(L)) - 6s_2(L)z^2 - 6A(L)^{-1}z\bar{z} + 12\log(\sigma(z;L)).$$

Now the theorem follows from Proposition 7.12. □

Definition 7.14. Define the Hecke L-functions associated to powers of $\bar{\psi}$ to be the analytic continuations of the Dirichlet series

$$L(\bar{\psi}^k, s) = \sum \frac{\bar{\psi}^k(\mathfrak{b})}{\mathbf{N}\mathfrak{b}^s},$$

summing over ideals $\mathfrak{b}$ of $\mathcal{O}$ prime to the conductor of $\bar{\psi}^k$. If $\mathfrak{m}$ is an ideal of $\mathcal{O}$ divisible by $\mathfrak{f}$ and $\mathfrak{c}$ is an ideal prime to $\mathfrak{m}$, we define the partial L-function $L_{\mathfrak{m}}(\bar{\psi}^k, s, \mathfrak{c})$ be the same formula, but with the sum restricted to ideals of K prime to $\mathfrak{m}$ such that $[\mathfrak{b}, K(\mathfrak{m})/K] = [\mathfrak{c}, K(\mathfrak{m})/K]$.

Recall that $\Omega \in \mathbf{C}^\times$ is such that $L = \Omega\mathcal{O}$.

Proposition 7.15. *Suppose $v \in KL/L$ has order $\mathfrak{m}$, where $\mathfrak{m}$ is divisible by $\mathfrak{f}$. Then for every $k \geq 1$,*

$$E_k(v; L) = v^{-k}\psi(\mathfrak{c})^k L_{\mathfrak{m}}(\bar{\psi}^k, k, \mathfrak{c})$$

where $\mathfrak{c} = \Omega^{-1}v\mathfrak{m}$.

Proof. Let μ be a generator of $\mathfrak{m}$, so that $v = \alpha\Omega/\mu$ for some $\alpha \in \mathcal{O}$ prime to $\mathfrak{m}$. For s large,

$$\sum_{\omega \in L} \frac{(\bar{v} + \bar{\omega})^k}{|v + \omega|^{2s}} = \frac{\mathbf{N}\mu^s}{\bar{\mu}^k} \frac{\bar{\Omega}^k}{|\Omega|^{2s}} \sum_{\beta \in \mathcal{O},\ \beta \equiv \alpha \ (\mathrm{mod}\ \mathfrak{m})} \frac{\bar{\beta}^k}{|\beta|^{2s}}.$$

By Corollary 5.16(i), if we define

$$\epsilon(\beta) = \psi(\beta\mathcal{O})/\beta$$

then ϵ is a multiplicative map from $\{\beta \in \mathcal{O} : \beta \text{ is prime to } \mathfrak{f}\}$ to $\mathcal{O}^\times$. By definition of the conductor, ϵ factors through $(\mathcal{O}/\mathfrak{f})^\times$. Thus if $\beta \equiv \alpha \pmod{\mathfrak{m}}$,

$$\bar{\beta} = \bar{\psi}(\beta\mathcal{O})\frac{\psi(\alpha\mathcal{O})}{\alpha}.$$

Therefore

$$\sum_{\beta \in \mathcal{O},\ \beta \equiv \alpha \ (\mathrm{mod}\ \mathfrak{m})} \frac{\bar{\beta}^k}{|\beta|^{2s}} = \frac{\psi(\alpha\mathcal{O})^k}{\alpha^k} \sum_{\mathfrak{b} \subset \mathcal{O},\, [\mathfrak{b}, K(\mathfrak{m})/K] = [\alpha\mathcal{O}, K(\mathfrak{m})/K]} \frac{\bar{\psi}(\mathfrak{b})^k}{\mathbf{N}\mathfrak{b}^s}$$

$$= \frac{\psi(\mathfrak{c})^k}{\alpha^k} L_{\mathfrak{m}}(\bar{\psi}^k, s, \sigma_{\mathfrak{c}})$$

and the proposition follows. □

Definition 7.16. Fix a generator f of $\mathfrak{f}$ and a set B of ideals of $\mathcal{O}$, prime to $\mathfrak{a}\mathfrak{f}$, such that the Artin map $\mathfrak{b} \mapsto [\mathfrak{b}, K(\mathfrak{f})/K]$ is a bijection from B to $\mathrm{Gal}(K(\mathfrak{f})/K)$. Let $u = \Omega/f \in \mathfrak{f}^{-1}L$ and define

$$\Lambda_{L,\mathfrak{a}}(z) = \Lambda_{L,\mathfrak{a},f}(z) = \Lambda_{E,\mathfrak{a},\xi(u)}(\xi(z)) = \prod_{\mathfrak{b} \in B} \Theta_{L,\mathfrak{a}}(\psi(\mathfrak{b})u + z).$$

By Proposition 7.8(ii), $\Lambda_{L,\mathfrak{a}} = \Lambda_{E,\mathfrak{a}} \circ \xi$.

Theorem 7.17. *For every $k \geq 1$,*

$$\left(\frac{d}{dz}\right)^k \log \Lambda_{L,\mathfrak{a}}(z) \mid_{z=0} = 12(-1)^{k-1}(k-1)! f^k (\mathbf{N}\mathfrak{a} - \psi(\mathfrak{a})^k) \Omega^{-k} L_{\mathfrak{f}}(\bar{\psi}^k, k).$$

Proof. By Theorem 7.13

$$\left(\frac{d}{dz}\right)^k \log \Lambda_{L,\mathfrak{a}}(z) \mid_{z=0} = \sum_{\mathfrak{b}\in B} \left(\frac{d}{dz}\right)^k \log \Theta_{L,\mathfrak{a}}(z) \mid_{z=\psi(\mathfrak{b})u}$$

$$= 12(-1)^{k-1}(k-1)! \left(\mathbf{N}\mathfrak{a} \sum_{\mathfrak{b}\in B} E_k(\psi(\mathfrak{b})u; L) - \sum_{\mathfrak{b}\in B} E_k(\psi(\mathfrak{b})u; \mathfrak{a}^{-1}L)\right).$$

By Proposition 7.15,

$$\sum_{\mathfrak{b}\in B} E_k(\psi(\mathfrak{b})u; L) = \sum_{\mathfrak{b}\in B} (\psi(\mathfrak{b})u)^{-k} \psi(\mathfrak{b})^k L_{\mathfrak{f}}(\bar{\psi}^k, k, \mathfrak{b}) = u^{-k} L_{\mathfrak{f}}(\bar{\psi}^k, k).$$

By inspection (and Corollary 5.16(i)) $E_k(z; \mathfrak{a}^{-1}L) = \psi(\mathfrak{a})^k E_k(\psi(\mathfrak{a})z; L)$, so

$$\sum_{\mathfrak{b}\in B} E_k(\psi(\mathfrak{b})u; \mathfrak{a}^{-1}L) = u^{-k} \psi(\mathfrak{a})^k L_{\mathfrak{f}}(\bar{\psi}^k, k).$$

□

Although we will not use it explicitly, the following theorem of Damerell is a corollary of this computation.

Corollary 7.18 (Damerell's Theorem). *For every $k \geq 1$,*

$$\Omega^{-k} L(\bar{\psi}^k, k) \in K.$$

Proof. By Proposition 7.8(i), $\Lambda_{L,\mathfrak{a}}(z)$ is a rational function of $\wp(z; L)$ and $\wp'(z; L)$ with coefficients in K. Differentiating the relation (from Theorem 2.3)

$$\wp'(z; L)^2 = 4\wp(z; L)^3 + 4a\wp(z; L) + 4b$$

shows that all derivatives $\wp^{(k)}(z; L)$ also belong to $K(\wp(z; L), \wp'(z; L))$, and hence $\Lambda_{L,\mathfrak{a}}^{(k)}$ does as well. Thus the corollary follows from Theorem 7.17. □

7.5 $\mathfrak{p}$-adic Expansions

Keep the notation of the previous sections. Fix a prime $\mathfrak{p}$ of K where E has good reduction, $\mathfrak{p} \nmid 6$. Suppose that our chosen Weierstrass model of E has good reduction at $\mathfrak{p}$ and that the auxiliary ideal $\mathfrak{a}$ is prime to $\mathfrak{p}$ as well as $6\mathfrak{f}$. Let $\hat{E}$ be the formal group attached to E over $\mathcal{O}_{\mathfrak{p}}$ as in §3.2, and $x(Z), y(Z) \in \mathcal{O}_{\mathfrak{p}}[[Z]]$ the power series of Theorem 3.7.

Definition 7.19. Let $\lambda_{\hat{E}}(Z) \in Z + Z^2 K_{\mathfrak{p}}[[Z]]$ be the logarithm map of $\hat{E}$ from Definition 3.10, so that $\lambda'_{\hat{E}}(Z) \in \mathcal{O}_{\mathfrak{p}}[[Z]]^\times$, and define an operator D on $\mathcal{O}_{\mathfrak{p}}[[Z]]$ by

$$D = \frac{1}{\lambda'_{\hat{E}}(Z)} \frac{d}{dZ}.$$

Proposition 7.20. *Identifying (x, y) both with $(\wp(z; L), \frac{1}{2}\wp'(z; L))$ and with $(x(Z), y(Z))$ leads to a commutative diagram*

$$\begin{array}{ccccccc}
K(\wp(z), \wp'(z)) & \xleftarrow{\sim} & K(E) & \xrightarrow{\sim} & K(x(Z), y(Z)) & \hookrightarrow & K_{\mathfrak{p}}((Z)) \\
\downarrow{\scriptstyle \frac{d}{dz}} & & \downarrow & & \downarrow{\scriptstyle D} & & \downarrow{\scriptstyle D} \\
K(\wp(z), \wp'(z)) & \xleftarrow{\sim} & K(E) & \xrightarrow{\sim} & K(x(Z), y(Z)) & \hookrightarrow & K_{\mathfrak{p}}((Z)).
\end{array}$$

Proof. Differentiating the relation $\wp'(z)^2 = 4\wp(z)^3 + 4a\wp(z) + 4b$ shows that

$$\wp''(z) = 6\wp(x)^2 + 2a \in K_{\mathfrak{p}}(\wp(z), \wp'(z)).$$

Thus, since both vertical maps are derivations, we need only check that $D(x(Z)) = 2y(Z)$ and $D(y(Z)) = 3x(Z)^2 + a$. (In fact, it would be enough to check either equality.) Both equalities are immediate from the definition (Definition 3.10) of $\hat{\omega}$ and $\lambda_{\hat{E}}$. □

Definition 7.21. Let $\Lambda_{\mathfrak{p},\mathfrak{a}}(Z)$ be the image of $\Lambda_{E,\mathfrak{a}}$ in $K_{\mathfrak{p}}((Z))$ under the map of Proposition 7.20.

Theorem 7.22. (i) $\Lambda_{\mathfrak{p},\mathfrak{a}}(Z) \in \mathcal{O}_{\mathfrak{p}}[[Z]]^\times$.
(ii) *For every $k \geq 1$,*

$$D^k \log(\Lambda_{\mathfrak{p},\mathfrak{a}}(Z))\,|_{Z=0} = 12(-1)^{k-1}(k-1)! f^k(\mathbf{N}\mathfrak{a} - \psi(\mathfrak{a})^k)\Omega^{-k} L(\bar{\psi}^k, k).$$

Proof. Fix an embedding $\bar{K} \hookrightarrow \bar{K}_{\mathfrak{p}}$ so that we can view $x(R) \in \bar{K}_{\mathfrak{p}}$ when $R \in E[\mathfrak{f}]$. Let $\mathcal{R}$ be the ring of integers of $\bar{K}_{\mathfrak{p}}$.

Consider one of the factors $x(\psi(\mathfrak{b})S + P) - x(Q)$ of $\Lambda_{E,\mathfrak{a}}(P)$, with $Q \in E[\mathfrak{a}] - O$. The explicit addition law for $x(P)$ ([Si] §III.2.3) shows that

$$x(\psi(\mathfrak{b})S + P) - x(Q) = \frac{(y(P) - y(\psi(\mathfrak{b})S))^2}{(x(P) - x(\psi(\mathfrak{b})S))^2} - x(P) - x(\psi(\mathfrak{b})S) - x(Q).$$

By Lemmas 7.3(ii) and 3.5, $x(\psi(\mathfrak{b})S), y(\psi(\mathfrak{b})S), x(Q) \in \mathcal{R}$. Substituting $x(Z)$ for $x(P)$, $y(Z)$ for $y(P)$ and using the expansions in Theorem 3.7 to show that

$$x(Z) \in Z^{-2} + Z\mathcal{O}_{\mathfrak{p}}[[Z]], \quad y(Z) \in -Z^{-3} + \mathcal{O}_{\mathfrak{p}}[[Z]]$$

gives

$$x(\psi(\mathfrak{b})S + P) - x(Q) \quad \mapsto \quad g_{\mathfrak{b},Q}(Z) \in \mathcal{R}[[Z]]$$

under the map of Proposition 7.20, where $g_{\mathfrak{b},Q}$ satisfies

$$g_{\mathfrak{b},Q}(0) = x(\psi(\mathfrak{b})S) - x(Q) \in \mathcal{R}^\times$$

by Lemma 7.3(iii), so $g_{\mathfrak{b},Q}(Z) \in \mathcal{R}[[Z]]^\times$. Also $\Delta(E), \alpha \in \mathcal{O}_\mathfrak{p}^\times$ since our Weierstrass equation has good reduction at $\mathfrak{p}$ and $\mathfrak{p} \nmid \mathfrak{a}$. Thus

$$\Lambda_{\mathfrak{p},\mathfrak{a}}(Z) = \Delta(E)^{(\mathbf{N}\mathfrak{a}-1)\#(B)}\alpha^{-12\#(B)} \prod_{\mathfrak{b},Q} g_{\mathfrak{b},Q}(Z)^{-6} \in \mathcal{R}[[Z]]^\times.$$

Since we already know $\Lambda_{\mathfrak{p},\mathfrak{a}} \in K_\mathfrak{p}((Z))$, this proves (i).

The second assertion is immediate from Theorem 7.17 and Proposition 7.20. □

8 Euler Systems

In this section we introduce Kolyvagin's concept of an Euler system (of which the elliptic units of §7 are an example) and we show how to use an Euler system to construct certain principal ideals in abelian extensions of K. In the next section we use these principal ideals (viewed as relations in ideal class groups) to bound the ideal class groups of abelian extensions of K.

As in the previous section, fix an imaginary quadratic field K and an elliptic curve E defined over K with complex multiplication by the ring of integers $\mathcal{O}$ of K. Let $\mathfrak{f}$ be the conductor of the Hecke character ψ of E, and fix a generator f of $\mathfrak{f}$.

Fix a prime $\mathfrak{p}$ of K not dividing $6\mathfrak{f}$, and for $n \geq 1$ let $K_n = K(E[\mathfrak{p}^n])$. Let p denote the rational prime below $\mathfrak{p}$. Fix a nontrivial ideal $\mathfrak{a}$ of $\mathcal{O}$ prime to $6\mathfrak{f}\mathfrak{p}$. Let $\mathcal{R} = \mathcal{R}(\mathfrak{a})$ denote the set of squarefree ideals of $\mathcal{O}$ prime to $6\mathfrak{f}\mathfrak{a}\mathfrak{p}$, and if $\mathfrak{r} \in \mathcal{R}$ let $K_n(\mathfrak{r}) = K_n(E[\mathfrak{r}]) = K(E[\mathfrak{r}\mathfrak{p}^n])$. The letter $\mathfrak{q}$ will always denote a prime of $\mathcal{R}$.

Also as in the previous section, fix a Weierstrass model of E which is minimal at $\mathfrak{p}$, let $L = \Omega\mathcal{O} \subset \mathbf{C}$ be the corresponding lattice given by Theorem 2.3(ii), and define $\xi = (\wp(\,\cdot\,;L), \wp'(\,\cdot\,;L)/2) : \mathbf{C}/L \xrightarrow{\sim} E(\mathbf{C})$.

8.1 The Euler System

Definition 8.1. If $\mathfrak{r} \in \mathcal{R}$ and $n \geq 0$ define

$$\eta_n(\mathfrak{r}) = \eta_n^{(\mathfrak{a})}(\mathfrak{r}) = \Lambda_{E,\mathfrak{a},\xi(\Omega/f)}(\xi(\psi(\mathfrak{p}^n\mathfrak{r})^{-1}\Omega)) = \Lambda_{L,\mathfrak{a}}(\psi(\mathfrak{p}^n\mathfrak{r})^{-1}\Omega).$$

where $\Lambda_{L,\mathfrak{a}}$ is as in Definition 7.16.

Proposition 8.2. *Suppose* $\mathfrak{r} \in \mathcal{R}$ *and* $n \geq 1$.

(i) $\eta_n(\mathfrak{r})$ *is a global unit in* $K_n(\mathfrak{r})$.

(ii) *If* $\mathfrak{q}$ *is a prime and* $\mathfrak{r}\mathfrak{q} \in \mathcal{R}$, *then*

$$\mathbf{N}_{K_n(\mathfrak{r}\mathfrak{q})/K_n(\mathfrak{r})}\eta_n(\mathfrak{q}\mathfrak{r}) = \eta_n(\mathfrak{r})^{1-\mathrm{Frob}_{\mathfrak{q}}^{-1}}.$$

(iii) $\mathbf{N}_{K_{n+1}(\mathfrak{r})/K_n(\mathfrak{r})}\eta_{n+1}(\mathfrak{r}) = \eta_n(\mathfrak{r})$.

Proof. Assertion (i) is just a restatement of Proposition 7.8(iii), and (ii) and (iii) are immediate from Corollary 7.7. □

8.2 Kolyvagin's Derivative Construction

Definition 8.3. Write $G_\mathfrak{r} = \mathrm{Gal}(K_n(\mathfrak{r})/K_n)$. By Corollary 5.20(ii), $G_\mathfrak{r}$ is independent of $n \geq 1$, and we have natural isomorphisms

$$\begin{array}{ccc} G_\mathfrak{r} & = & \prod_{\mathfrak{q}|\mathfrak{r}} G_\mathfrak{q} \\ \downarrow & & \downarrow \\ (\mathcal{O}/\mathfrak{r})^\times & = & \prod_{\mathfrak{q}|\mathfrak{r}} (\mathcal{O}/\mathfrak{q})^\times. \end{array}$$

If $\mathfrak{q} \mid \mathfrak{r}$ this allows us to view $G_\mathfrak{q}$ either as a quotient or a subgroup of $G_\mathfrak{r}$. By Corollary 5.20 if $\mathfrak{q}\mathfrak{r} \in \mathcal{R}$ then $K_n(\mathfrak{q}\mathfrak{r})/K_n(\mathfrak{r})$ is cyclic of degree $\mathbf{N}\mathfrak{q} - 1$, totally ramified at all primes above $\mathfrak{q}$ and unramified at all other primes.

For every $\mathfrak{r} \in \mathcal{R}$ define

$$N_\mathfrak{r} = \sum_{\sigma \in G_\mathfrak{r}} \sigma \in \mathbf{Z}[G_\mathfrak{r}]$$

so we clearly have

$$N_\mathfrak{r} = \prod_{\mathfrak{q}|\mathfrak{r}} N_\mathfrak{q}.$$

For every $n \geq 1$ and $\mathfrak{r} \in \mathcal{R}$, let $x_{n,\mathfrak{r}}$ be an indeterminate and define $X_{n,\mathfrak{r}}$ to be the $\mathrm{Gal}(K_n(\mathfrak{r})/K)$-module $Y_{n,\mathfrak{r}}/Z_{n,\mathfrak{r}}$ where

$$Y_{n,\mathfrak{r}} = \bigoplus_{\mathfrak{s}|\mathfrak{r}} \mathbf{Z}[\mathrm{Gal}(K_n(\mathfrak{s})/K)]x_{n,\mathfrak{s}},$$

$$Z_{n,\mathfrak{r}} = \sum_{\mathfrak{q}\mathfrak{s}|\mathfrak{r}} \mathbf{Z}[\mathrm{Gal}(K_n(\mathfrak{r})/K)]\left(N_\mathfrak{q}x_{n,\mathfrak{q}\mathfrak{s}} - (1 - \mathrm{Frob}_\mathfrak{q}^{-1})x_{n,\mathfrak{s}}\right) \subset Y_{n,\mathfrak{r}}.$$

In other words, $X_{n,\mathfrak{r}}$ is the quotient of the free $\mathbf{Z}[\mathrm{Gal}(K_n(\mathfrak{r})/K)]$-module on $\{x_{n,\mathfrak{s}} : \mathfrak{s} \mid \mathfrak{r}\}$ by the relations

- $G_{\mathfrak{r}/\mathfrak{s}}$ acts trivially on $x_{n,\mathfrak{s}}$, and
- $N_\mathfrak{q}x_{n,\mathfrak{q}\mathfrak{s}} = (1 - \mathrm{Frob}_\mathfrak{q}^{-1})x_{n,\mathfrak{s}}$ if $\mathfrak{q}\mathfrak{s} \mid \mathfrak{r}$.

For every prime $\mathfrak{q} \in \mathcal{R}$ fix once and for all a generator $\sigma_\mathfrak{q}$ of $G_\mathfrak{q}$ and define

$$D_\mathfrak{q} = \sum_{i=1}^{\mathbf{N}\mathfrak{q}-2} i\sigma_\mathfrak{q}^i \in \mathbf{Z}[G_\mathfrak{q}]$$

and for $\mathfrak{r} \in \mathcal{R}$

$$D_\mathfrak{r} = \prod_{\mathfrak{q}|\mathfrak{r}} D_\mathfrak{q} \in \mathbf{Z}[G_\mathfrak{r}].$$

If M is a power of p and $n \geq 1$ define $\mathcal{R}_{n,M} \subset \mathcal{R}$ to be the set of ideals $\mathfrak{r} \in \mathcal{R}$ such that for every prime $\mathfrak{q}$ dividing $\mathfrak{r}$,

- $\mathfrak{q}$ splits completely in K_n/K
- $\mathbf{N}\mathfrak{q} \equiv 1 \pmod{M}$.

Proposition 8.4. *Suppose M is a power of p, $n \geq 1$, and $\mathfrak{r} \in \mathcal{R}_{n,M}$.*

(i) *$X_{n,\mathfrak{r}}$ has no $\mathbf{Z}$-torsion.*
(ii) *$D_\mathfrak{r} x_{n,\mathfrak{r}} \in (X_{n,\mathfrak{r}}/MX_{n,\mathfrak{r}})^{G_\mathfrak{r}}$.*

Proof. For every prime $\mathfrak{q} \in \mathcal{R}$ and divisor $\mathfrak{s}$ of $\mathfrak{r}$, define

$$B_\mathfrak{q} = G_\mathfrak{q} - \{1\},$$
$$B_\mathfrak{s} = \prod_{\mathfrak{q}|\mathfrak{s}} B_\mathfrak{q} = \left\{ \prod_{\mathfrak{q}|\mathfrak{s}} g_\mathfrak{q} : g_\mathfrak{q} \in B_\mathfrak{q} \right\} \subset G_\mathfrak{r},$$
$$B = \cup_{\mathfrak{s}|\mathfrak{r}} B_\mathfrak{s} x_{n,\mathfrak{s}} \subset X_{n,\mathfrak{r}}$$

Then one can show by an easy combinatorial argument (see [Ru2] Lemma 2.1) that $X_{n,\mathfrak{r}}$ is a free $\mathbf{Z}$-module with basis B, which proves (i).

Note that

$$(\sigma_\mathfrak{q} - 1)D_\mathfrak{q} = \mathbf{N}\mathfrak{q} - 1 - N_\mathfrak{q}.$$

We will prove (ii) by induction on the number of primes dividing $\mathfrak{r}$. Suppose $\mathfrak{q} \mid \mathfrak{r}$, $\mathfrak{r} = \mathfrak{q}\mathfrak{s}$. Then

$$\begin{aligned}(\sigma_\mathfrak{q} - 1)D_\mathfrak{r} x_{n,\mathfrak{r}} &= (\sigma_\mathfrak{q} - 1)D_\mathfrak{q} D_\mathfrak{s} x_{n,\mathfrak{r}} \\ &= (\mathbf{N}\mathfrak{q} - 1)D_\mathfrak{s} x_{n,\mathfrak{r}} - (1 - \mathrm{Frob}_\mathfrak{q}^{-1})D_\mathfrak{s} x_{n,\mathfrak{s}}.\end{aligned}$$

Since $\mathfrak{q} \in \mathcal{R}_{n,M}$, $M \mid \mathbf{N}\mathfrak{q} - 1$ and $\mathrm{Frob}_\mathfrak{q} \in G_\mathfrak{s}$, so by the induction hypothesis

$$(\sigma_\mathfrak{q} - 1)D_\mathfrak{r} x_{n,\mathfrak{r}} \in MX_{n,\mathfrak{r}}.$$

Since the $\sigma_\mathfrak{q}$ generate $G_\mathfrak{r}$, this proves the proposition. □

Definition 8.5. An *Euler system* is a collection of global units

$$\{\eta(n,\mathfrak{r}) \in K_n(\mathfrak{r})^\times : n \geq 1, \ \mathfrak{r} \in \mathcal{R}\}$$

satisfying

$$\mathbf{N}_{K_n(\mathfrak{q}\mathfrak{r})/K_n(\mathfrak{r})}\eta(n,\mathfrak{q}\mathfrak{r}) = \eta(n,\mathfrak{r})^{1-\mathrm{Frob}_\mathfrak{q}^{-1}}, \tag{8}$$

$$\mathbf{N}_{K_{n+1}(\mathfrak{r})/K_n(\mathfrak{r})}\eta(n+1,\mathfrak{r}) = \eta(n,\mathfrak{r}). \tag{9}$$

Equivalently, an Euler system is a Galois equivariant map

$$\eta : \varinjlim_{n,\mathfrak{r}} X_{n,\mathfrak{r}} \longrightarrow \bigcup_{n,\mathfrak{r}} K_n(\mathfrak{r})^\times$$

such that $\eta(x_{n,\mathfrak{r}})$ is a global unit for every n and $\mathfrak{r}$. We will use these two definitions interchangeably.

For example, by Proposition 8.2 we can define an Euler system by

$$\eta(n,\mathfrak{r}) = \eta_n(\mathfrak{r}).$$

Proposition 8.6. *Suppose η is an Euler system and $\mathfrak{q} \in \mathcal{R}$ is a prime. Write $\mathbf{N}\mathfrak{q} - 1 = dp^k$ with d prime to p. Then for every $n \geq 1$ and every $\mathfrak{r} \in \mathcal{R}$ prime to $\mathfrak{q}$,*

$$\eta(n,\mathfrak{q}\mathfrak{r})^d \equiv \eta(n,\mathfrak{r})^{d\mathrm{Frob}_\mathfrak{q}^{-1}}$$

modulo every prime above $\mathfrak{q}$.

Proof. Suppose $m \geq n$, and let $G = \mathrm{Gal}(K_m(\mathfrak{q}\mathfrak{r})/K_n(\mathfrak{q}\mathfrak{r}))$. Fix a prime $\mathfrak{Q}$ of $K_m(\mathfrak{q}\mathfrak{r})$ above $\mathfrak{q}$, and let H be the decomposition group of $\mathfrak{q}$ in G. Let $H' \subset G$ be a set of coset representatives for G/H, and define

$$N_H = \sum_{\gamma\in H} \gamma, \quad N_{H'} = \sum_{\gamma \in H'} \gamma$$

so that $N_H N_{H'} = \sum_{\gamma\in G} \gamma$.

Since $\mathfrak{q}$ is totally ramified in $K_m(\mathfrak{q}\mathfrak{r})/K_m(\mathfrak{r})$, the Euler system distribution relation (8) reduces modulo $\mathfrak{Q}$ to

$$\eta(m,\mathfrak{q}\mathfrak{r})^{\mathbf{N}\mathfrak{q}-1} \equiv (\eta(m,\mathfrak{r})^{\mathrm{Frob}_\mathfrak{q}^{-1}})^{\mathbf{N}\mathfrak{q}-1} \pmod{\mathfrak{Q}}.$$

On the other hand, since H is generated by the Frobenius of $\mathfrak{q}$, if h denotes the degree of the residue field extension at $\mathfrak{q}$ in $K_n(\mathfrak{r})/K$ then (9) reduces to

$$\eta(n,\mathfrak{r}) = \eta(m,\mathfrak{r})^{N_{H'}N_H} \equiv (\eta(m,\mathfrak{r})^{N_{H'}})^t \pmod{\mathfrak{Q}}$$

and similarly $\eta(n,\mathfrak{q}\mathfrak{r}) \equiv (\eta(m,\mathfrak{q}\mathfrak{r})^{N_{H'}})^t \pmod{\mathfrak{Q}}$, where

$$t = \sum_{i=0}^{\#(H)-1} (\mathbf{N}\mathfrak{q}^h)^i \equiv \#(H) \pmod{\mathbf{N}\mathfrak{q} - 1}.$$

Recall that p^k is the highest power of p dividing $\mathbf{N}\mathfrak{q} - 1$. Since the decomposition group of $\mathfrak{q}$ in K_∞/K is infinite, for m sufficiently large we will have $p^k \mid t$, and then combining the congruences above proves the proposition. □

For the Euler system of elliptic units, one can prove directly, using Lemma 7.3, that the congruence of Proposition 8.6 holds with $d = 1$.

Definition 8.7. Suppose η is an Euler system, $n \geq 1$ and $\mathfrak{r} \in \mathcal{R}$. Using the map $X_{n,\mathfrak{r}} \to K_n(\mathfrak{r})^\times$ corresponding to η, we define a 1-cocycle $c = c_{\eta,n,\mathfrak{r}}$: $G_\mathfrak{r} \to K_n(\mathfrak{r})^\times$ by

$$c(\sigma) = \eta\left(\frac{(\sigma - 1)D_\mathfrak{r} x_{n,\mathfrak{r}}}{M}\right) \quad \text{for } \sigma \in G_\mathfrak{r}.$$

This is well defined by Proposition 8.4. Since $H^1(G_\mathfrak{r}, K_n(\mathfrak{r})^\times) = 0$, there is a $\beta \in K_n(\mathfrak{r})^\times$ such that $c(\sigma) = \beta^\sigma/\beta$ for every $\sigma \in G_\mathfrak{r}$. Then $\eta(x_{n,\mathfrak{r}})^{D_\mathfrak{r}}/\beta^M \in K_n^\times$ and we define

$$\kappa_{n,M}(\mathfrak{r}) = \eta(x_{n,\mathfrak{r}})^{D_\mathfrak{r}}/\beta^M \in K_n^\times/(K_n^\times)^M.$$

Since β is uniquely determined modulo $K_n^\times$, $\kappa_{n,M}(\mathfrak{r})$ is independent of the choice of β.

Remark 8.8. It is quite easy to show for every Euler system η, every n, and every $\mathfrak{r} \in \mathcal{R}_{n,M}$ that $\eta(n,\mathfrak{r})^{(\sigma-1)D_\mathfrak{r}}$ is an M-th power (Proposition 8.4(ii)). The reason for introducing the "universal Euler system" $X_{n,\mathfrak{r}}$ is to show that $\eta(n,\mathfrak{r})^{(\sigma-1)D_\mathfrak{r}}$ has a *canonical* M-th root, even when $K_n(\mathfrak{r})$ contains M-th roots of unity (Proposition 8.4(i)). This fact was used to construct the cocycle c above.

We next want to determine the ideal generated by $\kappa_{n,M}(\mathfrak{r})$ (modulo M-th powers).

Definition 8.9. Fix $n \geq 1$, a power M of p, and temporarily write $F = K_n$, $\mathcal{R}_{F,M} = \mathcal{R}_{n,M}$. Let $\mathcal{O}_F$ denote the ring of integers of F and

$$\mathcal{I}_F = \mathcal{I} = \oplus_\mathfrak{Q} \mathbf{Z}\mathfrak{Q}$$

the group of fractional ideals of F, written additively. For every prime $\mathfrak{q}$ of K let

$$\mathcal{I}_{F,\mathfrak{q}} = \mathcal{I}_\mathfrak{q} = \oplus_{\mathfrak{Q}|\mathfrak{q}} \mathbf{Z}\mathfrak{Q},$$

and if $y \in F^\times$ let $(y) \in \mathcal{I}$ denote the principal ideal generated by y, and $(y)_\mathfrak{q}$, $[y]$, and $[y]_\mathfrak{q}$ the projections of (y) to $\mathcal{I}_\mathfrak{q}$, $\mathcal{I}/M\mathcal{I}$, and $I_\mathfrak{q}/M\mathcal{I}_\mathfrak{q}$, respectively. Note that $[y]$ and $[y]_\mathfrak{q}$ are well defined for $y \in F^\times/(F^\times)^M$.

Suppose $\mathfrak{q} \in \mathcal{R}_{F,M}$, $\mathfrak{Q}$ is a prime of F above $\mathfrak{q}$, and $\bar{\mathfrak{Q}}$ is a prime of $\bar{K}$ above $\mathfrak{Q}$. Recall that $\mathfrak{Q}$ is completely split in F/K, and totally ramified of degree $\mathbf{N}\mathfrak{q} - 1 = \mathbf{N}\mathfrak{Q} - 1$ in $F(\mathfrak{q})/F$. Fix a lift $\sigma_\mathfrak{Q}$ of $\sigma_\mathfrak{q}$ to G_K so that $\sigma_\mathfrak{Q}$ belongs to the inertia group of $\bar{\mathfrak{Q}}$. Then there is an isomorphism

$$\mathbf{Z}/M\mathbf{Z} \xrightarrow{\sim} \mu_M$$

given by $a \mapsto (\pi^{a/M})^{1-\sigma_{\mathfrak{Q}}}$ where $\pi \in K$ is a generator of $\mathfrak{q}$. Let $\mathrm{Frob}_{\mathfrak{Q}} \in G_{F_{\mathfrak{Q}}}$ denote a Frobenius of $\mathfrak{Q}$ and define

$$\phi_{\mathfrak{Q}} : F_{\mathfrak{Q}}^{\times}/(F_{\mathfrak{Q}}^{\times})^M \to \mathbf{Z}/M\mathbf{Z}$$

to be the image of $\mathrm{Frob}_{\mathfrak{Q}}$ under the composition

$$G_{F_{\mathfrak{Q}}} \to \mathrm{Hom}(F_{\mathfrak{Q}}^{\times}, \mu_M) \xrightarrow{\sim} \mathrm{Hom}(F_{\mathfrak{Q}}^{\times}, \mathbf{Z}/M\mathbf{Z})$$

where the first map is the Kummer map and the second is induced by the isomorphism above. Concretely, since $\sigma_{\mathfrak{Q}}$ belongs to the inertia group, we have $\phi_{\mathfrak{Q}}(\alpha) = a$ where a is characterized by

$$(\alpha^{1/M})^{\mathrm{Frob}_{\mathfrak{Q}}-1} = (\pi^{a/M})^{1-\sigma_{\mathfrak{Q}}} \equiv (\beta^{1/M})^{1-\sigma_{\mathfrak{Q}}} \tag{10}$$

modulo the maximal ideal of $\bar{F_{\mathfrak{Q}}}$, where $\beta \in \bar{F_{\mathfrak{Q}}}^{\times}$ is an element satisfying $\mathrm{ord}_{\mathfrak{Q}}(\beta) = a$.

Finally, define

$$\phi_{\mathfrak{q}} : F^{\times}/(F^{\times})^M \to \mathcal{I}_{\mathfrak{q}}/M\mathcal{I}_{\mathfrak{q}}$$

by $\phi_{\mathfrak{q}}(\alpha) = \sum_{\mathfrak{Q}|\mathfrak{q}} \phi_{\mathfrak{Q}}(\alpha)\mathfrak{Q}$. It is not difficult to check that $\phi_{\mathfrak{q}}$ is $\mathrm{Gal}(F/K)$-equivariant, and that $\phi_{\mathfrak{q}}$ induces an isomorphism

$$\phi_{\mathfrak{q}} : (\mathcal{O}_F/\mathfrak{q}\mathcal{O}_F)^{\times}/((\mathcal{O}_F/\mathfrak{q}\mathcal{O}_F)^{\times})^M \xrightarrow{\sim} \mathcal{I}_{\mathfrak{q}}/M\mathcal{I}_{\mathfrak{q}}.$$

Proposition 8.10. *Suppose η is an Euler system, $n \geq 1$, $\mathfrak{r} \in \mathcal{R}_{n,M}$ and $\mathfrak{q}$ is a prime of K.*

(i) *If $\mathfrak{q} \nmid \mathfrak{r}$ then* $[\kappa_{n,M}(\mathfrak{r})]_{\mathfrak{q}} = 0$.
(ii) *If $\mathfrak{q} \mid \mathfrak{r}$ then* $[\kappa_{n,M}(\mathfrak{r})]_{\mathfrak{q}} = \phi_{\mathfrak{q}}(\kappa_{n,M}(\mathfrak{r}/\mathfrak{q}))$.

Proof. Suppose first that $\mathfrak{q} \nmid \mathfrak{r}$. Then $\mathfrak{q}$ is unramified in $K_n(\mathfrak{r})/K_n$, and by definition $\kappa_{n,M}(\mathfrak{r})$ is a global unit times an M-th power in $K_n(\mathfrak{r})^{\times}$, so $\mathrm{ord}_{\mathfrak{Q}}(\kappa_{n,M}(\mathfrak{r})) \equiv 0 \pmod{M}$ for every prime $\mathfrak{Q}$ of K_n above $\mathfrak{q}$. This proves (i).

Now suppose $\mathfrak{q} \mid \mathfrak{r}$, say $\mathfrak{r} = \mathfrak{q}\mathfrak{s}$. By definition

$$\kappa_{n,M}(\mathfrak{r}) = \eta(x_{n,\mathfrak{r}})^{D_{\mathfrak{r}}}/\beta_{\mathfrak{r}}^M, \quad \kappa_{n,M}(\mathfrak{s}) = \eta(x_{n,\mathfrak{s}})^{D_{\mathfrak{s}}}/\beta_{\mathfrak{s}}^M$$

where $\beta_{\mathfrak{r}} \in K_n(\mathfrak{r})^{\times}, \beta_{\mathfrak{s}} \in K_n(\mathfrak{s})^{\times}$ satisfy

$$\beta_{\mathfrak{r}}^{\sigma-1} = \eta((\sigma-1)D_{\mathfrak{r}}x_{n,\mathfrak{r}}/M), \quad \beta_{\mathfrak{s}}^{\sigma-1} = \eta((\sigma-1)D_{\mathfrak{s}}x_{n,\mathfrak{s}}/M)$$

for every $\sigma \in G_{\mathfrak{r}}$.

We will use (10) to evaluate $\phi_{\mathfrak{q}}(\kappa(\mathfrak{s}))$. Fix a prime $\mathfrak{Q}$ of K_n above $\mathfrak{q}$, let $\sigma_{\mathfrak{Q}}$ be as in Definition 8.9, and let d be the prime-to-p-part of $\mathbf{N}\mathfrak{q} - 1$ as in

Proposition 8.6. Modulo every prime above $\mathfrak{Q}$ we have

$$\begin{aligned}
(\kappa_{n,M}(\mathfrak{r})^{d/M})^{1-\sigma_{\mathfrak{Q}}} &= ((\eta(x_{n,\mathfrak{r}})^{D_{\mathfrak{r}}})^{1/M}/\beta_{\mathfrak{r}})^{d(1-\sigma_{\mathfrak{Q}})} \equiv \beta_{\mathfrak{r}}^{d(\sigma_{\mathfrak{q}}-1)} \\
&= \eta((\sigma_{\mathfrak{q}}-1)D_{\mathfrak{r}}x_{n,\mathfrak{r}}/M)^d \\
&= \eta((\mathbf{N}\mathfrak{q}-1-N_{\mathfrak{q}})D_{\mathfrak{s}}x_{n,\mathfrak{r}}/M)^d \\
&= \eta((\mathbf{N}\mathfrak{q}-1)D_{\mathfrak{s}}x_{n,\mathfrak{r}}/M)^d \eta((\mathrm{Frob}_{\mathfrak{q}}^{-1}-1)D_{\mathfrak{s}}x_{n,\mathfrak{s}}/M)^d \\
&= (\eta(x_{n,\mathfrak{r}})^{D_{\mathfrak{s}}})^{d(\mathbf{N}\mathfrak{q}-1)/M}/\beta_{\mathfrak{s}}^{d(1-\mathrm{Frob}_{\mathfrak{q}}^{-1})} \\
&= (\eta(x_{n,\mathfrak{r}})^{D_{\mathfrak{s}}})^{\mathrm{Frob}_{\mathfrak{q}}^{-1}d(\mathbf{N}\mathfrak{q}-1)/M}/\beta_{\mathfrak{s}}^{d(1-\mathrm{Frob}_{\mathfrak{q}}^{-1})} \\
&\equiv ((\eta(x_{n,\mathfrak{s}})^{D_{\mathfrak{s}}}/\beta_{\mathfrak{s}}^M)^{1/M})^{d(1-\mathrm{Frob}_{\mathfrak{Q}}^{-1})} \\
&\equiv (\kappa_{n,M}(\mathfrak{s})^{d/M})^{\mathrm{Frob}_{\mathfrak{Q}}-1}
\end{aligned}$$

using Proposition 8.6 for the second-to-last congruence. By (10) it follows that

$$d\,\phi_{\mathfrak{Q}}(\kappa_{n,M}(\mathfrak{s})) = d\,\mathrm{ord}_{\mathfrak{Q}}(\kappa_{n,M}(\mathfrak{r})),$$

and since d is prime to p, (ii) follows. □

9 Bounding Ideal Class Groups

In this section we describe Kolyvagin's method of using the Euler system of elliptic units, or rather the principal ideals deduced from elliptic units as in §8.2, to bound the size of certain ideal class groups. For a similar argument in the case of cyclotomic units and real abelian extensions of $\mathbf{Q}$, see [Ru1].

Keep the notation of the previous section. Let $F = K_1 = K(E[\mathfrak{p}])$ and let μ_F denote the roots of unity in F. Let $\Delta = \mathrm{Gal}(F/K)$, so $\Delta \cong (\mathcal{O}/\mathfrak{p})^\times$ is cyclic of order $p-1$ or p^2-1.

Since $\#(\Delta)$ is prime to p, the group ring $\mathbf{Z}_p[\Delta]$ is semisimple, i.e.,

$$\mathbf{Z}_p[\Delta] \cong \bigoplus_{\chi\in\Xi} R_\chi$$

where Ξ denotes the set of all irreducible $\mathbf{F}_p$-representations of Δ and R_χ denotes the corresponding direct summand of $\mathbf{Z}_p[\Delta]$. (We will also refer to elements of Ξ as irreducible $\mathbf{Z}_p$-representations of Δ.) Since $\#(\Delta)$ divides p^2-1, we have two cases:

- $\dim(\chi) = 1$, $R_\chi = \mathbf{Z}_p$,
- $\dim(\chi) = 2$, R_χ is the ring of integers of the unramified quadratic extension of $\mathbf{Q}_p$, and χ splits into two one-dimensional pieces over $\mathcal{O}_\mathfrak{p}$.

If $\chi \in \Xi$ and B is a $\mathbf{Z}[\Delta]$-module, we let $M^{(p)}$ denote the p-adic completion of M and

$$M^\chi = M^{(p)} \otimes_{\mathbf{Z}_p[\Delta]} R_\chi.$$

Then $M^{(p)} = \oplus_{\chi \in \Xi} M^{\chi}$, so we can view M^{χ} either as a quotient of M or a submodule of $M^{(p)}$. If $m \in M$ we write m^{χ} for the projection of m into M^{χ}.

Lemma 9.1. *For every nontrivial* $\chi \in \Xi$, $(\mathcal{O}_F^{\times}/\mu_F)^{\chi}$ *is free of rank one over* R_{χ}.

Proof. The Dirichlet unit theorem gives an exact sequence

$$0 \to \mathcal{O}_F^{\times} \otimes \mathbf{Q} \to \mathbf{Q}[\Delta] \to \mathbf{Q} \to 0$$

and the lemma follows by taking χ-components. □

Let A denote the ideal class group of F, and fix a $\chi \in \Xi$. We wish to bound the size of A^{χ}. Fix a power M of p, which we will later take to be large, and set $F_M = F(\mu_M)$.

Lemma 9.2. *The composition*

$$\mathrm{Hom}(A, \mathbf{Z}/M\mathbf{Z}) \to \mathrm{Hom}(G_F, \mathbf{Z}/M\mathbf{Z}) \to \mathrm{Hom}(G_{F_M}, \mathbf{Z}/M\mathbf{Z}),$$

given by class field theory and restriction to G_{F_M}, *is injective.*

Proof. The first map is clearly injective, and the kernel of the second is equal to $\mathrm{Hom}(\mathrm{Gal}(F_M/F), \mathbf{Z}/M\mathbf{Z})$. Thus to prove the lemma it suffices to show that there is no unramified p-extension of F in F_M. But the p-part of $\mathrm{Gal}(F_M/F)$ is $\mathrm{Gal}(F_M/F(\mu_p))$, which is totally ramified at all primes above p. This completes the proof. □

Lemma 9.3. *The map*

$$F^{\times}/(F^{\times})^M \to F_M{}^{\times}/(F_M{}^{\times})^M$$

is injective.

Proof. Kummer theory shows that $F^{\times}/(F^{\times})^M \cong H^1(F, \mu_M)$ and similarly for F_M, so the kernel of the map in the lemma is $H^1(F_M/F, \mu_M)$. Since $\mathrm{Gal}(F_M/F)$ is cyclic and acts faithfully on μ_M, and $p > 2$, it is easy to check that $H^1(F_M/F, \mu_M) = 0$. (See also Lemma 6.1.) □

Write $\mathcal{R}_{F,M}$ for $\mathcal{R}_{1,M}$, the set of primes of K defined in §8.

Proposition 9.4. *Suppose* $\kappa \in F^{\times}/(F^{\times})^M$ *and* $\alpha \in \mathrm{Hom}(A, \mathbf{Z}/M\mathbf{Z})$, $\alpha \neq 0$. *Then there is a prime* $\mathfrak{q} \in \mathcal{R}_{F,M}$ *and a prime* $\mathfrak{Q}$ *of* F *above* $\mathfrak{q}$ *such that*

(i) $\alpha(\mathfrak{c}) \neq 0$, *where* $\mathfrak{c}$ *denotes the class of* $\mathfrak{Q}$ *in* A,
(ii) $[\kappa]_{\mathfrak{q}} = 0$ *and for every* $d \in \mathbf{Z}$, $d\phi_{\mathfrak{q}}(\kappa) = 0 \Leftrightarrow \kappa^d \in (F^{\times})^M$.

Proof. Let t be the order of κ in $F^\times/(F^\times)^M$, and let $\rho \in \mathrm{Hom}(G_{F_M}, \mu_M)$ be the image of κ under the Kummer map. We view α as a map on G_{F_M} via the map of Lemma 9.2. Define two subgroups of G_{F_M}

$$H_\alpha = \{\gamma \in G_{F_M} : \alpha(\gamma) = 0\},$$
$$H_\kappa = \{\gamma \in G_{F_M} : \rho(\gamma) \text{ has order less than } t \text{ in } \mu_M\}.$$

Since $\alpha \neq 0$, Lemma 9.2 shows that $H_\alpha \neq G_{F_M}$. Similarly it follows from Lemma 9.3 that $H_\kappa \neq G_{F_M}$. Since a group cannot be a union of two proper subgroups, we can choose a $\gamma \in G_{F_M}$, $\gamma \notin H_\alpha \cup H_\kappa$. Let L be a finite Galois extension of F containing F_M such that both ρ and α are trivial on G_L. By the Cebotarev theorem we can choose a prime $\tilde{\mathfrak{Q}}$ of L, not dividing $6\mathfrak{a}\mathfrak{f}\mathfrak{p}$ and such that $[\kappa]_\mathfrak{q} = 0$, whose Frobenius in L/K is γ. Let $\mathfrak{Q}$ and $\mathfrak{q}$ denote the primes of F and K, respectively, below $\tilde{\mathfrak{Q}}$. We will show that these primes satisfy the conditions of the proposition.

First, the fact that γ fixes $F(\mu_M)$ means that $\mathfrak{q}$ splits completely in $F(\mu_M)$ and thus $\mathfrak{q} \in \mathcal{R}_{F,M}$.

The class field theory inclusion $\mathrm{Hom}(A, \mathbf{Z}/M\mathbf{Z}) \hookrightarrow \mathrm{Hom}(G_F, \mathbf{Z}/M\mathbf{Z})$ identifies $\alpha(\mathfrak{c})$ with $\alpha(\mathrm{Frob}_\mathfrak{Q}) = \alpha(\gamma)$, so (i) follows from the fact that $\gamma \notin H_\alpha$.

Since $\gamma \notin H_\kappa$, $(\kappa^{1/M})^{\mathrm{Frob}_\mathfrak{Q} - 1}$ is a primitive t-th root of unity. Therefore κ has order $t(\mathbf{N}\mathfrak{q} - 1)/M$ modulo $\mathfrak{Q}$, and hence has order at least t (and hence exactly t) in $(\mathcal{O}_F/\mathfrak{q}\mathcal{O}_F)^\times/((\mathcal{O}_F/\mathfrak{q}\mathcal{O}_F)^\times)^M$. Since $\phi_\mathfrak{q}$ is an isomorphism on $(\mathcal{O}_F/\mathfrak{q}\mathcal{O}_F)^\times/((\mathcal{O}_F/\mathfrak{q}\mathcal{O}_F)^\times)^M$, this proves (ii). □

Suppose η is an Euler system as defined in Definition 8.5. Define $\mathcal{C} = \mathcal{C}_\eta \subset \mathcal{O}_F^\times$ to be the group generated over $\mathbf{Z}[\Delta]$ by μ_F and $\eta(1, \mathcal{O})$.

Theorem 9.5. *With notation as above, if η is an Euler system and χ is an irreducible $\mathbf{Z}_p$-representation of Δ then*

$$\#(A^\chi) \leq \#((\mathcal{O}_F^\times/\mathcal{C}_\eta)^\chi).$$

Proof. If χ is the trivial character then A^χ is the p-part of the ideal class group of K, which is zero. Hence we may assume that $\chi \neq 1$.

By Lemma 9.1

$$(\mathcal{O}_F^\times/\mathcal{C})^\chi \cong R_\chi/mR_\chi$$

for some $m \in R_\chi$. If $m = 0$ then there is nothing to prove, so we may assume $m \neq 0$. Choose M large enough so that M/m annihilates A. For $\mathfrak{r} \in \mathcal{R}_{F,M}$ we will write $\kappa(\mathfrak{r})$ for the element $\kappa_{1,M}(\mathfrak{r}) \in F^\times/(F^\times)^M$ constructed in Definition 8.7.

Number the elements of $\mathrm{Hom}(A^\chi, \mathbf{Z}/M\mathbf{Z}) \subset \mathrm{Hom}(A, \mathbf{Z}/M\mathbf{Z})$ so that

$$\mathrm{Hom}(A^\chi, \mathbf{Z}/M\mathbf{Z}) = \{\alpha_1, \ldots, \alpha_k\}.$$

Using Proposition 9.4 we choose inductively a sequence of primes $\mathfrak{q}_1, \ldots, \mathfrak{q}_k \in \mathcal{R}_{F,M}$ and $\mathfrak{Q}_i$ of F above $\mathfrak{q}_i$ such that, if $\mathfrak{c}_i$ denotes the class of $\mathfrak{Q}_i$ in A and

$\mathfrak{r}_i = \prod_{j\le i} \mathfrak{q}_j$ for $0 \le i \le k$,

$$\alpha_i(\mathfrak{c}_i) \neq 0, \tag{11}$$

$$d\phi_{\mathfrak{q}_i}(\kappa(\mathfrak{r}_{i-1})^\chi) = 0 \Leftrightarrow (\kappa(\mathfrak{r}_{i-1})^\chi)^d = 0 \in F^\times/(F^\times)^M \tag{12}$$

(just apply Proposition 9.4 with $\kappa = \kappa(\mathfrak{r}_{i-1})^\chi$ and $\alpha = \alpha_i$ to produce $\mathfrak{q}_i$ and $\mathfrak{Q}_i$).

First we claim that the classes $\{\mathfrak{c}_i^\chi\}$ generate A^χ. For if not, then there is an $\alpha \in \mathrm{Hom}(A^\chi, \mathbf{Z}/M\mathbf{Z})$ such that $\alpha(\mathfrak{c}_j) = 0$ for every j. But $\alpha = \alpha_i$ for some i, so (11) shows this is not the case.

If $1 \le i \le k$ let s_i denote the order of $\mathfrak{c}_i^\chi$ in $A^\chi/\langle \mathfrak{c}_1^\chi, \ldots, \mathfrak{c}_{i-1}^\chi\rangle$. Since the $\mathfrak{c}_i^\chi$ generate A^χ we have

$$\#(A^\chi) = \prod_{i=1}^{k} [R_\chi : s_i R_\chi].$$

If $0 \le i \le k-1$ let t_i denote the order of $\kappa(\mathfrak{r}_i)^\chi$ in $F^\times/(F^\times)^M$. By (12) and Proposition 8.10(ii), for $i \ge 1$ the order of $[\kappa(\mathfrak{r}_i)^\chi]_{\mathfrak{q}_i}$ is t_{i-1}. In particular it follows that $t_{i-1} \mid t_i$. Since $\kappa(\mathfrak{r}_0)$ is the image of $\eta(1, \mathcal{O})$ in $\mathcal{O}_F^\times/(\mathcal{O}_F^\times)^M$, the exact sequence

$$0 \to R_\chi\kappa(\mathfrak{r}_0)/\mu_F \cap R_\chi\kappa(\mathfrak{r}_0) \to (\mathcal{O}_F^\times/\mu_F(\mathcal{O}_F^\times)^M)^\chi \to (\mathcal{O}_F^\times/\mathcal{C})^\chi \to 0$$

shows that $M \mid t_0 m$.

For each i we can choose $\nu_i \in F^\times/(F^\times)^M$ such that $\nu_i^{M/t_i} = \kappa(\mathfrak{r}_i)^\chi\zeta$ with $\zeta \in \mu_F$. In particular

$$(M/t_i)[\nu_i]_{\mathfrak{q}_i} = [\kappa(\mathfrak{r}_i)^\chi]_{\mathfrak{q}_i}$$

so $[\nu_i]_{\mathfrak{q}_i}$ has order $t_{i-1}M/t_i$ in $(\mathcal{I}_\mathfrak{q}/M\mathcal{I}_\mathfrak{q})^\chi \cong R_\chi/MR_\chi$. Thus, using Proposition 8.10(i), there is a unit $u \in R_\chi^\times$ such that

$$(\nu_i) \equiv u(t_i/t_{i-1})\mathfrak{q}_i^\chi \pmod{\mathcal{I}_{\mathfrak{q}_1}, \ldots, \mathcal{I}_{\mathfrak{q}_{i-1}}, t_i\mathcal{I}}.$$

We know that $t_0 \mid t_i$ and $(M/m) \mid t_0$. Thus by our choice of M, t_i annihilates A and we conclude that

$$(t_i/t_{i-1})\mathfrak{c}_i^\chi = 0 \quad \text{in } A^\chi/\langle \mathfrak{c}_1^\chi, \ldots, \mathfrak{c}_{i-1}^\chi\rangle.$$

Therefore $s_i \mid (t_i/t_{i-1})$ for every $i \ge 1$, so

$$\#(A^\chi) = \prod_{i=1}^{k} [R_\chi : s_i R_\chi] \text{ divides } [t_0 R_\chi : t_k R_\chi].$$

Since $t_k \mid M$ and $M \mid t_0 m$, this index divides $[R_\chi : mR_\chi]$. This proves the theorem. □

Corollary 9.6. *Let $C_{\mathfrak{a}}$ denote the group of (elliptic) units of F generated over $\mathbf{Z}_p[\Delta]$ by μ_F and by $\eta_1^{(\mathfrak{a})}$. If χ is an irreducible $\mathbf{Z}_p$-representation of Δ then*

$$\#(A^\chi) \quad \textit{divides} \quad \#((\mathcal{O}_F^\times/C_{\mathfrak{a}})^\chi).$$

Proof. Apply Theorem 9.5 with the Euler system $\eta(n,\mathfrak{r}) = \eta_n^{(\mathfrak{a})}(\mathfrak{r})$. □

Remark 9.7. If C_F denotes the full group of elliptic units of F (see for example [Ru2] §1), then one can combine Theorem 9.5 with a well-known argument using the analytic class number formula to prove that for every χ,

$$\#(A^\chi) = \#((\mathcal{O}_F^\times/C_F)^\chi).$$

See Theorem 3.3 of [Ru2].

Corollary 9.8. *With notation as above, if* $(\eta_1^{(\mathfrak{a})})^\chi \notin \mu_F^\chi((\mathcal{O}_F^\times)^\chi)^p$ *then* $A^\chi = 0$.

Proof. Immediate from Corollary 9.6 and Lemma 9.1. □

10 The Theorem of Coates and Wiles

Keep the notation of the previous sections. In this section we will prove the following theorem.

Theorem 10.1 (Coates-Wiles [CW1]). *If $L(\bar{\psi},1) \neq 0$ then $E(K)$ is finite.*

Suppose for the rest of this section that $\mathfrak{p}$ is a prime of K not dividing $\mathfrak{f}$, of residue characteristic $p > 7$ (see remark 10.3 below). As in §9 we let $F = K(E[\mathfrak{p}])$, $\Delta = \mathrm{Gal}(F/K)$ and A is the ideal class group of F.

Lemma 10.2. *There is an ideal $\mathfrak{a}$ of $\mathcal{O}$, prime to $6\mathfrak{p}\mathfrak{f}$, such that* $\mathbf{N}\mathfrak{a} \not\equiv \psi(\mathfrak{a}) \pmod{\mathfrak{p}}$.

Proof. By Corollary 5.18, $E[\bar{\mathfrak{p}}] \not\subset E(K)$. Choose a prime $\mathfrak{q}$ of K, not dividing $6p\mathfrak{f}$, such that $[\mathfrak{q}, K(E[\bar{\mathfrak{p}}])/K] \neq 1$. By Corollary 5.16(ii) we deduce that $\psi(\mathfrak{q}) \not\equiv 1 \pmod{\bar{\mathfrak{p}}}$, and so $\bar{\psi}(\mathfrak{q}) \not\equiv 1 \pmod{\mathfrak{p}}$. Since $\psi(\mathfrak{q})\bar{\psi}(\mathfrak{q}) = \mathbf{N}\mathfrak{q}$, the lemma is satisfied with $\mathfrak{a} = \mathfrak{q}$. □

Remark 10.3. Lemma 10.2 is not in general true without the assumption $p > 7$, since for small p it may happen that $E[\bar{\mathfrak{p}}] \subset E(K)$.

By Corollary 5.20(iv), F/K is totally ramified at $\mathfrak{p}$. Let $\mathcal{P}$ denote the prime of F above $\mathfrak{p}$. By Lemma 3.6 and Corollary 5.16 $E[\mathfrak{p}] \subset E_1(F_{\mathcal{P}})$, so the isomorphism of Corollary 3.8 restricts to an isomorphism

$$E[\mathfrak{p}] \xrightarrow{\sim} \hat{E}[\mathfrak{p}] \subset \hat{E}(F_{\mathcal{P}})$$

where $\hat{E}$ is the formal group attached to E. Let $\mathcal{O}_{F,\mathcal{P}}$ denote the completion of $\mathcal{O}_F$ at $\mathcal{P}$.

Lemma 10.4. *The map*

$$E[\mathfrak{p}] \xrightarrow{\sim} \hat{E}[\mathfrak{p}] \xrightarrow{1+} (1+\mathcal{P}\mathcal{O}_{F,\mathcal{P}})/(1+\mathcal{P}^2\mathcal{O}_{F,\mathcal{P}})$$

is a Δ-equivariant isomorphism.

Proof. The map in question is a well-defined homomorphism, and by Lemma 7.3 it is injective. Both groups have order $\mathbf{N}\mathfrak{p}$, so it is an isomorphism. The Δ-equivariance is clear. □

Now fix an ideal $\mathfrak{a}$ satisfying Lemma 10.2, a generator Ω of the period lattice of E as in §7.4, and a generator f of the conductor $\mathfrak{f}$. With these choices define the elliptic units $\eta_n(\mathfrak{r})$ as in §8.1. Let $\eta = \eta_1(\mathcal{O})$, a global (elliptic) unit of F which depends on the choice of $\mathfrak{a}$.

Definition 10.5. Define

$$\delta : \mathcal{O}_{F,\mathcal{P}}^\times \to E[\mathfrak{p}]$$

to be the composition of the natural projection

$$\mathcal{O}_{F,\mathcal{P}}^\times \twoheadrightarrow (1+\mathcal{P}\mathcal{O}_{F,\mathcal{P}})/(1+\mathcal{P}^2\mathcal{O}_{F,\mathcal{P}})$$

with the inverse of the isomorphism of Lemma 10.4.

Recall that by Corollary 7.18, $L(\bar{\psi},1)/\Omega \in K$.

Proposition 10.6. *$L(\bar{\psi},1)/\Omega$ is integral at $\mathfrak{p}$, and*

$$L(\bar{\psi},1)/\Omega \equiv 0 \pmod{\mathfrak{p}} \;\Leftrightarrow\; \delta(\eta) = 0.$$

Proof. Let $P = (\wp(\Omega/\psi(\mathfrak{p}); \Omega\mathcal{O}), \wp'(\Omega/\psi(\mathfrak{p}); \Omega\mathcal{O})/2) \in E[\mathfrak{p}]$ and

$$z = -x(P)/y(P) \in \mathcal{P},$$

the image of P in $\hat{E}[\mathfrak{p}]$. Then $\eta = \Lambda_{\mathfrak{p},\mathfrak{a}}(z)$, where $\Lambda_{\mathfrak{p},\mathfrak{a}}$ is the power series of Definition 7.21.

By Theorem 7.22, $\Lambda_{\mathfrak{p},\mathfrak{a}}(0) \in \mathcal{O}_\mathfrak{p}^\times$, $12f(\mathbf{N}\mathfrak{a} - \psi(\mathfrak{a}))(L(\bar{\psi},1)/\Omega) \in \mathcal{O}_\mathfrak{p}$, and

$$\eta \equiv \Lambda_{\mathfrak{p},\mathfrak{a}}(0)(1 + 12f(\mathbf{N}\mathfrak{a} - \psi(\mathfrak{a}))(L(\bar{\psi},1)/\Omega)z) \pmod{\mathcal{P}^2}.$$

Thus

$$\delta(\eta) = 12f(\mathbf{N}\mathfrak{a} - \psi(\mathfrak{a}))(L(\bar{\psi},1)/\Omega)P$$

and with our choice of $\mathfrak{a}$, $12f(\mathbf{N}\mathfrak{a} - \psi(\mathfrak{a})) \in \mathcal{O}_\mathfrak{p}^\times$. This proves the proposition. □

Definition 10.7. Let χ_E denote the representation of Δ on $E[\mathfrak{p}]$; by Corollary 5.20 χ_E is $\mathbf{F}_\mathfrak{p}$-irreducible. Then in the notation of §9 we have $E[\mathfrak{p}] \cong R_{\chi_E}/\mathfrak{p}R_{\chi_E}$ as Δ-modules.

Theorem 10.8. *Suppose $L(\bar{\psi},1)/\Omega$ is a unit at $\mathfrak{p}$. Then*

$$A^{\chi_E} = 0.$$

Proof. Since the map δ is Δ-equivariant,

$$\delta(\eta^{\chi_E}) = \delta(\eta)^{\chi_E} = \delta(\eta) \neq 0$$

by Proposition 10.6. Hence

$$\eta^{\chi_E} \notin ((\mathcal{O}_F^\times)^{\chi_E})^p.$$

The Weil pairing (see [Si] Proposition III.8.1) gives a Galois-equivariant isomorphism

$$E[p] \cong \mathrm{Hom}(E[p], \mu_p).$$

If $\mu_F^{\chi_E}$ were nontrivial, then $E[p]^{G_K}$ would be nontrivial, and this is impossible by Corollary 5.18. Now the theorem follows from Corollary 9.8. □

Lemma 10.9. *Suppose p splits into two primes in K and $\mathrm{Tr}_{K/\mathbf{Q}}\psi(\mathfrak{p}) \neq 1$. Then*

(i) $\mu_p \not\subset F_{\mathcal{P}}$,
(ii) $(\mathcal{O}_{F,\mathcal{P}}^\times)^{\chi_E}$ *is free of rank one over R_{χ_E}.*

Proof. By Theorem 5.15(ii), $[\psi(\mathfrak{p}), F_{\mathcal{P}}/K_{\mathfrak{p}}] = 1$. On the other hand, class field theory over $\mathbf{Q}$ shows that $[p, \mathbf{Q}_p(\mu_p)/\mathbf{Q}_p] = 1$. Thus we have (again using Theorem 5.15(ii))

$$\begin{aligned}\mu_p \subset F_{\mathcal{P}} \Rightarrow F_{\mathcal{P}} = K_{\mathfrak{p}}(\mu_p) \Rightarrow [p/\psi(\mathfrak{p}), F_{\mathcal{P}}/K_{\mathfrak{p}}] &= 1 \\ \Rightarrow p/\psi(\mathfrak{p}) &\equiv 1 \pmod{\mathfrak{p}} \\ \Rightarrow \mathrm{Tr}_{K/\mathbf{Q}}\psi(\mathfrak{p}) &\equiv 1 \pmod{\mathfrak{p}} \\ \Rightarrow \mathrm{Tr}_{K/\mathbf{Q}}\psi(\mathfrak{p}) &= 1,\end{aligned}$$

the last implication because $|\mathrm{Tr}_{K/\mathbf{Q}}\psi(\mathfrak{p})| \leq 2\sqrt{p} < p-1$. This proves (i).

We have isomorphisms

$$\mathcal{O}_{F,\mathcal{P}}^\times \otimes \mathbf{Q}_p \xrightarrow{\sim} \mathcal{O}_{F,\mathcal{P}} \otimes \mathbf{Q}_p \cong K_{\mathfrak{p}}[\Delta],$$

the first one given by the $\mathfrak{p}$-adic logarithm map. Together with (i) this proves (ii). □

Theorem 10.10. *Suppose $L(\bar{\psi},1)/\Omega$ is a unit at $\mathfrak{p}$, p splits into two primes in K, and $\mathrm{Tr}_{K/\mathbf{Q}}\psi(\mathfrak{p}) \neq 1$. Then the natural (injective) map*

$$(\mathcal{O}_F^\times)^{\chi_E} \to (\mathcal{O}_{F,\mathcal{P}}^\times)^{\chi_E}$$

is surjective.

Proof. As in the proof of Theorem 10.8, $\delta(\eta^{\chi_E}) = \delta(\eta)$. Thus by Proposition 10.6 $(\mathcal{O}_F^\times)^{\chi_E} \not\subset ((\mathcal{O}_{F,\mathcal{P}}^\times)^{\chi_E})^p$, so the Theorem follows from Lemma 10.9. □

Proof (of the Coates-Wiles Theorem 10.1). Using the Cebotarev theorem we can find infinitely many primes $\mathfrak{p}$ which split in K and such that $\mathrm{Tr}_{K/\mathbf{Q}}\psi(\mathfrak{p}) \neq 1$. Choose one which does not divide $6\mathfrak{f}$ or $L(\bar{\psi},1)/\Omega$. Then by Theorems 10.8 and 10.10 and Corollary 6.10, the Selmer group $S_{\psi(\mathfrak{p})}(E) = 0$. In particular $E(K)/\mathfrak{p}E(K) = 0$, so (using the Mordell-Weil Theorem 4.6), $E(K)$ is finite. □

Remark 10.11. This proof also shows that for primes $\mathfrak{p}$ satisfying the hypotheses of Theorem 10.10, the $\mathfrak{p}$-part of the Tate-Shafarevich group $\text{Ш}(E)$ is trivial.

Using the Explicit Reciprocity Law of Wiles ([Wil] or [dS] §I.4) one can show that $\delta = -\delta_1$ where δ_1 is the map of Lemma 6.8. Together with Proposition 10.6, Theorem 10.8 and Corollary 6.10, this shows that $S_{\psi(\mathfrak{p})}(E) = 0$ for every $\mathfrak{p}$ not dividing $2 \cdot 3 \cdot 5 \cdot 7 \cdot \mathfrak{f} \cdot (L(\bar{\psi},1)/\Omega)$. We will prove a stronger version of this (Corollary 12.13 and Theorem 12.19) in §12.

11 Iwasawa Theory and the "Main Conjecture"

In order to study the Selmer group under more general conditions than in §10, we need to prove Iwasawa-theoretic versions (Theorem 11.7 and Corollary 11.8 below) of Theorem 9.5 and Remark 9.7. As in the previous sections, we fix an elliptic curve E defined over an imaginary quadratic field K, with $\mathrm{End}_K(E) = \mathcal{O}$, the ring of integers of K. We fix a prime $\mathfrak{p}$ of K where E has good reduction, and for simplicity we still assume that $p > 7$ (in order to apply Lemma 10.2).

Write $K_n = K(E[\mathfrak{p}^n])$, $n = 0, 1, 2, \ldots, \infty$, and let $G_\infty = \mathrm{Gal}(K_\infty/K)$. By Corollary 5.20(ii), we have

$$G_\infty \cong \mathcal{O}_\mathfrak{p}^\times \cong \Delta \times \Gamma$$

where

$$\Delta \cong \mathrm{Gal}(K_1/K) \cong (\mathcal{O}/\mathfrak{p})^\times$$

is the prime-to-p part of G_∞ and

$$\Gamma = \mathrm{Gal}(K_\infty/K_1) \cong 1 + \mathfrak{p}\mathcal{O}_\mathfrak{p} \cong \mathbf{Z}_p^{[K_\mathfrak{p}:\mathbf{Q}_p]}$$

is the p-part.

11.1 The Iwasawa Algebra

Define the *Iwasawa algebra*

$$\Lambda = \mathbf{Z}_p[[G_\infty]] = \varprojlim_n \mathbf{Z}_p[\mathrm{Gal}(K_n/K)] = \varprojlim_n \mathbf{Z}_p[\Delta][\mathrm{Gal}(K_n/K_1)].$$

Then

$$\Lambda = \bigoplus_{\chi \in \Xi} \Lambda_\chi$$

where Ξ is the set of irreducible $\mathbf{Z}_p$-representations of Δ as in §9 and

$$\Lambda_\chi = \Lambda \otimes_{\mathbf{Z}_p[\Delta]} R_\chi \cong R_\chi[[\Gamma]].$$

The following algebraic properties of the Iwasawa algebra and its modules are well-known. For proofs, see for example [Iw] and [Se].

For every irreducible $\mathbf{Z}_p$-representation χ of Δ, Λ_χ is a complete local noetherian ring, noncanonically isomorphic to a power series ring in $[K_\mathfrak{p} : \mathbf{Q}_p]$ variables over R_χ. In particular Λ is not an integral domain, but rather is a direct sum of local integral domains. Let $\mathcal{M}$ denote the (finite) intersection of all maximal ideals of Λ, i.e., $\mathcal{M}$ is the kernel of the natural map $\Lambda \twoheadrightarrow \mathbf{F}_p[\Delta]$.

A Λ-module M will be called a torsion Λ-module if it is annihilated by a non-zero-divisor in Λ. A Λ-module will be called pseudo-null if it is annihilated by an ideal of height at least two in Λ. If $\Gamma \cong \mathbf{Z}_p$ then a module is pseudo-null if and only if it is finite.

If M is a finitely generated torsion Λ-module, then there is an injective Λ-module homomorphism

$$\bigoplus_{i=1}^{r} \Lambda/f_i\Lambda \hookrightarrow M$$

with pseudo-null cokernel, where the elements $f_i \in \Lambda$ can be chosen to satisfy $f_{i+1} \mid f_i$ for $1 \le i \le r$. The elements f_i are not uniquely determined, but the ideal $\prod_i f_i\Lambda$ is. We call the ideal $\prod_i f_i\Lambda$ the *characteristic ideal* $\mathrm{char}(M)$ of the torsion Λ-module M. The characteristic ideal is multiplicative in exact sequences: if $0 \to M' \to M \to M'' \to 0$ is an exact sequence of torsion Λ-modules then $\mathrm{char}(M) = \mathrm{char}(M')\mathrm{char}(M'')$.

11.2 The Iwasawa Modules

Define

$A_n =$ the p-part of the ideal class group of K_n,
$U_n =$ the p-adic completion of the local units of $K_n \otimes K_\mathfrak{p}$ (equivalently, the 1-units of $K_n \otimes K_\mathfrak{p}$),
$\mathcal{E}_n =$ the global units of K_n,
$\bar{\mathcal{E}}_n =$ the p-adic completion of $\mathcal{E}_n$ (equivalently, since Leopoldt's conjecture holds for K_n, the closure of the image of $\mathcal{E}_n$ in U_n),
$\mathcal{C}_n =$ the elliptic units of K_n, the subgroup of $\mathcal{E}_n$ generated over the group ring $\mathbf{Z}[\mathrm{Gal}(K_n/K)]$ by the $\eta_n^{(\mathfrak{a})} = \eta_n^{(\mathfrak{a})}(\mathcal{O})$ (see Definition 8.1) for *all* choices of ideal $\mathfrak{a}$ prime to $6\mathfrak{p}\mathfrak{f}$, and the roots of unity in K_n,
$\bar{\mathcal{C}}_n =$ the p-adic completion of $\mathcal{C}_n$ (equivalently, the closure of the image of $\mathcal{C}_n$ in U_n),

and

$$A_\infty = \varprojlim_n A_n, \quad U_\infty = \varprojlim_n U_n, \quad \mathcal{E}_\infty = \varprojlim_n \bar{\mathcal{E}}_n, \quad \mathcal{C}_\infty = \varprojlim_n \bar{\mathcal{C}}_n,$$

inverse limits with respect to norm maps. Also define $X_\infty = \mathrm{Gal}(M_\infty/K_\infty)$ where M_∞ is the maximal abelian p-extension of K_∞ unramified outside of the prime above $\mathfrak{p}$.

Class field theory identifies A_∞ with $\mathrm{Gal}(L_\infty/K_\infty)$, where L_∞ is the maximal everywhere-unramified abelian p-extension of K_∞, and identifies the inertia group in X_∞ of the unique prime above $\mathfrak{p}$ with $U_\infty/\mathcal{E}_\infty$. Thus there is an exact sequence of Λ-modules

$$0 \to \mathcal{E}_\infty/\mathcal{C}_\infty \to U_\infty/\mathcal{C}_\infty \to X_\infty \to A_\infty \to 0. \tag{13}$$

For every $n \geq 0$, let $\Lambda_n = \mathbf{Z}_p[\mathrm{Gal}(K_n/K)]$ and let $\mathcal{J}_n \subset \Lambda$ denote the kernel of the restriction map $\Lambda \to \Lambda_n$. In particular $\mathcal{J}_0$ is the augmentation ideal of Λ.

Lemma 11.1. *For every $n \geq 1$, the natural map*

$$A_\infty/\mathcal{J}_n A_\infty \to A_n$$

is an isomorphism.

Proof. When $\Gamma = \mathbf{Z}_p$ this is a standard argument going back to Iwasawa [Iw], using the fact that only one prime of K ramifies in K_∞ and it is totally ramified.

For the general case, consider the diagram of fields below, where L_n is the maximal unramified abelian p-extension of K_n, and L'_n is the fixed field of $\mathcal{J}_n A_\infty$ in L_∞. Since K_∞/K_n is totally ramified above $\mathfrak{p}$, $K_\infty \cap L_n = K_n$, and so $\mathrm{Gal}(K_\infty L_n/K_\infty) = A_n$ and the map $A_\infty/\mathcal{J}_n A_\infty \to A_n$ is just the restriction map. We will show that $K_\infty L_n = L'_n$, and the lemma will follow.

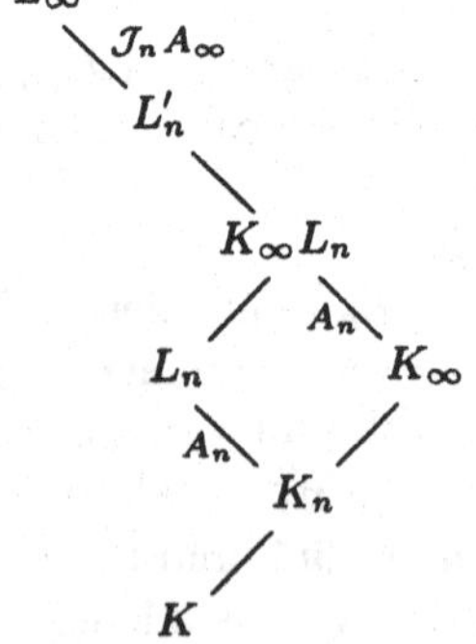

Since $\mathrm{Gal}(K_\infty/K_n)$ acts on $\mathrm{Gal}(L_\infty/K_\infty)$ by conjugation, $\mathcal{J}_n A_\infty$ is generated by commutators

$$\mathrm{Gal}(L_\infty/L'_n) = [\mathrm{Gal}(L_\infty/K_n), \mathrm{Gal}(L_\infty/K_\infty)].$$

Such commutators are trivial on L_n, so $K_\infty L_n \subset L'_n$.

On the other hand, only the unique prime above $\mathfrak{p}$ ramifies in the abelian extension L'_n/K_n, and it is totally ramified in K_∞/K_n. If we write $\mathcal{I}$ for the inertia group of this prime in $\mathrm{Gal}(L'_n/K_n)$, the inverse of the projection isomorphism $\mathcal{I} \xrightarrow{\sim} \mathrm{Gal}(K_\infty/K_n)$ gives a splitting of the exact sequence

$$0 \to A_\infty/\mathcal{J}_n A_\infty \to \mathrm{Gal}(L'_n/K_n) \to \mathrm{Gal}(K_\infty/K_n) \to 0.$$

It follows that ${L'_n}^{\mathcal{I}}$ is an abelian, everywhere-unramified p-extension of K_n, and hence ${L'_n}^{\mathcal{I}} \subset L_n$ and so $L'_n = K_\infty L_n$. □

Proposition 11.2. *A_∞ is a finitely-generated torsion Λ-module.*

Proof. By Lemma 11.1, $A_\infty/\mathcal{J}_n A_\infty$ is finite for every n, and the proposition follows. □

Proposition 11.3. (i) *X_∞ is a finitely-generated Λ-module and for every χ*

$$\mathrm{rank}_{\Lambda_\chi} X_\infty^\chi = [K_\mathfrak{p} : \mathbf{Q}_p] - 1.$$

(In particular if $K_\mathfrak{p} = \mathbf{Q}_p$ then X_∞ is a finitely-generated torsion Λ-module.)

(ii) *X_∞ has no nonzero pseudo-null submodules.*

Proof. See [Gr]. □

Proposition 11.4. *U_∞ is a finitely-generated, torsion-free Λ-module, and for every χ*

$$\mathrm{rank}_{\Lambda_\chi}(U_\infty^\chi) = [K_\mathfrak{p} : \mathbf{Q}_p].$$

Further, if $[K_\mathfrak{p} : \mathbf{Q}_p] = 2$ then $U_\infty^{\chi_E}$ is free of rank 2 over Λ_χ.

Proof. See [Iw] §12 or [Win]. □

Proposition 11.5. *$\mathcal{E}_\infty$ is a finitely-generated Λ-module, and for every χ*

$$\mathrm{rank}_{\Lambda_\chi}(\mathcal{E}_\infty^\chi) = 1.$$

Proof. The natural map $\mathcal{E}_\infty \to U_\infty$ is injective, so the proposition follows from (13) and Propositions 11.2, 11.3 and 11.4. □

Proposition 11.6. *$C_\infty^{\chi_E}$ is free of rank one over Λ_{χ_E}.*

Proof. Choose an ideal $\mathfrak{a}$ of $\mathcal{O}$ such that $\psi(\mathfrak{a}) \not\equiv \mathbf{N}\mathfrak{a} \pmod{\mathfrak{p}}$ (Lemma 10.2). We will show that $\mathcal{C}_\infty^{\chi_E}$ is generated over Λ_{χ_E} by $\{(\eta_n^{(\mathfrak{a})})^{\chi_E}\}_n$ with this choice of $\mathfrak{a}$. Suppose $\mathfrak{b}$ is some other ideal of $\mathcal{O}$ prime to $6\mathfrak{p}\mathfrak{f}$. It follows from Theorem 7.4(ii) and Lemma 7.10 that for every n

$$(\eta_n^{(\mathfrak{a})})^{\sigma_\mathfrak{b}-\mathbf{N}\mathfrak{b}} = (\eta_n^{(\mathfrak{b})})^{\sigma_\mathfrak{a}-\mathbf{N}\mathfrak{a}}$$

where $\sigma_\mathfrak{a} = [\mathfrak{a}, K_n/K]$, $\sigma_\mathfrak{b} = [\mathfrak{b}, K_n/K]$. Since $\psi(\mathfrak{a}) \not\equiv \mathbf{N}\mathfrak{a} \pmod{\mathfrak{p}}$, and $\sigma_\mathfrak{a}$ acts as $\psi(\mathfrak{a})$ on $E[\mathfrak{p}]$ (Corollary 5.16(ii)), we see that $\sigma_\mathfrak{a} - \mathbf{N}\mathfrak{a}$ acts bijectively on $E[\mathfrak{p}]$. But $E[\mathfrak{p}] \cong \Lambda_{\chi_E}/\mathcal{M}_{\chi_E}$ where $\mathcal{M}_{\chi_E}$ denotes the maximal ideal of the local ring Λ_{χ_E}. Therefore $\sigma_\mathfrak{a} - \mathbf{N}\mathfrak{a}$ is invertible in Λ_{χ_E}, so

$$\{(\eta_n^{(\mathfrak{b})})^{\chi_E}\}_n \in \Lambda_{\chi_E}\{(\eta_n^{(\mathfrak{a})})^{\chi_E}\}_n$$

as claimed. Since U_∞ is torsion-free (Proposition 11.4), $\mathcal{C}_\infty^{\chi_E}$ must be free of rank 1. □

11.3 Application of the Euler System of Elliptic Units

Theorem 11.7. $\mathrm{char}(A_\infty)$ *divides* $\mathrm{char}(\mathcal{E}_\infty/\mathcal{C}_\infty)$.

The rest of this section will be devoted to a proof of this theorem. The techniques are similar to those of the proof of Theorem 9.5, but messier and more technically complicated because one needs to study modules over $\mathbf{Z}_p[\mathrm{Gal}(K_n/K)]$ rather than $\mathbf{Z}_p[\Delta]$. See [Ru2] for the details which are not included below, and see [Ru1] for the analogous result for cyclotomic fields.

We also record, but will not prove, the following corollary. With a better definition of elliptic units, it would hold for more representations χ of Δ. See [Ru2] Theorem 4.1 for a precise statement and [Ru2] §10 (see also [dS] §III.2) for the proof, which is an application of the analytic class number formula.

Corollary 11.8. $\mathrm{char}(A_\infty^{\chi_E}) = \mathrm{char}(\mathcal{E}_\infty^{\chi_E}/\mathcal{C}_\infty^{\chi_E})$.

Definition 11.9. Since A_∞ is a torsion Λ-module, we can fix once and for all an injective Λ-module map with pseudo-null cokernel

$$\bigoplus_{i=1}^{r} \Lambda/f_i\Lambda \hookrightarrow A_\infty$$

with $f_i \in \Lambda$, $f_{i+1} \mid f_i$ for $1 \le i \le r$. Let A_∞^0 denote the image of this map, so

$$A_\infty^0 = \oplus_{i=1}^r \Lambda y_i \subset A_\infty$$

where $y_i \in A_\infty$ is the image of $1 \in \Lambda/f_i\Lambda$. Then A_∞^0 is an "elementary" submodule of A_∞ and A_∞/A_∞^0 is pseudo-null.

Let $\Omega = K_\infty(\boldsymbol{\mu}_{p^\infty})$. If $\sigma \in G_\Omega$ we write $[\sigma] \in A_\infty$ for the restriction of σ to L_∞. Note that if $K_\mathfrak{p} \ne \mathbf{Q}_p$ then the Weil pairing (see [Si] Proposition

III.8.1) shows that $\Omega = K_\infty$, and if $K_\mathfrak{p} = \mathbf{Q}_p$ then Ω/K_∞ is totally ramified at the prime $p/\mathfrak{p}$. Thus in either case the map $G_\Omega \to A_\infty$ is surjective.

If $0 \le k \le r$, a *Frobenius sequence* $\boldsymbol{\sigma}$ of length k is a k-tuple $(\sigma_1, \ldots, \sigma_k)$ of elements of G_Ω satisfying

$$[\sigma_i] - y_i \in \mathcal{M}A_\infty^0$$

for $1 \le i \le k$, where $\mathcal{M}$ is as defined in §11.1, the intersection of all maximal ideals of Λ.

Suppose $n \ge 1$ and M is a power of p. Recall the subset $\mathcal{R}_{n,M}$ of $\mathcal{R}$ defined in §8.2. For $0 \le k \le r$ we call a k-tuple $(\tilde{\pi}_1, \ldots, \tilde{\pi}_k)$ of primes of K_n a *Kolyvagin sequence* (for n and M) if

- the $\tilde{\pi}_i$ lie above distinct primes of K belonging to $\mathcal{R}_{n,M}$, and
- there is a Frobenius sequence $\boldsymbol{\sigma} = (\sigma_1, \ldots, \sigma_k)$ such that for $1 \le i \le k$,

$$\mathrm{Frob}_{\tilde{\pi}_i} = \sigma_i \quad \text{on } L_n$$

  where L_n is the maximal unramified abelian p-extension of K_n.

If $\boldsymbol{\pi}$ is a Kolyvagin sequence of length k we will write π_i for the prime of K below $\tilde{\pi}_i$ and we define

$$\mathfrak{r}(\boldsymbol{\pi}) = \prod_{i=1}^{k} \pi_i \in \mathcal{R}_{n,M}.$$

Let $\Pi(k, n, M)$ be the set of all Kolyvagin sequences of length k for n and M.

Fix an ideal $\mathfrak{a}$ so that $\{(\eta_n^{(\mathfrak{a})})^{\chi_E}\}_n$ generates $C_\infty^{\chi_E}$, as in the proof of Proposition 11.6. Using the Euler system of elliptic units $\eta_n^{(\mathfrak{a})}(\mathfrak{r})$, for $\mathfrak{r} \in \mathcal{R}_{n,M}$ we obtain the Kolyvagin derivative classes

$$\kappa_{n,M}(\mathfrak{r}) \in K_n^\times/(K_n^\times)^M$$

as in Definition 8.7. For every n recall that $\Lambda_n = \mathbf{Z}_p[\mathrm{Gal}(K_n/K)]$ and let $\Lambda_{n,M} = (\mathbf{Z}/M\mathbf{Z})[\mathrm{Gal}(K_n/K)] = \Lambda_n/M\Lambda_n$. If $0 \le k \le r$ define $\Psi(k, n, M)$ to be the ideal of $\Lambda_{n,M}$ generated by

$$\{\psi(\kappa_{n,M}(\mathfrak{r}(\boldsymbol{\pi}))) : \boldsymbol{\pi} \in \Pi(k,n,M), \psi \in \mathrm{Hom}_{\Lambda_n}(\Lambda_{n,M}\kappa_{n,M}(\mathfrak{r}(\boldsymbol{\pi})), \Lambda_{n,M})\}$$

When $k = 0$, $\Pi(k, n, M)$ has a single element (the empty sequence) and

$$\Psi(0, n, M) \supset \{\psi(\eta_n^{(\mathfrak{a})}) \pmod{M} : \psi \in \mathrm{Hom}_{\Lambda_n}(\mathcal{E}_n, \Lambda_n)\} \tag{14}$$

It follows from Lemma 11.1 that $A_\infty/\mathcal{J}_n A_\infty$ is finite for every n. From this it is not difficult to show that $\Lambda_n/\mathrm{char}(A_\infty)\Lambda_n$ is also finite for every n. For every n define N_n to be the product of $\#(A_n)$ and the smallest power of p which annihilates $\Lambda_n/\mathrm{char}(A_\infty)\Lambda_n$.

The following proposition is the key to the proof of Theorem 11.7.

Proposition 11.10. *There is an ideal $\mathcal{B}$ of height at least two in Λ such that for every $n \geq 1$, power M of p, and $0 \leq k < r$,*

$$\mathcal{B}\Psi(k, n, MN_n)\Lambda_{n,M} \subset f_{k+1}\Psi(k+1, n, M).$$

We will first show how to complete the proof of Theorem 11.7 assuming Proposition 11.10, and then we will prove Proposition 11.10.

Lemma 11.11. *Suppose G is a finite abelian group and B is finitely generated $\mathbf{Z}_p[G]$-module with no p-torsion. If $f \in \mathbf{Z}_p[G]$ is not a zero-divisor, $b \in B$, and*

$$\{\psi(b) : \psi \in \mathrm{Hom}_{\mathbf{Z}_p[G]}(B, \mathbf{Z}_p[G])\} \subset f\mathbf{Z}_p[G],$$

then $b \in fB$.

Proof. Let $B' = \mathbf{Z}_p[G]b + fB$. Since f is not a zero-divisor, we have a commutative diagram

$$\begin{array}{ccccc}
\mathrm{Hom}_{\mathbf{Z}_p[G]}(B', f\mathbf{Z}_p[G]) & \xleftarrow{f} & \mathrm{Hom}_{\mathbf{Z}_p[G]}(B', \mathbf{Z}_p[G]) & \xrightarrow{\sim} & \mathrm{Hom}_{\mathbf{Z}_p}(B', \mathbf{Z}_p) \\
\downarrow & & \downarrow & & \downarrow \\
\mathrm{Hom}_{\mathbf{Z}_p[G]}(fB, f\mathbf{Z}_p[G]) & \xleftarrow{f} & \mathrm{Hom}_{\mathbf{Z}_p[G]}(fB, \mathbf{Z}_p[G]) & \xrightarrow{\sim} & \mathrm{Hom}_{\mathbf{Z}_p}(fB, \mathbf{Z}_p)
\end{array}$$

in which the horizontal maps are all isomorphisms.

Choose $\bar{\varphi} \in \mathrm{Hom}_{\mathbf{Z}_p[G]}(fB, f\mathbf{Z}_p[G])$. Since B has no p-torsion and f is not a zero-divisor, $\bar{\varphi}$ extends uniquely to a map $\varphi : B \to \mathbf{Z}_p[G]$, and by our assumption, $\varphi \in \mathrm{Hom}_{\mathbf{Z}_p[G]}(B', f\mathbf{Z}_p[G])$. Thus all the vertical maps in the diagram above are isomorphisms. Since B' and fB are free $\mathbf{Z}_p$-modules, the surjectivity of the right-hand vertical map shows that $B' = fB$, which proves the lemma. □

Proof (of Theorem 11.7, assuming Proposition 11.10). Fix $n \geq 1$ and $\psi \in \mathrm{Hom}_{\Lambda_n}(\bar{\mathcal{E}}_n, \Lambda_n)$, and let $\mathcal{B} \subset \Lambda$ be an ideal of height at least two satisfying Proposition 11.10. We will show that, for every choice of $\mathfrak{a}$,

$$\mathcal{B}^r\psi(\eta_n^{(\mathfrak{a})}) \subset \mathrm{char}(A_\infty)\Lambda_n. \tag{15}$$

Assuming this, Lemma 11.11 applied with $B = \bar{\mathcal{E}}_n/(\bar{\mathcal{E}}_n)_{\mathrm{tors}}$ shows that

$$\mathcal{B}^r\eta_n^{(\mathfrak{a})} \subset \mathrm{char}(A_\infty)\bar{\mathcal{E}}_n + (\bar{\mathcal{E}}_n)_{\mathrm{tors}}.$$

Since $\mathcal{E}_\infty$ has no Λ-torsion (Lemma 11.5), it follows that

$$\mathcal{B}^r\{\eta_n^{(\mathfrak{a})}\}_n \subset \mathrm{char}(A_\infty)\mathcal{E}_\infty.$$

Thus $\mathcal{B}^r\mathcal{C}_\infty \subset \mathrm{char}(A_\infty)\mathcal{E}_\infty$, and since $\mathcal{B}^r$ is an ideal of height at least two the theorem follows.

It remains to prove (15). Suppose $0 \le k < r$ and M is a power of p. Proposition 11.10 shows that

$$\mathcal{B}\Psi(k,n,MN_n^{r-k})\Lambda_{n,M} \subset f_{k+1}\Psi(k+1,n,MN_n^{r-k-1})\Lambda_{n,M},$$

so by induction we conclude that

$$\mathcal{B}^r\Psi(0,n,MN_n^r)\Lambda_{n,M} \subset \left(\prod_{i=1}^r f_i\right)\Psi(r,n,M) \subset \mathrm{char}(A_\infty)\Lambda_{n,M}. \qquad (16)$$

Using (14) it follows that

$$\mathcal{B}^r\psi(\eta_n^{(a)})\Lambda_{n,M} \subset \mathrm{char}(A_\infty)\Lambda_{n,M},$$

and since this holds for every M, it proves (15). This completes the proof of Theorem 11.7. □

The rest of this section is devoted to proving Proposition 11.10. If $\sigma = (\sigma_1,\ldots,\sigma_k)$ is a Frobenius sequence define

$$A_\sigma = \sum_{i=1}^k \Lambda[\sigma_i] \subset A_\infty^0.$$

Lemma 11.12. *If σ is a Frobenius sequence of length k then A_σ is a direct summand of A_∞^0 and $A_\sigma = \oplus_{i=1}^k \Lambda/f_i\Lambda$.*

Proof. Recall that $A_\infty^0 = \oplus_{i=1}^r \Lambda y_i$. Define $Y_k = \sum_{i=k+1}^r \Lambda y_i$. The image of $A_\sigma + Y_k$ in $A_\infty^0/\mathcal{M}A_\infty^0$ contains all the y_i, so by Nakayama's Lemma, $A_\sigma + Y_k = A_\infty^0$. We will show that $A_\sigma \cap Y_k = 0$, and thus $A_\infty^0 = A_\sigma \oplus Y_k$ and

$$A_\sigma \cong A_\infty^0/Y_k \cong \oplus_{i=1}^k \Lambda/f_i\Lambda.$$

For $1 \le i \le k$ write

$$[\sigma_i] = y_i + v_i + w_i$$

where $v_i \in \mathcal{M}(\oplus_{i\le k}\Lambda y_i)$ and $w_i \in \mathcal{M}Y_k$. Suppose

$$\sum_{i=1}^k a_i[\sigma_i] \in Y_k$$

with $a_i \in \Lambda$. Then we must have $\sum_{i=1}^k a_i(y_i + v_i) = 0$. We can write this in matrix form, using the basis $y_1,\ldots,y_k$ of $\oplus_{i\le k}\Lambda y_i$, as

$$(a_1,\ldots,a_k)B \in (f_1\Lambda,\ldots,f_k\Lambda)$$

where B is a $k\times k$ matrix with entries in Λ, congruent to the identity matrix modulo $\mathcal{M}$. Therefore B is invertible, and, since $f_k \mid f_i$ for every $i \le k$, we conclude that each a_i is divisible by f_k. But f_k annihilates Y_k, so we deduce that

$$\sum_{i=1}^{k} a_i[\sigma_i] = \sum_{i=1}^{k} a_i w_i = 0.$$

This completes the proof of the lemma. □

Recall $\Omega = K_\infty(\mu_{p^\infty})$.

Proposition 11.13. *Suppose W is a finite subgroup of $K_n^\times/(K_n^\times)^M$ for some n and M. Then G_K acts trivially on the cokernel of the natural Kummer map*

$$G_\Omega \to \mathrm{Hom}(W, \mu_M).$$

Proof. Let $\bar{W}$ denote the image of W in $\Omega^\times/(\Omega^\times)^M$. The map in question factors

$$G_\Omega \to \mathrm{Hom}(\bar{W}, \mu_M) \to \mathrm{Hom}(W, \mu_M).$$

where the first (Kummer) map is surjective and the cokernel of the second is $\mathrm{Hom}(V, \mu_M)$ with

$$V = \ker(W \to \bar{W}) \subset \ker(H^1(K_n, \mu_M) \to H^1(\Omega, \mu_M)) = H^1(\Omega/K_n, \mu_M).$$

Since Ω is abelian over K, G_K acts on V via the cyclotomic character, and hence G_K acts trivially on $\mathrm{Hom}(V, \mu_M)$. The proposition follows. □

Let $\mathcal{A}$ denote the annihilator in Λ of A_∞/A_∞^0, so $\mathcal{A}$ is an ideal of height at least two.

Lemma 11.14. *Suppose $n \geq 0$, M is a power of p, $k < r$, and $\pi = \{\tilde{\pi}_1, \ldots, \tilde{\pi}_{k+1}\} \in \Pi(k+1, n, MN_n)$. Let $\mathfrak{Q} = \tilde{\pi}_{k+1}$, $\mathfrak{q} = \pi_{k+1}$ and $\mathfrak{r} = \mathfrak{q}^{-1}\mathfrak{r}(\pi)$. If $\rho \in \mathcal{A}$ then there is a Galois-equivariant homomorphism*

$$\psi : \Lambda_{n,M}\kappa_{n,M}(\mathfrak{r}\mathfrak{q}) \to \Lambda_{n,M}$$

such that

$$\rho\phi_{\mathfrak{q}}(\kappa_{n,M'}(\mathfrak{r})) \equiv f_{k+1}\psi(\kappa_{n,M}(\mathfrak{r}\mathfrak{q}))\mathfrak{Q} \pmod{M}$$

where $\phi_{\mathfrak{q}} : \Lambda_{n,MN_n}\kappa_{n,MN_n}(\mathfrak{r}) \to \Lambda_{n,MN_n}\mathfrak{Q}$ is the map of Definition 8.9.

Proof. Write $M' = MN_n$, and let σ be a Frobenius sequence corresponding to π. Let $\bar{A}_n$ denote the quotient of A_n by the Λ_n-submodule generated by the classes of $\tilde{\pi}_1, \ldots, \tilde{\pi}_k$, and let $[\mathfrak{Q}]$ denote the class of $\mathfrak{Q}$ in $\bar{A}_n$. Since the Frobenius of $\mathfrak{Q}$ on the Hilbert class field of K_n is σ, $[\mathfrak{Q}]$ is the projection of $[\sigma]$ to $\bar{A}_n$. By Lemma 11.12 the annihilator of $[\sigma]$ in A_∞^0/A_σ is $f_{k+1}\Lambda$ and A_σ is a direct summand of A_∞^0, so the annihilator of $[\sigma]$ in $(A_\infty^0/A_\sigma) \otimes \Lambda_n$ is $f_{k+1}\Lambda_n$. By Lemma 11.1, $\bar{A}_n = A_\infty/(A_\sigma + \mathcal{J}_n A_\infty)$ so the kernel of the natural map $(A_\infty^0/A_\sigma)\otimes\Lambda_n \to \bar{A}_n$ is annihilated by $\mathcal{A}$. Therefore if $\mathcal{A}' \subset \Lambda_n$ is the annihilator of $[\mathfrak{Q}]$ in $\bar{A}_n$, then

$$\mathcal{A}\mathcal{A}' \subset f_{k+1}\Lambda_n.$$

Since $\#(A_n)$ divides N_n, Proposition 8.10(i) shows that $[\kappa_{n,M'}(\mathfrak{r}\mathfrak{q})]_\mathfrak{q}$ is 0 in $\bar{A}_n$. Therefore $[\kappa_{n,M'}(\mathfrak{r}\mathfrak{q})]_\mathfrak{q} \in \mathcal{A}'\Lambda_{n,M'}\mathfrak{Q}$, so if $\rho \in \mathcal{A}$ then

$$\rho[\kappa_{n,M'}(\mathfrak{r}\mathfrak{q})]_\mathfrak{q} \in f_{k+1}\Lambda_{n,M'}\mathfrak{Q}.$$

Since f_{k+1} divides f_1 and f_1 divides N_n in Λ_n, the map

$$f_{k+1}^{-1} : \Lambda_{n,M'} \to \Lambda_{n,M}$$

is well-defined, and we will define $\psi : \Lambda_{n,M}\kappa_{n,M}(\mathfrak{r}\mathfrak{q}) \to \Lambda_{n,M}\mathfrak{Q}$ by

$$\psi(\kappa_{n,M}(\mathfrak{r}\mathfrak{q}))\mathfrak{Q} = f_{k+1}^{-1}\rho[\kappa_{n,M'}(\mathfrak{r}\mathfrak{q})]_\mathfrak{q}.$$

If we can show that ψ is well-defined, then by Proposition 8.10(ii) we will have

$$\rho\phi_\mathfrak{q}(\kappa_{n,M'}(\mathfrak{r})) = \rho[\kappa_{n,M'}(\mathfrak{r}\mathfrak{q})]_\mathfrak{q} \equiv f_{k+1}\psi(\kappa_{n,M}(\mathfrak{r}\mathfrak{q}))\mathfrak{Q} \pmod{M}$$

as desired.

We need to show that ψ is well-defined, i.e., if $\eta \in \Lambda_n$ and $\kappa_{n,M}(\mathfrak{r}\mathfrak{q})^\eta \in (K_n^\times)^M$ then $\eta\rho[\kappa_{n,M'}(\mathfrak{r}\mathfrak{q})]_\mathfrak{q} \in f_{k+1}M\Lambda_{n,M'}\mathfrak{Q}$. But this is essentially the same argument as above. If η annihilates $\kappa_{n,M}(\mathfrak{r}\mathfrak{q})$ then $\kappa_{n,M'}(\mathfrak{r}\mathfrak{q})^\eta = \alpha^M$ for some $\alpha \in K_n^\times$. Again using Proposition 8.10(i), $[\alpha]_\mathfrak{q}$ is 0 in $\bar{A}_n$, so $\rho[\alpha]_\mathfrak{q} \in f_{k+1}\Lambda_{n,N_n}\mathfrak{Q}$ and the desired inclusion follows. □

Proof (of Proposition 11.10). Let $\Lambda_\Omega = \mathbf{Z}_p[[\mathrm{Gal}(\Omega/K)]]$ and denote by ϵ both the cyclotomic character $\mathrm{Gal}(\Omega/K) \to \mathbf{Z}_p^\times$ and the induced map $\Lambda_\Omega \to \mathbf{Z}_p$. Define

$$\mathrm{tw}_\epsilon : \Lambda_\Omega \to \Lambda_\Omega$$

to be the homomorphism induced by $\gamma \mapsto \epsilon(\gamma)\gamma^{-1}$ for $\gamma \in \mathrm{Gal}(\Omega/K)$.

Recall that $\mathcal{A}$ is the annihilator of A_∞/A_∞^0, and define

$$\mathcal{B} = \begin{cases} \mathcal{A} & \text{if } K_\mathfrak{p} = \mathbf{Q}_p \\ \mathcal{A}\,\mathrm{tw}_\epsilon(\mathcal{M}\mathcal{A}\mathcal{J}_0) & \text{if } K_\mathfrak{p} \neq \mathbf{Q}_p \end{cases}$$

(recall that $\Omega = K_\infty$ if $K_\mathfrak{p} \neq \mathbf{Q}_p$). Then $\mathcal{B}$ is an ideal of height at least two, and we will show that Proposition 11.10 holds with this choice of $\mathcal{B}$.

Fix n and M, and write $M' = MN_n$. Fix a Kolyvagin sequence $\pi \in \Pi(k,n,M')$, let $\mathfrak{r} = \mathfrak{r}(\pi)$, and suppose $\psi : \Lambda_{n,M'}\kappa_{n,M'}(\mathfrak{r}) \to \Lambda_{n,M'}$. We need to show that

$$\mathcal{B}\psi(\kappa_{n,M'}(\mathfrak{r}))\Lambda_{n,M} \subset f_{k+1}\Psi(k+1,n,M).$$

We will do this by constructing suitable Kolyvagin sequences of length $k+1$ extending π.

There is a $\mathbf{Z}_p$-module isomorphism

$$\iota : \mathrm{Hom}_{\Lambda_n}(\Lambda_{n,M'}\kappa_{n,M'}(\mathfrak{r}), \Lambda_{n,M'}) \otimes \mu_{M'} \xrightarrow{\sim} \mathrm{Hom}_{\mathbf{Z}_p}(\Lambda_{n,M'}\kappa_{n,M'}(\mathfrak{r}), \mu_{M'})$$

induced by

$$(\sum_{\gamma} a_\gamma \gamma) \otimes \zeta \mapsto \zeta^{a_1}.$$

One can check that if $\phi \in \mathrm{Hom}_{\Lambda_n}(\Lambda_{n,M'}\kappa_{n,M'}(\mathfrak{r}), \Lambda_{n,M'})$, $\zeta \in \mu_{M'}$, and $\rho \in \Lambda_\Omega$ then

$$\rho\iota(\phi \otimes \zeta) = \iota((\mathrm{tw}_\epsilon(\rho)\phi) \otimes \zeta).$$

Suppose $\tau \in G_K$, fix a primitive M'-th root of unity $\zeta_{M'}$, and let $\mathrm{Kum}_{M'}$ denote the Kummer map $G_\Omega \to \mathrm{Hom}(K_n^\times, \mu_{M'})$. By Proposition 11.13 there is a $\gamma_0 \in G_\Omega$ such that

$$\mathrm{Kum}_{M'}(\gamma_0) = (\tau - 1)\iota(\psi \otimes \zeta_{M'}) \quad \text{on } \Lambda_{n,M'}\kappa_{n,M'}(\mathfrak{r}).$$

Choose $\rho \in \Lambda_\Omega$ such that the projection of ρ to Λ lies in $\mathcal{M}\mathcal{A}$ and let $\gamma \in G_\Omega$ be such that $\gamma = \gamma_0^\rho$ on Ω^{ab} (we view $\mathrm{Gal}(\Omega/K)$ as acting on $\mathrm{Gal}(\Omega^{\mathrm{ab}}/\Omega)$ in the usual way).

Let σ be a Frobenius sequence corresponding to π. We define two Frobenius sequences σ' and σ'' of length $k+1$ extending σ as follows. Let σ'_{k+1} be an element of G_Ω such that $[\sigma'_{k+1}] = y_{k+1}$, and let $\sigma''_{k+1} = \sigma'_{k+1}\gamma$ with γ as above.

Since $[\gamma] = \rho[\gamma_0] \in \mathcal{M}\mathcal{A}^0_\infty$, both σ' and σ'' are Frobenius sequences.

Let $\mathfrak{q}'$ and $\mathfrak{q}''$ be primes of K whose Frobenius elements (for some choice of primes "upstairs") in $H_n(\mu_{M'}, (\Lambda_n\kappa_{n,M'}(\mathfrak{r}))^{1/M'})/K$ are the restrictions of σ' and σ'', respectively, where H_n is the Hilbert class field of K_n. Let $\mathfrak{Q}'$ and $\mathfrak{Q}''$ be primes of K_n above $\mathfrak{q}'$ and $\mathfrak{q}''$ with these Frobenius elements. It follows from Definition 8.9 that there are integers a' and a'' such that

$$\iota(\bar\phi_{\mathfrak{q}'} \otimes \zeta_{M'}^{a'}) = \mathrm{Kum}_{M'}(\sigma'), \quad \iota(\bar\phi_{\mathfrak{q}''} \otimes \zeta_{M'}^{a''}) = \mathrm{Kum}_{M'}(\sigma'')$$

where $\phi_{\mathfrak{q}'} : \Lambda_{n,M'}\kappa_{n,M'}(\mathfrak{r}) \to \Lambda_{n,M'}\mathfrak{Q}'$ is the map of Definition 8.9, $\bar\phi_{\mathfrak{q}'} \in \mathrm{Hom}(\Lambda_{n,M'}\kappa_{n,M'}(\mathfrak{r}), \Lambda_{n,M'})$ is defined by $\phi_{\mathfrak{q}'} = \bar\phi_{\mathfrak{q}'}\mathfrak{Q}'$, and similarly for $\mathfrak{q}''$. Now

$$\begin{aligned}\iota(\mathrm{tw}_\epsilon(\rho(\tau-1))\psi) \otimes \zeta_{M'} &= \rho\mathrm{Kum}_{M'}(\gamma_0) \\ &= \mathrm{Kum}_{M'}(\sigma'') - \mathrm{Kum}_{M'}(\sigma') \\ &= \iota(a''\bar\phi_{\mathfrak{q}''} \otimes \zeta_{M'} - a'\bar\phi_{\mathfrak{q}'} \otimes \zeta_{M'})\end{aligned}$$

and so finally

$$\mathrm{tw}_\epsilon(\rho(\tau-1))\psi = a''\bar\phi_{\mathfrak{q}''} - a'\bar\phi_{\mathfrak{q}'}. \tag{17}$$

If $K_\mathfrak{p} \neq \mathbf{Q}_p$, then the $\mathrm{tw}_\epsilon(\rho)(\epsilon(\tau)\tau - 1)$, with our choices of τ and ρ, generate $\mathrm{tw}_\epsilon(\mathcal{MAJ}_0)$. If $K_\mathfrak{p} = \mathbf{Q}_p$, then $\mu_p \not\subset K_\infty$ (since K_∞/K is unramified at $p/\mathfrak{p}$) and so we can choose τ and ρ so that $\mathrm{tw}_\epsilon(\rho(\tau - 1))$ projects to a unit in Λ. Now Lemma 11.14 completes the proof of Proposition 11.10. □

12 Computing the Selmer Group

In this section we compute the order of the p-power Selmer group for primes $p > 7$ of good reduction, and thereby prove assertion (ii) of the theorem of the introduction. The computation divides naturally into two cases depending on whether p splits in K or not.

Keep the notation of the previous section. In particular E is an elliptic curve defined over an imaginary quadratic field K, with complex multiplication by the full ring of integers of K, and $\mathfrak{p}$ is a prime of K of residue characteristic greater than 7 where E has good reduction.

Definition 12.1. Let $\pi = \psi(\mathfrak{p})$ and recall that the π-adic Tate module of E is defined by

$$T_\pi(E) = \varprojlim_n E[\mathfrak{p}^n],$$

inverse limit with respect to multiplication by π. For every n let $\delta_n : U_n \to E[\mathfrak{p}^n]$ be the map of Lemma 6.8. It is clear from the definition that we have commutative diagrams

$$\begin{array}{ccc} U_{n+1} & \xrightarrow{\delta_{n+1}} & E[\mathfrak{p}^{n+1}] \\ {\scriptstyle \mathbf{N}_{K_{n+1}/K_n}}\downarrow & & \downarrow{\scriptstyle \pi} \\ U_n & \xrightarrow{\delta_n} & E[\mathfrak{p}^n], \end{array}$$

and we define

$$\delta_\infty = \varprojlim_n \delta_n : U_\infty \longrightarrow T_\pi(E).$$

Recall the Selmer group $\mathcal{S}_{\pi^n}(E)$ of Definition 4.1 and the extended Selmer group $\mathcal{S}'_{\pi^n}(E)$ of Definition 6.3. Define

$$\mathcal{S}_{\mathfrak{p}^\infty} = \varinjlim_n \mathcal{S}_{\pi^n}(E), \quad \mathcal{S}'_{\mathfrak{p}^\infty} = \varinjlim_n \mathcal{S}'_{\pi^n}(E).$$

Thus there is an exact sequence

$$0 \to E(K) \otimes \mathbf{Q}_p/\mathbf{Z}_p \to \mathcal{S}_{\mathfrak{p}^\infty} \to \text{Ш}(E)_{\mathfrak{p}^\infty} \to 0. \tag{18}$$

Proposition 12.2. (i) $\mathcal{S}'_{\mathfrak{p}^\infty} = \mathrm{Hom}(X_\infty, E[\mathfrak{p}^\infty])^{G_\infty} = \mathrm{Hom}(X_\infty^{\chi_E}, E[\mathfrak{p}^\infty])^{\Gamma}$.
(ii) $\mathcal{S}_{\mathfrak{p}^\infty}$ *is the kernel of the composition*

$$\mathcal{S}'_{\mathfrak{p}^\infty} \xrightarrow{\sim} \mathrm{Hom}(X_\infty^{\chi_E}, E[\mathfrak{p}^\infty])^{\Gamma} \to \mathrm{Hom}(\ker(\delta_\infty), E[\mathfrak{p}^\infty])$$

induced by (i) *and local class field theory.*

Proof. The first assertion is just a restatement of Proposition 6.5, and the second follows from Theorem 6.9. □

Theorem 12.3 (Wiles' explicit reciprocity law [Wil]). *Suppose* $\mathbf{x}$ *is an* $\mathcal{O}_\mathfrak{p}$*-generator of* $T_\pi(E)$, $\mathbf{z} = (z_n)$ *is the corresponding generator of* $T_\pi(\hat{E})$, $\mathbf{u} = (u_n) \in U_\infty$, *and* $f(Z) \in \mathcal{O}_\mathfrak{p}[[Z]]$ *is such that* $f(z_n) = u_n$ *for every* n. *Then*

$$\delta_\infty(\mathbf{u}) = (\psi(\mathfrak{p}) - 1)\frac{f'(0)}{f(0)}\mathbf{x}.$$

See [Wil] or [dS] Theorem I.4.2 for the proof.

Corollary 12.4. $\delta_\infty(\mathcal{C}_\infty) = \dfrac{L(\bar{\psi},1)}{\Omega}T_\pi(E)$.

Proof. Using Theorem 12.3, we see that $\delta_\infty(\mathcal{C}_\infty)$ is the ideal of $\mathcal{O}_\mathfrak{p}$ generated by the values $(\Lambda'_{\mathfrak{p},\mathfrak{a}}(0)/\Lambda_{\mathfrak{p},\mathfrak{a}}(0))$ where $\Lambda_{\mathfrak{p},\mathfrak{a}}$ is defined in Definition 7.21, and we allow the ideal $\mathfrak{a}$ to vary. The corollary now follows from Theorem 7.22(ii) and Lemma 10.2. □

Remark 12.5. In fact, for every $\mathbf{u} \in U_\infty$ there is a power series $f_\mathbf{u} \in \mathcal{O}_\mathfrak{p}[[X]]$ such that $f_\mathbf{u}(z_n) = u_n$ for every n as in Theorem 12.3. See [Col] or [dS] §I.2.

Definition 12.6. Let $\rho_E : G_\infty \to \mathcal{O}_\mathfrak{p}^\times$ be the character giving the action of G_∞ on $E[\mathfrak{p}^\infty]$. We can also view ρ_E as a homomorphism from Λ to $\mathcal{O}_\mathfrak{p}$, and we define $\mathcal{A}_E \subset \Lambda$ to be the kernel of this homomorphism.

If $a, b \in K_\mathfrak{p}$ we will write $a \sim b$ to mean that $a/b \in \mathcal{O}_\mathfrak{p}^\times$.

12.1 Determination of the Selmer Group when $K_\mathfrak{p} = \mathbf{Q}_p$

For this subsection we suppose (in addition to our other assumptions) that $K_\mathfrak{p} = \mathbf{Q}_p$.

If M is a Λ-module we will write

$$M^{\mathcal{A}_E=0} = \{m \in M : \mathcal{A}_E m = 0\}.$$

Proposition 12.7. *Suppose that* M *is a finitely-generated torsion* Λ *module.*

(i) $\mathrm{Hom}(M, E[\mathfrak{p}^\infty])^{G_\infty}$ *is finite* $\Leftrightarrow \rho_E(\mathrm{char}(M)) \neq 0 \Leftrightarrow M^{\mathcal{A}_E=0}$ *is finite.*
(ii) $\#(\mathrm{Hom}(M, E[\mathfrak{p}^\infty])^{G_\infty}) \sim \rho_E(\mathrm{char}(M))\#(M^{\mathcal{A}_E=0})$.

Proof. Fix an exact sequence of Λ-modules

$$0 \to \bigoplus_{i=1}^{k} \Lambda/f_i\Lambda \to M \to Z \to 0$$

with pseudo-null (in this case, finite) cokernel Z. Fix a topological generator γ of $G_\infty = \Delta \times \Gamma$. Then $\mathcal{A}_E = (\gamma - \rho_E(\gamma))\Lambda$, so multiplication by $\gamma - \rho_E(\gamma)$ leads to a snake lemma exact sequence of kernels and cokernels

$$0 \to \bigoplus_{i=1}^{k} (\Lambda/f_i\Lambda)^{\mathcal{A}_E=0} \to M^{\mathcal{A}_E=0} \to Z^{\mathcal{A}_E=0}$$
$$\to \bigoplus_{i=1}^{k} \Lambda/(f_i\Lambda + \mathcal{A}_E) \to M/\mathcal{A}_E M \to Z/\mathcal{A}_E \to 0.$$

Also

$$\mathrm{Hom}(M, E[\mathfrak{p}^\infty])^{G_\infty} = \mathrm{Hom}(M/\mathcal{A}_E M, E[\mathfrak{p}^\infty]).$$

The map ρ_E induces an isomorphism $\Lambda/(f_i\Lambda + \mathcal{A}_E) \xrightarrow{\sim} \mathbf{Z}_p/\rho_E(f_i)\mathbf{Z}_p$, and

$$(\Lambda/f_i\Lambda)^{\mathcal{A}_E=0} = \{g \in \Lambda, g\mathcal{A}_E \subset f_i\Lambda\}/f_i\Lambda$$

so since $\mathcal{A}_E$ is a prime ideal,

$$(\Lambda/f_i\Lambda)^{\mathcal{A}_E=0} \neq 0 \Leftrightarrow f_i \in \mathcal{A}_E \Leftrightarrow (\Lambda/f_i\Lambda)^{\mathcal{A}_E=0} \text{ is infinite.}$$

Since Z is finite, the exact sequence

$$0 \to Z^{\mathcal{A}_E=0} \to Z \xrightarrow{\gamma-\rho_E(\gamma)} Z \to Z/\mathcal{A}_E Z \to 0$$

shows that $\#(Z^{\mathcal{A}_E=0}) = \#(Z/\mathcal{A}_E Z)$. Since $\mathrm{char}(M) = \prod_i f_i\Lambda$, the lemma follows. □

Theorem 12.8. $\#(\mathcal{S}'_{\mathfrak{p}^\infty}) = [\mathbf{Z}_p : \rho_E(\mathrm{char}(X_\infty))]$.

Proof. This is immediate from Propositions 12.2(i), 11.3, and 12.7 (note that if $\rho_E(\mathrm{char}(X_\infty)) \neq 0$ then $X_\infty^{\mathcal{A}_E=0}$ is finite by Proposition 12.7 and hence zero by Proposition 11.3). □

Theorem 12.9. $\mathrm{char}(X_\infty^{\chi_E}) = \mathrm{char}(U_\infty^{\chi_E}/C_\infty^{\chi_E})$.

Proof. Immediate from Corollary 11.8 and (13). □

Theorem 12.10 (Coates and Wiles). *Let $\mathcal{D}$ denote the ring of integers of the completion of the maximal unramified extension of $\mathbf{Q}_p$. Then there is a $\mathfrak{p}$-adic period $\Omega_\mathfrak{p} \in \mathcal{D}^\times$ such that* $\mathrm{char}(U_\infty/C_\infty)\mathcal{D}[[G_\infty]]$ *has a generator $\mathcal{L}_E$ satisfying*

$$\rho_E^k(\mathcal{L}_E) = \Omega_\mathfrak{p}^k(1 - \psi(\mathfrak{p}^k)/p)\frac{L(\bar{\psi}^k, k)}{\Omega^k}$$

for every $k \geq 1$.

Proof. See [CW2] or [dS] Corollary III.1.5. □

Corollary 12.11. $\#(\mathcal{S}'_{\mathfrak{p}^\infty}) \sim (1 - \psi(\mathfrak{p})/p)\dfrac{L(\bar{\psi}, 1)}{\Omega}$.

Proof. Immediate from Theorem 12.8, Theorem 12.9, and Theorem 12.10. □

Proposition 12.12. $[\mathcal{S}'_{\mathfrak{p}^\infty} : \mathcal{S}_{\mathfrak{p}^\infty}] \sim (1 - \psi(\mathfrak{p})/p)$.

For a proof see [PR] Proposition II.8 or [Co] Proposition 2 and Lemma 3.

Corollary 12.13. *Suppose* $\mathfrak{p} \nmid \mathfrak{f}$, $p > 7$, *and* $K_{\mathfrak{p}} = \mathbf{Q}_p$.

(i) *If* $L(\bar{\psi}, 1) = 0$ *then* $\mathcal{S}_{\mathfrak{p}^\infty}$ *is infinite.*
(ii) *If* $L(\bar{\psi}, 1) \neq 0$ *then*

$$\#(\text{Ш}(E)_{\mathfrak{p}^\infty}) \sim \frac{L(\bar{\psi}, 1)}{\Omega}.$$

Proof. This is immediate from Corollary 12.11 and Proposition 12.12. (For (ii), we also use (18).) □

12.2 Determination of the Selmer Group when $[K_{\mathfrak{p}} : \mathbf{Q}_p] = 2$

For this subsection we suppose that $[K_{\mathfrak{p}} : \mathbf{Q}_p] = 2$, so $\Gamma \cong \mathbf{Z}_p^2$ and $E[\mathfrak{p}^\infty] \cong K_{\mathfrak{p}}/\mathcal{O}_{\mathfrak{p}}$ has $\mathbf{Z}_p$-corank 2.

Lemma 12.14. *There is a decomposition*

$$U_\infty^{\chi_E} = V_1 \oplus V_2$$

where V_1 *and* V_2 *are free of rank one over* Λ_{χ_E}, $\delta_\infty(V_2) = 0$, *and* $\mathcal{E}_\infty^{\chi_E} \not\subset V_2$.

Proof. By Proposition 11.4, $U_\infty^{\chi_E}$ is free of rank two over Λ_{χ_E}. Fix a splitting $U_\infty^{\chi_E} = \Lambda_{\chi_E} v_1 \oplus \Lambda_{\chi_E} v_2$. By Corollary 5.20(ii), ρ_E is surjective, and it follows that $\delta_\infty(\Lambda_{\chi_E} v_1)$ and $\delta_\infty(\Lambda_{\chi_E} v_2)$ are $\mathcal{O}_{\mathfrak{p}}$-submodules of $T_\pi(E)$. Since δ_∞ is surjective (Lemma 6.8) and $\delta_\infty(U_\infty) = \delta_\infty(U_\infty^{\chi_E})$, it follows that either $\delta_\infty(\Lambda_{\chi_E} v_1) = T_\pi(E)$ or $\delta_\infty(\Lambda_{\chi_E} v_2) = T_\pi(E)$.

Thus, by renumbering if necessary, we may assume that $\delta_\infty(\Lambda_{\chi_E} v_1) = T_\pi(E)$. In particular we can choose $g \in \Lambda_{\chi_E}$ so that $\delta_\infty(v_2) = \delta_\infty(g v_1)$, and (by adjusting g if necessary by an element of the kernel of ρ_E) we may assume that $\mathcal{E}_\infty \not\subset \Lambda_{\chi_E}(v_2 - g v_1)$. Now the lemma is satisfied with

$$V_1 = \Lambda_{\chi_E} v_1, \quad V_2 = \Lambda_{\chi_E}(v_2 - g v_1).$$

□

Definition 12.15. Fix a decomposition of $U_\infty^{\chi_E}$ as in Lemma 12.14 and define

$$\tilde{U} = U_\infty^{\chi_E}/V_2, \quad \tilde{X} = X_\infty^{\chi_E}/\text{image}(V_2)$$

where image(V_2) denotes the image of V_2 in X_∞ under the Artin map of local class field theory.

Lemma 12.16. (i) $\tilde{X}$ *is a torsion* Λ_{χ_E}*-module with no nonzero pseudo-null submodules.*
(ii) $\mathrm{char}(\tilde{X}) = \mathrm{char}(\tilde{U}/\mathrm{image}(\mathcal{C}_\infty^{\chi_E}))$.

Proof. Since $\mathcal{E}_\infty \not\subset V_2$ and V_2 is free, (i) follows from Proposition 11.3. Also, the exact sequence (13) induces an exact sequence

$$0 \to \mathcal{E}_\infty^{\chi_E}/\mathcal{C}_\infty^{\chi_E} \to \tilde{U}/\mathrm{image}(\mathcal{C}_\infty^{\chi_E}) \to \tilde{X} \to A_\infty^{\chi_E} \to 0,$$

so (ii) follows from Corollary 11.8. □

Proposition 12.17. $\mathcal{S}_{\mathfrak{p}^\infty} = \mathrm{Hom}(\tilde{X}, E[\mathfrak{p}^\infty])^\Gamma$.

Proof. By our choice of V_1 and V_2 (Proposition 12.14), we see that $\ker(\delta_\infty) = \mathcal{A}_E V_1 + V_2$. Thus

$$\tilde{X}/\mathcal{A}_E\tilde{X} = X_\infty/(\mathcal{A}_E X_\infty + \mathrm{image}(V_2)) = X_\infty/(\mathcal{A}_E X_\infty + \mathrm{image}(\ker(\delta_\infty)))$$

and so by Proposition 12.2(ii)

$$\mathcal{S}_{\mathfrak{p}^\infty} = \mathrm{Hom}(X_\infty/(\mathcal{A}_E X_\infty + \mathrm{image}(\ker(\delta_\infty))), E[\mathfrak{p}^\infty]) = \mathrm{Hom}(\tilde{X}, E[\mathfrak{p}^\infty])^\Gamma.$$

□

Proposition 12.18. *Suppose M is a finitely-generated torsion Λ_{χ_E}-module and F is a $\mathbf{Z}_p$-extension of K_1 in K_∞ satisfying*

(i) *M has no nonzero pseudo-null submodules,*
(ii) *If γ generates* $\mathrm{Gal}(K_\infty/F)$ *then char$(M) \not\subset (\gamma - 1)\Lambda_{\chi_E}$, char$(M) \not\subset (\gamma - \rho_E(\gamma))\Lambda_{\chi_E}$, and $M/(\gamma-1)M$ has no nonzero finite submodules.*

Then

$$\#(\mathrm{Hom}(M, E[\mathfrak{p}^\infty])^\Gamma) = [\mathcal{O}_\mathfrak{p} : \rho_E(\mathrm{char}(M))].$$

Proof (sketch). For a complete proof see [Ru2], Lemmas 6.2 and 11.15.

Let $\hat{T}_\pi = \mathrm{Hom}(T_\pi, \mathcal{O}_\mathfrak{p})$, let $\Lambda_F = \mathbf{Z}_p[[\mathrm{Gal}(F/K)]]$, and let $\bar{M}$ denote the $\Lambda_F^{\chi_E}$-module $(M \otimes \hat{T}_\pi)/(\gamma - 1)(M \otimes \hat{T}_\pi)$. Using the hypotheses on M and F it is not difficult to show (see [Ru2] Lemma 11.15) that $\bar{M}$ has no nonzero finite submodules. Therefore exactly as in Proposition 12.7,

$$\begin{aligned}\#(\mathrm{Hom}(M, E[\mathfrak{p}^\infty])^\Gamma) &= \#(\mathrm{Hom}(M \otimes \hat{T}_\pi, \mathcal{O}_\mathfrak{p})^\Gamma)\\ &= \#(\mathrm{Hom}(\bar{M}, \mathcal{O}_\mathfrak{p})^{\mathrm{Gal}(F/K)})\\ &= [\mathcal{O}_\mathfrak{p} : \mathbb{1}(\mathrm{char}_F(\bar{M}))]\end{aligned}$$

where $\mathbb{1}$ denotes the trivial character and $\mathrm{char}_F(\bar{M})$ is the characteristic ideal of $\bar{M}$ as a $\Lambda_F^{\chi_E}$-module.

By an argument similar to the proof of Proposition 12.7, one can show that $\mathrm{char}_F(\bar{M}) = \mathrm{char}(M \otimes \hat{T}_\pi)\Lambda_F^{\chi_E}$. Therefore

$$\#(\mathrm{Hom}(M, E[\mathfrak{p}^\infty])^\Gamma) = [\mathcal{O}_\mathfrak{p} : \mathbb{1}(\mathrm{char}(M \otimes \hat{T}_\pi))] = [\mathcal{O}_\mathfrak{p} : \rho_E(\mathrm{char}(M))].$$

□

Theorem 12.19. *Suppose* $\mathfrak{p} \nmid \mathfrak{f}$, $p > 7$, *and* $K_\mathfrak{p} \neq \mathbf{Q}_p$.

(i) *If* $L(\bar{\psi}, 1) = 0$ *then* $\mathcal{S}_{\mathfrak{p}^\infty}$ *is infinite.*
(ii) *If* $L(\bar{\psi}, 1) \neq 0$ *then*

$$\#(\text{Ш}(E)_{\mathfrak{p}^\infty}) = [\mathcal{O}_\mathfrak{p} : (L(\bar{\psi}, 1)/\Omega)\mathcal{O}_\mathfrak{p}].$$

Proof. Lemma 12.16(i) shows that $\tilde{X}$ satisfies the first hypothesis of Proposition 12.18, and the same argument with K_∞ replaced by F verifies the second hypothesis for all but finitely many choices of F. Also, $\tilde{U}/\text{image}(\mathcal{C}_\infty^{\chi_E})$ satisfies the hypotheses of Proposition 12.18 since it is a quotient of one free Λ_{χ_E}-module by another (Proposition 11.6). Therefore by Proposition 12.18 and Lemma 12.16(ii),

$$\#(\text{Hom}(\tilde{X}, E[\mathfrak{p}^\infty])^\Gamma) = \#(\text{Hom}(\tilde{U}/\text{image}(\mathcal{C}_\infty^{\chi_E}), E[\mathfrak{p}^\infty])^\Gamma).$$

The left-hand side of this equality is $\#(\mathcal{S}_{\mathfrak{p}^\infty})$ by Proposition 12.17. On the other hand,

$$\text{Hom}(\tilde{U}/\text{image}(\mathcal{C}_\infty^{\chi_E}), E[\mathfrak{p}^\infty])^\Gamma = \text{Hom}(\tilde{U}/(\text{image}(\mathcal{C}_\infty^{\chi_E} + \mathcal{A}_E\tilde{U})), E[\mathfrak{p}^\infty]),$$

and $\delta_\infty : \tilde{U}/\mathcal{A}_E\tilde{U} \to T_\pi(E)$ is an isomorphism (Lemmas 6.8 and 12.14). Therefore

$$\#(\text{Hom}(\tilde{U}/\text{image}(\mathcal{C}_\infty^{\chi_E}), E[\mathfrak{p}^\infty])^\Gamma) = \#(T_\pi/\delta_\infty(\mathcal{C}_\infty))$$

and the Theorem follows from Wiles' explicit reciprocity law (Corollary 12.4) and (18). □

12.3 Example

We conclude with one example. Let E be the elliptic curve $y^2 = x^3 - x$. The map $(x, y) \mapsto (-x, iy)$ is an automorphism of order 4 defined over $K = \mathbf{Q}(i)$, so $\text{End}_K(E) = \mathbf{Z}[i]$. Let $\mathfrak{p}_2$ denote the prime $(1+i)$ above 2.

Clearly $E(\mathbf{Q})_{\text{tors}} \supset E[2] = \{O, (0,0), (1,0), (-1,0)\}$. With a bit more effort one checks that $E(K)$ contains the point $(-i, 1+i)$ of order $\mathfrak{p}_2^3$, and using the Theorem of Nagell and Lutz ([Si] Corollary VIII.7.2) or Corollary 5.18 one can show that in fact $E(K)_{\text{tors}} = E[\mathfrak{p}_2^3]$.

The discriminant of E is 64, so E has good reduction at all primes of K different from $\mathfrak{p}_2$. Since $E[\mathfrak{p}_2^3] \subset E(K)$, if we write ψ_E for the Hecke character of K attached to E, Corollary 5.16 shows that $\psi_E(\mathfrak{a}) \equiv 1 \pmod{\mathfrak{p}_2^3}$ for every ideal $\mathfrak{a}$ prime to $\mathfrak{p}_2$. But every such ideal has a *unique* generator congruent to 1 modulo $\mathfrak{p}_2^3$, so this characterizes ψ_E and shows that its conductor is $\mathfrak{p}_2^3$.

Standard computational techniques now show that

$$L(\bar{\psi}, 1) = .6555143885\ldots$$
$$\Omega = 2.622057554\ldots$$

Therefore by the Coates-Wiles theorem (Theorem 10.1), $E(K) = E[\mathfrak{p}_2^3]$ and $E(\mathbf{Q}) = E[2]$. Further, $L(\bar{\psi}, 1)/\Omega$ is approximately $1/4$. By Proposition 10.6 $L(\bar{\psi}, 1)/\Omega$ is integral at all primes $\mathfrak{p}$ of residue characteristic greater than 7. In fact the same techniques show that $L(\bar{\psi}, 1)/\Omega$ is integral at all primes $\mathfrak{p} \neq \mathfrak{p}_2$, and give a bound on the denominator at $\mathfrak{p}_2$ from which we can conclude that $L(\bar{\psi}, 1)/\Omega = 1/4$.

Therefore by Corollary 12.13 and Theorem 12.19, $\mathcal{S}_{\mathfrak{p}^\infty} = Ш(E_{/K})_{\mathfrak{p}^\infty} = 0$ for all primes $\mathfrak{p}$ of residue characteristic greater than 7, and again the same proof works for all $\mathfrak{p} \neq \mathfrak{p}_2$. It follows easily from this that $Ш(E_{/\mathbf{Q}})_p = 0$ for all odd rational primes. Fermat did the 2-descent necessary to show that $Ш(E_{/\mathbf{Q}})_2 = 0$ (see [We] Chap. II), so in fact $Ш(E_{/\mathbf{Q}}) = 0$. Together with the fact that the Tamagawa factor at 2 is equal to 4, this shows that the full Birch and Swinnerton-Dyer conjecture holds for E over $\mathbf{Q}$.

References

[Ca] Cassels, J.W.S., Diophantine equations with special reference to elliptic curves, *J. London Math. Soc.* **41** (1966) 193–291.

[Co] Coates, J., Infinite descent on elliptic curves with complex multiplication. In: Arithmetic and Geometry, M. Artin and J. Tate, eds., *Prog. in Math.* **35**, Boston: Birkhäuser (1983) 107–137.

[CW1] Coates, J., Wiles, A., On the conjecture of Birch and Swinnerton-Dyer, *Inventiones math.* **39** (1977) 223–251.

[CW2] Coates, J., Wiles, A., On p-adic L-functions and elliptic units, *J. Austral. Math. Soc. (ser. A)* **26** (1978) 1–25.

[Col] Coleman, R., Division values in local fields, *Inventiones math.* **53** (1979) 91–116.

[dS] de Shalit, E., Iwasawa theory of elliptic curves with complex multiplication, *Perspectives in Math.* **3**, Orlando: Academic Press (1987).

[GS] Goldstein, C., Schappacher, N., Séries d'Eisenstein et fonctions L de courbes elliptiques à multiplication complexe, *J. für die reine und angew. Math.* **327** (1981) 184–218.

[Gr] Greenberg, R., On the structure of certain Galois groups, *Inventiones math.* **47** (1978) 85–99.

[Iw] Iwasawa, K., On $\mathbf{Z}_\ell$-extensions of algebraic number fields, *Annals of Math.* (2) **98** (1973) 246–326.

[Ko] Kolyvagin, V. A., Euler systems. In: The Grothendieck Festschrift (Vol. II), P. Cartier et al., eds., *Prog. in Math* **87**, Boston: Birkhäuser (1990) 435–483.

[La] Lang, S., Elliptic Functions, Reading: Addison Wesley (1973).

[PR] Perrin-Riou, B., Arithmétique des courbes elliptiques et théorie d'Iwasawa, *Bull. Soc. Math. France, Mémoire* Nouvelle série **17** 1984.

[Ro] Robert, G., Unités elliptiques, *Bull. Soc. Math. France, Mémoire* **36** (1973).

[Ru1] Rubin, K., The main conjecture. Appendix to: Cyclotomic fields I and II, S. Lang, *Graduate Texts in Math.* **121**, New York: Springer-Verlag (1990) 397–419.

[Ru2] Rubin, K., The "main conjectures" of Iwasawa theory for imaginary quadratic fields, *Inventiones Math.* **103** (1991) 25–68.

[Se] Serre, J.-P., Classes des corps cyclotomiques (d'après K. Iwasawa), Séminaire Bourbaki exposé 174, December 1958. In: Séminaire Bourbaki vol. 5, Paris: Société de France (1995) 83–93.

[ST] Serre, J.-P., Tate, J., Good reduction of abelian varieties, *Ann. of Math.* **88** (1968) 492–517.

[Sh] Shimura, G., Introduction to the arithmetic theory of automorphic functions, Princeton: Princeton Univ. Press (1971).

[Si] Silverman, J., The arithmetic of elliptic curves, *Graduate Texts in Math.* **106**, New York: Springer-Verlag (1986).

[Ta] Tate, J., Algorithm for determining the type of a singular fiber in an elliptic pencil. In: Modular functions of one variable IV, *Lecture Notes in Math.* **476**, New York: Springer-Verlag (1975) 33–52.

[We] Weil, A., Number theory. An approach through history. From Hammurapi to Legendre, Boston: Birkhäuser (1984).

[Wil] Wiles, A., Higher explicit reciprocity laws, *Annals of Math.* **107** (1978) 235–254.

[Win] Wintenberger, J.-P., Structure galoisienne de limites projectives d'unités locales, *Comp. Math.* **42** (1981) 89–103.

LIST OF C.I.M.E. SEMINARS — Publisher

1954 -	1. Analisi funzionale	C.I.M.E.
	2. Quadratura delle superficie e questioni connesse	"
	3. Equazioni differenziali non lineari	"
1955 -	4. Teorema di Riemann-Roch e questioni connesse	"
	5. Teoria dei numeri	"
	6. Topologia	"
	7. Teorie non linearizzate in elasticità, idrodinamica,aerodinamica	"
	8. Geometria proiettivo-differenziale	"
1956 -	9. Equazioni alle derivate parziali a caratteristiche reali	"
	10. Propagazione delle onde elettromagnetiche	"
	11. Teoria della funzioni di più variabili complesse e delle funzioni automorfe	"
1957 -	12. Geometria aritmetica e algebrica (2 vol.)	"
	13. Integrali singolari e questioni connesse	"
	14. Teoria della turbolenza (2 vol.)	"
1958 -	15. Vedute e problemi attuali in relatività generale	"
	16. Problemi di geometria differenziale in grande	"
	17. Il principio di minimo e le sue applicazioni alle equazioni funzionali	"
1959 -	18. Induzione e statistica	"
	19. Teoria algebrica dei meccanismi automatici (2 vol.)	"
	20. Gruppi, anelli di Lie e teoria della coomologia	"
1960 -	21. Sistemi dinamici e teoremi ergodici	"
	22. Forme differenziali e loro integrali	"
1961 -	23. Geometria del calcolo delle variazioni (2 vol.)	"
	24. Teoria delle distribuzioni	"
	25. Onde superficiali	"
1962 -	26. Topologia differenziale	"
	27. Autovalori e autosoluzioni	"
	28. Magnetofluidodinamica	"

1963 - 29. Equazioni differenziali astratte "
30. Funzioni e varietà complesse "
31. Proprietà di media e teoremi di confronto in Fisica Matematica "

1964 - 32. Relatività generale "
33. Dinamica dei gas rarefatti "
34. Alcune questioni di analisi numerica "
35. Equazioni differenziali non lineari "

1965 - 36. Non-linear continuum theories "
37. Some aspects of ring theory "
38. Mathematical optimization in economics "

1966 - 39. Calculus of variations Ed. Cremonese, Firenze
40. Economia matematica "
41. Classi caratteristiche e questioni connesse "
42. Some aspects of diffusion theory "

1967 - 43. Modern questions of celestial mechanics "
44. Numerical analysis of partial differential equations "
45. Geometry of homogeneous bounded domains "

1968 - 46. Controllability and observability "
47. Pseudo-differential operators "
48. Aspects of mathematical logic "

1969 - 49. Potential theory "
50. Non-linear continuum theories in mechanics and physics and their applications "
51. Questions of algebraic varieties "

1970 - 52. Relativistic fluid dynamics "
53. Theory of group representations and Fourier analysis "
54. Functional equations and inequalities "
55. Problems in non-linear analysis "

1971 - 56. Stereodynamics "
57. Constructive aspects of functional analysis (2 vol.) "
58. Categories and commutative algebra "

1972 -	59. Non-linear mechanics		"
	60. Finite geometric structures and their applications		"
	61. Geometric measure theory and minimal surfaces		"
1973 -	62. Complex analysis		"
	63. New variational techniques in mathematical physics		"
	64. Spectral analysis		"
1974 -	65. Stability problems		"
	66. Singularities of analytic spaces		"
	67. Eigenvalues of non linear problems		"
1975 -	68. Theoretical computer sciences		"
	69. Model theory and applications		"
	70. Differential operators and manifolds		"
1976 -	71. Statistical Mechanics		Ed Liguori, Napoli
	72. Hyperbolicity		"
	73. Differential topology		"
1977 -	74. Materials with memory		"
	75. Pseudodifferential operators with applications		"
	76. Algebraic surfaces		"
1978 -	77. Stochastic differential equations		"
	78. Dynamical systems		Ed Liguori, Napoli and Birhäuser Verlag
1979 -	79. Recursion theory and computational complexity		"
	80. Mathematics of biology		"
1980 -	81. Wave propagation		"
	82. Harmonic analysis and group representations		"
	83. Matroid theory and its applications		"
1981 -	84. Kinetic Theories and the Boltzmann Equation	(LNM 1048)	Springer-Verlag
	85. Algebraic Threefolds	(LNM 947)	"
	86. Nonlinear Filtering and Stochastic Control	(LNM 972)	"
1982 -	87. Invariant Theory	(LNM 996)	"
	88. Thermodynamics and Constitutive Equations	(LN Physics 228)	"
	89. Fluid Dynamics	(LNM 1047)	"

1983 -	90.	Complete Intersections	(LNM 1092)	Springer-Verlag
	91.	Bifurcation Theory and Applications	(LNM 1057)	"
	92.	Numerical Methods in Fluid Dynamics	(LNM 1127)	"
1984 -	93.	Harmonic Mappings and Minimal Immersions	(LNM 1161)	"
	94.	Schrödinger Operators	(LNM 1159)	"
	95.	Buildings and the Geometry of Diagrams	(LNM 1181)	"
1985 -	96.	Probability and Analysis	(LNM 1206)	"
	97.	Some Problems in Nonlinear Diffusion	(LNM 1224)	"
	98.	Theory of Moduli	(LNM 1337)	"
1986 -	99.	Inverse Problems	(LNM 1225)	"
	100.	Mathematical Economics	(LNM 1330)	"
	101.	Combinatorial Optimization	(LNM 1403)	"
1987 -	102.	Relativistic Fluid Dynamics	(LNM 1385)	"
	103.	Topics in Calculus of Variations	(LNM 1365)	"
1988 -	104.	Logic and Computer Science	(LNM 1429)	"
	105.	Global Geometry and Mathematical Physics	(LNM 1451)	"
1989 -	106.	Methods of nonconvex analysis	(LNM 1446)	"
	107.	Microlocal Analysis and Applications	(LNM 1495)	"
1990 -	108.	Geometric Topology: Recent Developments	(LNM 1504)	"
	109.	H_{∞} Control Theory	(LNM 1496)	"
	110.	Mathematical Modelling of Industrial Processes	(LNM 1521)	"
1991 -	111.	Topological Methods for Ordinary Differential Equations	(LNM 1537)	"
	112.	Arithmetic Algebraic Geometry	(LNM 1553)	"
	113.	Transition to Chaos in Classical and Quantum Mechanics	(LNM 1589)	"
1992 -	114.	Dirichlet Forms	(LNM 1563)	"
	115.	D-Modules, Representation Theory, and Quantum Groups	(LNM 1565)	"
	116.	Nonequilibrium Problems in Many-Particle Systems	(LNM 1551)	"

1993 -	117. Integrable Systems and Quantum Groups	(LNM 1620)	Springer-Verlag
	118. Algebraic Cycles and Hodge Theory	(LNM 1594)	
	119. Phase Transitions and Hysteresis	(LNM 1584)	"
1994 -	120. Recent Mathematical Methods in Nonlinear Wave Propagation	(LNM 1640)	"
	121. Dynamical Systems	(LNM 1609)	"
	122. Transcendental Methods in Algebraic Geometry	(LNM 1646)	"
1995 -	123. Probabilistic Models for Nonlinear PDE's	(LNM 1627)	"
	124. Viscosity Solutions and Applications	(LNM 1660)	"
	125. Vector Bundles on Curves. New Directions	(LNM 1649)	"
1996 -	126. Integral Geometry, Radon Transforms and Complex Analysis	(LNM 1684)	"
	127. Calculus of Variations and Geometric Evolution Problems	LNM 1713	"
	128. Financial Mathematics	LNM 1656	"
1997 -	129. Mathematics Inspired by Biology	LNM 1714	"
	130. Advanced Numerical Approximation of Nonlinear Hyperbolic Equations	LNM 1697	"
	131. Arithmetic Theory of Elliptic Curves	LNM 1716	"
	132. Quantum Cohomology	to appear	"
1998 -	133. Optimal Shape Design	to appear	"
	134. Dynamical Systems and Small Divisors	to appear	"
	135. Mathematical Problems in Semiconductor Physics	to appear	"
	136. Stochastic PDE's and Kolmogorov Equations in Infinite Dimension	LNM 1715	"
	137. Filtration in Porous Media and Industrial Applications	to appear	"
1999 -	138. Computional Mathematics driven by Industrual Applicationa	to appear	"
	139. Iwahori-Hecke Algebras and Representation Theory	to appear	"
	140. Theory and Applications of Hamiltonian Dynamics	to appear	"

141. Global Theory of Minimal Surfaces in Flat Spaces — to appear — "

142. Direct and Inverse Methods in Solving Nonlinear Evolution Equations — to appear — "

FONDAZIONE C.I.M.E.
CENTRO INTERNAZIONALE MATEMATICO ESTIVO
INTERNATIONAL MATHEMATICAL SUMMER CENTER

"Computational Mathematics driven by Industrial Applications"

is the subject of the first 1999 C.I.M.E. Session.

The session, sponsored by the Consiglio Nazionale delle Ricerche (C.N.R.), the Ministero dell'Università e della Ricerca Scientifica e Tecnologica (M.U.R.S.T.) and the European Community, will take place, under the scientific direction of Professors Vincenzo CAPASSO (Università di Milano), Heinz W. ENGL (Johannes Kepler Universitaet, Linz) and Doct. Jacques PERIAUX (Dassault Aviation) at the Ducal Palace of Martina Franca (Taranto), from 21 to 27 June, 1999.

Courses

a) Paths, trees and flows: graph optimisation problems with industrial applications (5 lectures in English) Prof. Rainer BURKARD (Technische Universität Graz)

Abstract

Graph optimisation problems play a crucial role in telecommunication, production, transportation, and many other industrial areas. This series of lectures shall give an overview about exact and heuristic solution approaches and their inherent difficulties. In particular the essential algorithmic paradigms such as greedy algorithms, shortest path computation, network flow algorithms, branch and bound as well as branch and cut, and dynamic programming will be outlined by means of examples stemming from applications.

References

1) R. K. Ahuja, T. L. Magnanti & J. B. Orlin, *Network Flows: Theory, Algorithms and Applications*, Prentice Hall, 1993

2) R. K. Ahuja, T. L. Magnanti, J.B.Orlin & M. R. Reddy, *Applications of Network Optimization*. Chapter 1 in: Network Models (Handbooks of Operations Research and Management Science, Vol. 7), ed. by M. O. Ball et al., North Holland 1995, pp. 1-83

3) R. E. Burkard & E. Cela, *Linear Assignment Problems and Extensions*, Report 127, June 1998 (to appear in Handbook of Combinatorial Optimization, Kluwer, 1999).

Can be downloaded by anonymous ftp from

ftp.tu-graz.ac.at, directory/pub/papers/math

4) R. E. Burkard, E. Cela, P. M. Pardalos & L. S. Pitsoulis, *The Quadratic Assignment Problem*, Report 126 May 1998 (to appear in Handbook of Combinatorial Optimization, Kluwer, 1999). Can be downloaded by anonymous ftp from ftp.tu-graz.ac.at, directory /pub/papers/math.

5) E. L. Lawler, J. K. Lenstra, A. H. G.Rinnooy Kan & D. B. Shmoys (Eds.), *The Travelling Salesman Problem*, Wiley, Chichester, 1985.

b) New Computational Concepts, Adaptive Differential Equations Solvers and Virtual Labs (5 lectures in English) Prof. Peter DEUFLHARD (Konrad Zuse Zentrum, Berlin).

Abstract

The series of lectures will address computational mathematical projects that have been tackled by the speaker and his group. In all the topics to be presented novel mathematical modelling, advanced algorithm developments. and efficient visualisation play a joint role to solve problems of practical relevance. Among the applications to be exemplified are:

1) Adaptive multilevel FEM in clinical cancer therapy planning;

2) Adaptive multilevel FEM in optical chip design;

3) Adaptive discrete Galerkin methods for countable ODEs in polymer chemistry;

4) Essential molecular dynamics in RNA drug design.

References

1) P. Deuflhard & A Hohmann, *Numerical Analysis. A first Course in Scientific Computation*, Verlag de Gruyter, Berlin, 1995

2) P. Deuflhard et al *A nonlinear multigrid eigenproblem solver for the complex Helmoltz equation*, Konrad Zuse Zentrum Berlin SC 97-55 (1997)

3) P. Deuflhard et al. *Recent developments in chemical computing,* Computers in Chemical Engineering, 14, (1990),pp.1249-1258.

4) P. Deuflhard et al. (eds) *Computational molecular dynamics: challenges, methods, ideas,* Lecture Notes in Computational Sciences and Engineering, vol.4 Springer Verlag, Heidelberg, 1998.

5) P.Deuflhard & M. Weiser, *Global inexact Newton multilevel FEM for nonlinear elliptic problems,* Konrad Zuse Zentrum SC 96-33, 1996.

c) Computational Methods for Aerodynamic Analysis and Design. (5 lectures in English) Prof. Antony JAMESON (Stanford University, Stanford).

Abstract

The topics to be discussed will include: - Analysis of shock capturing schemes, and fast solution algorithms for compressible flow; - Formulation of aerodynamic shape optimisation based on control theory; - Derivation of the adjoint equations for compressible flow modelled by the potential Euler and Navies-Stokes equations; - Analysis of alternative numerical search procedures; - Discussion of geometry control and mesh perturbation methods; - Discussion of numerical implementation and practical applications to aerodynamic design.

d) Mathematical Problems in Industry (5 lectures in English) Prof. Jacques-Louis LIONS (Collège de France and Dassault Aviation, France).

Abstract

1. Interfaces and scales. The industrial systems are such that for questions of reliability, safety, cost no subsystem can be underestimated. Hence the need to address problems of scales, both in space variables and in time and the crucial importance of modelling and numerical methods.

2. Examples in Aerospace Examples in Aeronautics and in Spatial Industries. Optimum design.

3. Comparison of problems in Aerospace and in Meteorology. Analogies and differences

4 Real time control. Many methods can be thought of. Universal decomposition methods will be presented.

References

1) J. L. Lions, *Parallel stabilization hyperbolic and Petrowsky systems*, WCCM4 Conference, CDROM Proceedings, Buenos Aires, June 29- July 2, 1998.

2) W. Annacchiarico & M. Cerolaza, *Structural shape optimization of 2-D finite elements models using Beta-splines and genetic algorithms*, WCCM4 Conference, CDROM Proceedings, Buenos Aires, June 29- July 2, 1998.

3) J. Periaux, M. Sefrioui & B. Mantel, *Multi-objective strategies for complex optimization problems in aerodynamics using genetic algorithms*, ICAS '98 Conference, Melbourne, September '98, ICAS paper 98-2.9.1

e) Wavelet transforms and Cosine Transform in Signal and Image Processing (5 lectures in English) Prof. Gilbert STRANG (MIT, Boston).

Abstract

In a series of lectures we will describe how a linear transform is applied to the sampled data in signal processing, and the transformed data is compressed (and quantized to a string of bits). The quantized signal is transmitted and then the inverse transform reconstructs a very good approximation to the original signal. Our analysis concentrates on the construction of the transform. There are several important constructions and we emphasise two: 1) the discrete cosine transform (DCT); 2) discrete wavelet transform (DWT). The DCT is an orthogonal transform (for which we will give a new proof). The DWT may be orthogonal, as for the Daubechies family of wavelets. In other cases it may be biorthogonal - so the reconstructing transform is the inverse but not the transpose of the analysing transform. The reason for this possibility is that orthogonal wavelets cannot also be symmetric, and symmetry is essential property in image processing (because our visual system objects to lack of symmetry). The wavelet construction is based on a "bank" of filters - often a low pass and high pass filter. By iterating the low pass filter we decompose the input space into "scales" to produce a multiresolution. An infinite iteration yields in the limit the scaling function and a wavelet: the crucial equation for the theory is the refinement equation or dilatation equation that yields the scaling function. We discuss the mathematics of the refinement equation: the existence and the smoothness of the solution, and the construction by the cascade algorithm. Throughout these lectures we will be developing the mathematical ideas, but always for a purpose. The insights of wavelets have led to new bases for function spaces and there is no doubt that other ideas are waiting to be developed. This is applied mathematics.

References

1) I. Daubechies, *Ten lectures on wavelets*, SIAM, 1992.

2) G. Strang & T. Nguyen, *Wavelets and filter banks*, Wellesley-Cambridge, 1996.

3) Y. Meyer, *Wavelets: Algorithms and Applications*, SIAM, 1993.

Seminars

Two hour seminars will be held by the Scientific Directors and Professor R. Mattheij.

1) **Mathematics of the crystallisation process of polymers.** Prof. Vincenzo CAPASSO (Un. di Milano).

2) **Inverse Problems: Regularization methods, Application in Industry.** Prof. H. W. ENGL (Johannes Kepler Un., Linz).

3) **Mathematics of Glass.** Prof. R. MATTHEIJ (TU Eindhoven).

4) **Combining game theory and genetic algorithms for solving multi-objective shape optimization problems in Aerodynamics Engineering.** Doct. J. PERIAUX (Dassault Aviation).

Applications

Those who want to attend the Session should fill in an application to C.I.M.E Foundation at the address below, **not later than April 30, 1999**. An important consideration in the acceptance of applications is the scientific relevance of the Session to the field of interest of the applicant. Applicants are requested, therefore, to submit, along with their application, a scientific curriculum and a letter of recommendation. Participation will only be allowed to persons who have applied in due time and have had their application accepted. CIME will be able to partially support some of the youngest participants. Those who plan to apply for support have to mention it explicitely in the application form.

Attendance

No registration fee is requested. Lectures will be held at Martina Franca on June 21, 22, 23, 24, 25, 26, 27. Participants are requested to register on June 20, 1999.

Site and lodging

Martina Franca is a delightful baroque town of white houses of Apulian spontaneous architecture. Martina Franca is the major and most aristocratic centre of the "Murgia dei Trulli" standing on an hill which dominates the well known Itria valley spotted with "Trulli" conical dry stone houses which go back to the 15th century. A masterpiece of baroque architecture is the Ducal palace where the workshop will be hosted. Martina Franca is part of the province of Taranto, one of the major centres of Magna Grecia, particularly devoted to mathematics. Taranto houses an outstanding museum of Magna Grecia with fabulous collections of gold manufactures.

Lecture Notes

Lecture notes will be published as soon as possible after the Session.

Arrigo CELLINA
CIME Director

Vincenzo VESPRI
CIME Secretary

Fondazione C.I.M.E. c/o Dipartimento di Matematica ?U. Dini? Viale Morgagni, 67/A - 50134 FIRENZE (ITALY) Tel. +39-55-434975 / +39-55-4237123 FAX +39-55-434975 / +39-55-4222695 E-mail CIME@UDINI.MATH.UNIFI.IT

Information on CIME can be obtained on the system World-Wide-Web on the file HTTP: //WWW.MATH.UNIFI.IT/CIME/WELCOME.TO.CIME

FONDAZIONE C.I.M.E.
CENTRO INTERNAZIONALE MATEMATICO ESTIVO
INTERNATIONAL MATHEMATICAL SUMMER CENTER

"Iwahori-Hecke Algebras and Representation Theory"

is the subject of the second 1999 C.I.M.E. Session.

The session, sponsored by the Consiglio Nazionale delle Ricerche (C.N.R.), the Ministero dell'Università e della Ricerca Scientifica e Tecnologica (M.U.R.S.T.) and the European Community, will take place, under the scientific direction of Professors Velleda BALDONI (Università di Roma "Tor Vergata") and Dan BARBASCH (Cornell University) at the Ducal Palace of Martina Franca (Taranto), from June 28 to July 6, 1999.

Courses

a) Double HECKE algebras and applications (6 lectures in English)
Prof. Ivan CHEREDNIK (Un. of North Carolina at Chapel Hill, USA)
Abstract:
The starting point of many theories in the range from arithmetic and harmonic analysis to path integrals and matrix models is the formula:

$$\Gamma(k+1/2) = 2\int_0^\infty e^{-x^2}x^{2k}dx.$$

Recently a q-generalization was found based on the Hecke algebra technique, which completes the 15 year old Macdonald program.

The course will be about applications of the double affine Hecke algebras (mainly one-dimensional) to the Macdonald polynomials, Verlinde algebras, Gauss integrals and sums. It will be understandable for those who are not familiar with Hecke algebras and (hopefully) interesting to the specialists.

1) *q-Gauss integrals.* We will introduce a q-analogue of the classical integral formula for the gamma-function and use it to generalize the Gaussian sums at roots of unity.

2) *Ultraspherical polynomials.* A connection of the q-ultraspherical polynomials (the Rogers polynomials) with the one-dimensional double affine Hecke algebra will be established.

3) *Duality.* The duality for these polynomials (which has no classical counterpart) will be proved via the double Hecke algebras in full details.

4) *Verlinde algebras.* We will study the polynomial representation of the 1-dim. DHA at roots of unity, which leads to a generalization and a simplification of the Verlinde algebras.

5) *$PSL_2(\mathbf{Z})$-action.* The projective action of the $PSL_2(\mathbf{Z})$ on DHA and the generalized Verlinde algebras will be considered for A_1 and arbitrary root systems.

6) *Fourier transform of the q-Gaussian.* The invariance of the q-Gaussian with respect to the q-Fourier transform and some applications will be discussed.

References:

1) *From double Hecke algebra to analysis*, Proceedings of ICM98, Documenta Mathematica (1998).

2) *Difference Macdonald–Mehta conjecture*, IMRN:10, 449–467 (1997).

3) *Lectures on Knizhnik-Zamolodchikov equations and Hecke algebras*, MSJ Memoirs (1997).

b) Representation theory of affine Hecke algebras

Prof. Gert HECKMAN (Catholic Un., Nijmegen, Netherlands)

Abstract.

1. The Gauss hypergeometric equation.
2. Algebraic aspects of the hypergeometric system for root systems.
3. The hypergeometric function for root systems.
4. The Plancherel formula in the hypergeometric context.
5. The Lauricella hypergeometric function.
6. A root system analogue of 5.

I will assume that the audience is familiar with the classical theory of ordinary differential equations in the complex plane, in particular the concept of regular singular points and monodromy (although in my first lecture I will give a brief review of the Gauss hypergeometric function). This material can be found in many text books, for example E.L. Ince, Ordinary differential equations, Dover Publ, 1956. E.T. Whittaker and G.N. Watson, A course of modern analysis, Cambridge University Press, 1927.

I will also assume that the audience is familiar with the theory of root systems and reflection groups, as can be found in N. Bourbaki, Groupes et algèbres de Lie, Ch. 4,5 et 6, Masson, 1981. J. E. Humphreys, Reflection groups and Coxeter groups, Cambridge University Press, 1990. or in one of the text books on semisimple groups.

For the material covered in my lectures references are W.J. Couwenberg, Complex reflection groups and hypergeometric functions, Thesis Nijmegen, 1994. G.J. Heckman, Dunkl operators, Sem Bourbaki no 828, 1997. E.M. Opdam, Lectures on Dunkl operators, preprint 1998.

c) Representations of affine Hecke algebras.

Prof. George LUSZTIG (MIT, Cambridge, USA)

Abstract

Affine Hecke algebras appear naturally in the representation theory of p-adic groups. In these lectures we will discuss the representation theory of affine Hecke algebras and their graded version using geometric methods such as equivariant K-theory or perverse sheaves.

References.

1. V. Ginzburg, *Lagrangian construction of representations of Hecke algebras*, Adv. in Math. 63 (1987), 100-112.

2. D. Kazhdan and G. Lusztig, *Proof of the Deligne-Langlands conjecture for Hecke algebras.*, Inv. Math. 87 (1987), 153-215.

3. G. Lusztig, *Cuspidal local systems and graded Hecke algebras, I*, IHES Publ. Math. 67 (1988),145-202; II, in "Representation of groups" (ed. B. Allison and G. Cliff), Conf. Proc. Canad. Math. Soc.. 16, Amer. Math. Soc. 1995, 217-275.

4. G. Lusztig, *Bases in equivariant K-theory, Represent. Th.*, 2 (1998).

d) Affine-like Hecke Algebras and p-adic representation theory

Prof. Roger HOWE (Yale Un., New Haven, USA)

Abstract

Affine Hecke algebras first appeared in the study of a special class of representations (the spherical principal series) of reductive groups with coefficients in p-adic fields. Because of their connections with this and other topics, the structure and representation theory of affine Hecke algebras has been intensively studied by a variety of authors. In the meantime, it has gradually emerged that affine Hecke algebras, or slight generalizations of them, allow one to understand far more of the representations of p-adic groups than just the spherical principal series. Indeed, it seems possible that such algebras will allow one to understand all representations of p-adic groups. These lectures will survey progress in this approach to p-adic representation theory.

Topics:

1) Generalities on spherical function algebras on p-adic groups.
2) Iwahori Hecke algebras and generalizations.
3) - 4) Affine Hecke algebras and harmonic analysis
5) - 8) Affine-like Hecke algebras and representations of higher level.

References:

J. Adler, *Refined minimal K-types and supercuspidal representations*, Ph.D. Thesis, University of Chicago.

D. Barbasch, *The spherical dual for p-adic groups*, in Geometry and Representation Theory of Real and p-adic Groups, J. Tirao, D. Vogan, and J. Wolf, eds, Prog. In Math. 158, Birkhauser Verlag, Boston, 1998, 1 - 20.

D. Barbasch and A. Moy, *A unitarity criterion for p-adic groups*, Inv. Math. 98 (1989), 19 - 38.

D. Barbasch and A. Moy, *Reduction to real infinitesimal character in affine Hecke algebras*, J. A. M. S.6 (1993), 611- 635.

D. Barbasch, *Unitary spherical spectrum for p-adic classical groups*, Acta. Appl. Math. 44 (1996), 1 - 37.

C. Bushnell and P. Kutzko, *The admissible dual of GL(N) via open subgroups*, Ann. of Math. Stud. 129, Princeton University Press, Princeton, NJ, 1993.

C. Bushnell and P. Kutzko, *Smooth representations of reductive p-adic groups*: Structure theory via types, D. Goldstein, *Hecke algebra isomorphisms for tamely ramified characters*, R. Howe and A. Moy, *Harish-Chandra Homomorphisms for p-adic Groups*, CBMS Reg. Conf. Ser. 59, American Mathematical Society, Providence, RI, 1985.

R. Howe and A. Moy, *Hecke algebra isomorphisms for GL(N) over a p-adic field*, J. Alg. 131 (1990), 388 - 424.

J-L. Kim, *Hecke algebras of classical groups over p-adic fields and supercuspidal representations,I, II, III*, preprints, 1998.

G. Lusztig, *Classification of unipotent representations of simple p-adic groups*, IMRN 11 (1995), 517 - 589.

G. Lusztig, *Affine Hecke algebras and their graded version*, J. A. M. S. 2 (1989), 599 - 635.

L. Morris, *Tamely ramified supercuspidal representations of classical groups, I, II*, Ann. Ec. Norm. Sup 24, (1991) 705 - 738; 25 (1992), 639 - 667.

L. Morris, *Tamely ramified intertwining algebras*, Inv. Math. 114 (1994), 1 - 54.

A. Roche, *Types and Hecke algebras for principal series representations of split reductive p-adic groups*, preprint, (1996).

J-L. Waldspurger, *Algebres de Hecke et induites de representations cuspidales pour GLn*, J. reine u. angew. Math. 370 (1986), 27 - 191.

J-K. Yu, *Tame construction of supercuspidal representations*, preprint, 1998.

Applications

Those who want to attend the Session should fill in an application to the Director of C.I.M.E at the address below, **not later than** April 30, 1999.

An important consideration in the acceptance of applications is the scientific relevance of the Session to the field of interest of the applicant.

Applicants are requested, therefore, to submit, along with their application, a scientific curriculum and a letter of recommendation.

Participation will only be allowed to persons who have applied in due time and have had their application accepted.

CIME will be able to partially support some of the youngest participants. Those who plan to apply for support have to mention it explicitely in the application form.

Attendance

No registration fee is requested. Lectures will be held at Martina Franca on June 28, 29, 30, July 1, 2, 3, 4, 5, 6. Participants are requested to register on June 27, 1999.

Site and lodging

Martina Franca is a delightful baroque town of white houses of Apulian spontaneous architecture. Martina Franca is the major and most aristocratic centre of the Murgia dei Trulli standing on an hill which dominates the well known Itria valley spotted with Trulli conical dry stone houses which go back to the 15th century. A masterpiece of baroque architecture is the Ducal palace where the workshop will be hosted. Martina Franca is part of the province of Taranto, one of the major centres of Magna Grecia, particularly devoted to mathematics. Taranto houses an outstanding museum of Magna Grecia with fabulous collections of gold manufactures.

Lecture Notes

Lecture notes will be published as soon as possible after the Session.

Arrigo CELLINA
CIME Director

Vincenzo VESPRI
CIME Secretary

Fondazione C.I.M.E. c/o Dipartimento di Matematica U. Dini Viale Morgagni, 67/A - 50134 FIRENZE (ITALY) Tel. +39-55-434975 / +39-55-4237123 FAX +39-55-434975 / +39-55-4222695 E-mail CIME@UDINI.MATH.UNIFI.IT

Information on CIME can be obtained on the system World-Wide-Web on the file HTTP: //WWW.MATH.UNIFI.IT/CIME/WELCOME.TO.CIME.

FONDAZIONE C.I.M.E.
CENTRO INTERNAZIONALE MATEMATICO ESTIVO
INTERNATIONAL MATHEMATICAL SUMMER CENTER

"Theory and Applications of Hamiltonian Dynamics"

is the subject of the third 1999 C.I.M.E. Session.

The session, sponsored by the Consiglio Nazionale delle Ricerche (C.N.R.), the Ministero dell'Università e della Ricerca Scientifica e Tecnologica (M.U.R.S.T.) and the European Community, will take place, under the scientific direction of Professor Antonio GIORGILLI (Un. di Milano), at Grand Hotel San Michele,Cetraro (Cosenza), from July 1 to July 10, 1999.

Courses

a) Physical applications of Nekhoroshev theorem and exponential estimates (6 lectures in English)

Prof. Giancarlo BENETTIN (Un. di Padova, Italy)

Abstract

The purpose of the lectures is to introduce exponential estimates (i.e., construction of normal forms up to an exponentially small remainder) and Nekhoroshev theorem (exponential estimates plus geometry of the action space) as the key to understand the behavior of several physical systems, from the Celestial mechanics to microphysics.

Among the applications of the exponential estimates, we shall consider problems of adiabatic invariance for systems with one or two frequencies coming from molecular dynamics. We shall compare the traditional rigorous approach via canonical transformations, the heuristic approach of Jeans and of Landau–Teller, and its possible rigorous implementation via Lindstet series. An old conjecture of Boltzmann and Jeans, concerning the possible presence of very long equilibrium times in classical gases (the classical analog of "quantum freezing") will be reconsidered. Rigorous and heuristic results will be compared with numerical results, to test their level of optimality.

Among the applications of Nekhoroshev theorem, we shall study the fast rotations of the rigid body, which is a rather complete problem, including degeneracy and singularities. Other applications include the stability of elliptic equilibria, with special emphasis on the stability of triangular Lagrangian points in the spatial restricted three body problem.

References:

For a general introduction to the subject, one can look at chapter 5 of V.I. Arnold, VV. Kozlov and A.I. Neoshtadt, in *Dynamical Systems III*, V.I. Arnold Editor (Springer, Berlin 1988). An introduction to physical applications of Nekhorshev theorem and exponential estimates is in the proceeding of the Noto School "Non-Linear Evolution and Chaotic Phenomena", G. Gallavotti and P.W. Zweifel Editors (Plenum Press, New York, 1988), see the contributions by G. Benettin, L. Galgani and A. Giorgilli.

General references on Nekhoroshev theorem and exponential estimates: N.N. Nekhoroshev, Usp. Mat. Nauk. **32**:6, 5-66 (1977) [Russ. Math. Surv. **32**:6, 1-65

(1977)]; G. Benettin, L. Galgani, A. Giorgilli, Cel. Mech. **37**, 1 (1985); A. Giorgilli and L. Galgani, Cel. Mech. **37**, 95 (1985); G. Benettin and G. Gallavotti, Journ. Stat. Phys. **44**, 293-338 (1986); P. Lochak, Russ. Math. Surv. **47**, 57-133 (1992); J. Pöschel, Math. Z. **213**, 187–216 (1993).

Applications to statistical mechanics: G. Benettin, in: *Boltzmann's legacy 150 years afrer his birth*, Atti Accad. Nazionale dei Lincei **131**, 89-105 (1997); G. Benettin, A. Carati and P. Sempio, Journ. Stat. Phys. **73**, 175–192 (1993); G. Benettin, A. Carati and G. Gallavotti, Nonlinearity **10**, 479-505 (1997); G. Benettin, A. Carati e F. Fassò, Physica D **104**, 253-268 (1997); G. Benettin, P. Hjorth and P. Sempio, *Exponentially long equilibrium times in a one dimensional collisional model of a classical gas*, in print in Journ. Stat. Phys.

Applications to the rigid body: G. Benettin and F. Fassò, Nonlinearity **9**, 137-186 (1996); G. Benettin, F. Fassò e M. Guzzo, Nonlinearity **10**, 1695-1717 (1997).

Applications to elliptic equilibria (recent nonisochronous approach): F. Fassò, M. Guzzo e G. Benettin, Comm. Math. Phys. **197**, 347-360 (1998); L. Niederman, *Nonlinear stability around an elliptic equilibrium point in an Hamiltonian system*, preprint (1997). M. Guzzo, F. Fasso' e G. Benettin, Math. Phys. Electronic Journal, Vol. **4**, paper 1 (1998); G. Benettin, F. Fassò e M. Guzzo, *Nekhoroshev–stability of L4 and L5 in the spatial restricted three–body problem*, in print in Regular and Chaotic Dynamics.

b) KAM-theory (6 lectures in English)

Prof. Hakan ELIASSON (Royal Institute of Technology, Stockholm, Sweden)

Abstract

Quasi-periodic motions (or invariant tori) occur naturally when systems with periodic motions are coupled. The perturbation problem for these motions involves small divisors and the most natural way to handle this difficulty is by the quadratic convergence given by Newton's method. A basic problem is how to implement this method in a particular perturbative situation. We shall describe this difficulty, its relation to linear quasi-periodic systems and the way given by KAM-theory to overcome it in the most generic case. Additional difficulties occur for systems with elliptic lower dimensional tori and even more for systems with weak non-degeneracy.

We shall also discuss the difference between initial value and boundary value problems and their relation to the Lindstedt and the Poincaré-Lindstedt series.

The classical books Lectures in Celestial Mechanics by Siegel and Moser (Springer 1971) and Stable and Random Motions in Dynamical Systems by Moser (Princeton University Press 1973) are perhaps still the best introductions to KAM-theory. The development up to middle 80's is described by Bost in a Bourbaki Seminar (no. 6 1986). After middle 80's a lot of work have been devoted to elliptic lower dimensional tori, and to the study of systems with weak non-degeneracy starting with the work of Cheng and Sun (for example "*Existence of KAM-tori in Degenerate Hamiltonian systems*", J. Diff. Eq. 114, 1994). Also on linear quasi-periodic systems there has been some progress which is described in my article "*Reducibility and point spectrum for quasi-periodic skew-products*", Proceedings of the ICM, Berlin volume II 1998.

c) The Adiabatic Invariant in Classical Dynamics: Theory and applications (6 lectures in English).

Prof. Jacques HENRARD (Facultés Universitaires Notre Dame de la Paix, Namur, Belgique).

Abstract

The adiabatic invariant theory applies essentially to oscillating non-autonomous Hamiltonian systems when the time dependance is considerably slower than the oscillation periods. It describes "easy to compute" and "dynamicaly meaningful" quasi-invariants by which on can predict the approximate evolution of the system on very large time scales. The theory makes use and may serve as an illustration of several classical results of Hamiltonian theory.

1) Classical Adiabatic Invariant Theory (Including an introduction to angle-action variables)

2) Classical Adiabatic Invariant Theory (continued) and some applications (including an introduction to the "magnetic bottle")

3) Adiabatic Invariant and Separatrix Crossing (Neo-adiabatic theory)

4) Applications of Neo-Adiabatic Theory: Resonance Sweeping in the Solar System

5) The chaotic layer of the "Slowly Modulated Standard Map"

References:

J.R. Cary, D.F. Escande, J.L. Tennison: Phys.Rev. A, 34, 1986, 3256-4275

J. Henrard, in "*Dynamics reported*" (n=B02- newseries), Springer Verlag 1993; pp 117-235)

J. Henrard: in "*Les méthodes moderne de la mécanique céleste*" (Benest et Hroeschle eds), Edition Frontieres, 1990, 213-247

J. Henrard and A. Morbidelli: Physica D, 68, 1993, 187-200.

d) Some aspects of qualitative theory of Hamiltonian PDEs (6 lectures in English).

Prof. Sergei B. KUKSIN (Heriot-Watt University, Edinburgh, and Steklov Institute, Moscow)

Abstract.

I) Basic properties of Hamiltonian PDEs. Symplectic structures in scales of Hilbert spaces, the notion of a Hamiltonian PDE, properties of flow-maps, etc.

II) Around Gromov's non-squeezing property. Discussions of the finite-dimensional Gromov's theorem, its version for PDEs and its relevance for mathematical physics, infinite-dimensional symplectic capacities.

III) Damped Hamiltonian PDEs and the turbulence-limit. Here we establish some qualitative properties of PDEs of the form <non-linear Hamiltonian PDE>+<small linear damping> and discuss their relations with theory of decaying turbulence

Parts I)-II) will occupy the first three lectures, Part III - the last two.

References

[1] S.K., *Nearly Integrable Infinite-dimensional Hamiltonian Systems.* LNM 1556, Springer 1993.

[2] S.K., *Infinite-dimensional symplectic capacities and a squeezing theorem for Hamiltonian PDE's.* Comm. Math. Phys. 167 (1995), 531-552.

[3] Hofer H., Zehnder E., *Symplectic invariants and Hamiltonian dynamics.* Birkhauser, 1994.

[4] S.K. *Oscillations in space-periodic nonlinear Schroedinger equations.* Geometric and Functional Analysis 7 (1997), 338-363.

For I) see [1] (Part 1); for II) see [2,3]; for III) see [4]."

e) An overview on some problems in Celestial Mechanics (6 lectures in English)

Prof. Carles SIMO' (Universidad de Barcelona, Spagna)

Abstract

1. Introduction. The N-body problem. Relative equilibria. Collisions.

2. The 3D restricted three-body problem. Libration points and local stability analysis.

3. Periodic orbits and invariant tori. Numerical and symbolical computation.

4. Stability and practical stability. Central manifolds and the related stable/unstable manifolds. Practical confiners.

5. The motion of spacecrafts in the vicinity of the Earth-Moon system. Results for improved models. Results for full JPL models.

References:

C. Simò, *An overview of some problems in Celestial Mechanics*, available at http://www-ma1.upc.es/escorial .

Click of "curso completo" of Prof. Carles Simó

Applications

Deadline for application: **May 15, 1999.**

Applicants are requested to submit, along with their application, a scientific curriculum and a letter of recommendation.

CIME will be able to partially support some of the youngest participants. Those who plan to apply for support have to mention it explicitely in the application form.

Attendance

No registration fee is requested. Lectures will be held at Cetraro on July 1, 2, 3, 4, 5, 6, 7, 8, 9, 10. Participants are requested to register on June 30, 1999.

Site and lodging

The session will be held at Grand Hotel S. Michele at Cetraro (Cosenza), Italy. Prices for full board (bed and meals) are roughly 150.000 italian liras p.p. day in a single room, 130.000 italian liras in a double room. Cheaper arrangements for multiple lodging in a residence are avalaible. More detailed information may be obtained from the Direction of the hotel (tel. +39-098291012, Fax +39-098291430, email: sanmichele@antares.it.

Further information on the hotel at the web page www.sanmichele.it

Arrigo CELLINA
CIME Director

Vincenzo VESPRI
CIME Secretary

Fondazione C.I.M.E. c/o Dipartimento di Matematica U. Dini Viale Morgagni, 67/A - 50134 FIRENZE (ITALY) Tel. +39-55-434975 / +39-55-4237123 FAX +39-55-434975 / +39-55-4222695 E-mail CIME@UDINI.MATH.UNIFI.IT

Information on CIME can be obtained on the system World-Wide-Web on the file HTTP: //WWW.MATH.UNIFI.IT/CIME/WELCOME.TO.CIME.

FONDAZIONE C.I.M.E.
CENTRO INTERNAZIONALE MATEMATICO ESTIVO
INTERNATIONAL MATHEMATICAL SUMMER CENTER

"Global Theory of Minimal Surfaces in Flat Spaces"

is the subject of the fourth 1999 C.I.M.E. Session.

The session, sponsored by the Consiglio Nazionale delle Ricerche (C.N.R.), the Ministero dell'Università e della Ricerca Scientifica e Tecnologica (M.U.R.S.T.) and the European Community, will take place, under the scientific direction of Professor Gian Pietro PIROLA (Un. di Pavia), at Ducal Palace of Martina Franca (Taranto), from July 7 to July 15, 1999.

Courses

a) Asymptotic geometry of properly embedded minimal surfaces (6 lecture in English)

Prof. William H. MEEKS, III (Un. of Massachusetts, Amherst, USA).

Abstract:

In recent years great progress has been made in understanding the asymptotic geometry of properly embedded minimal surfaces. The first major result of this type was the solution of the generalized Nitsch conjecture by P. Collin, based on earlier work by Meeks and Rosenberg. It follows from the resolution of this conjecture that whenever M is a properly embedded minimal surface with more than one end and $E \subset M$ is an annular end representative, then E has finite total curvature and is asymptotic to an end of a plan or catenoid. Having finite total curvature in the case of an annular end is equivalent to proving the end has quadratic area growth with respect to the radial function r. Recently Collin, Kusner, Meeks and Rosenberg have been able to prove that any middle end of M, even one with infinite genus, has quadratic area growth. It follows from this result that middle ends are never limit ends and hence M can only have one or two limit ends which must be top or bottom ends. With more work it is shown that the middle ends of M stay a bounded distance from a plane or an end of a catenoid.

The goal of my lectures will be to introduce the audience to the concepts in the theory o f properly embedded minimal surfaces needed to understand the above results and to understand some recent classification theorems on proper minimal surfaces of genus 0 in flat three-manifolds.

References

1) H. Rosenberg, *Some recent developments in the theory of properly embedded minimal surfaces in E*, Asterisque **206**, (19929, pp. 463-535;

2) W. Meeks & H. Rosenberg, *The geometry and conformal type of properly embedded minimal surfaces in E*, Invent.Math. **114**, (1993), pp. 625-639;

3) W. Meeks, J. Perez & A. Ros, *Uniqueness of the Riemann minimal examples*, Invent. Math. **131**, (1998), pp. 107-132;

4) W. Meeks & H. Rosenberg, *The geometry of periodic minimal surfaces*, Comm. Math. Helv. **68**, (1993), pp. 255-270;

5) P. Collin, *Topologie et courbure des surfaces minimales proprement plongees dans E*, Annals of Math. **145**, (1997), pp. 1-31;

6) H. Rosenberg, *Minimal surfaces of finite type*, Bull. Soc. Math. France **123**, (1995), pp. 351-359;

7) Rodriquez & H. Rosenberg, *Minimal surfaces in E with one end and bounded curvature*, Manusc. Math. **96**, (1998), pp. 3-9.

b) Properly embedded minimal surfaces with finite total curvature (6 lectures in English)

Prof. Antonio ROS (Universidad de Granada, Spain)

Abstact:

Among properly embedded minimal surfaces in Euclidean 3-space, those that have finite total curvature form a natural and important subclass. These surfaces have finitely many ends which are all parallel and asymptotic to planes or catenoids. Although the structure of the space $\mathcal{M}$ of surfaces of this type which have a fixed topology is not well understood, we have a certain number of partial results and some of them will be explained in the lectures we will give.

The first nontrivial examples, other than the plane and the catenoid, were constructed only ten years ago by Costa, Hoffman and Meeks. Schoen showed that if the surface has two ends, then it must be a catenoid and López and Ros proved that the only surfaces of genus zero are the plane and the catenoid. These results give partial answers to an interesting open problem: decide which topologies are supported by this kind of surfaces. Ros obtained certain compactness properties of $\mathcal{M}$. In general this space is known to be noncompact but he showed that $\mathcal{M}$ is compact for some fixed topologies. Pérez and Ros studied the local structure of $\mathcal{M}$ around a nondegenerate surface and they proved that around these points the moduli space can be naturally viewed as a Lagrangian submanifold of the complex Euclidean space.

In spite of that analytic and algebraic methods compete to solve the main problems in this theory, at this moment we do not have a satisfactory idea of the behaviour of the moduli space $\mathcal{M}$. Thus the above is a good research field for young geometers interested in minimal surfaces.

References

1) C. Costa, *Example of a compete minimal immersion in* $\mathbf{R}^3$ *of genus one and three embedded ends*, Bull. SOc. Bras. Math. **15, (1984), pp. 47-54;**

2) D. Hoffman & H. Karcher, *Complete embedded minimal surfaces of finite total curvature*, R. Osserman ed., Encyclopedia of Math., vol. of Minimal Surfaces, **5-90**, Springer 1997;

3) D. Hoffman & W. H. Meeks III, *Embedded minimal surfaces of finite topology*, Ann. Math. **131**, (1990), pp. 1-34;

4) F. J. Lòpez & A. Ros, *On embedded minimal surfaces of genus zero*, J. Differential Geometry **33**, (1991), pp. 293-300;

5) J. P. Perez & A. Ros, *Some uniqueness and nonexistence theorems for embedded minimal surfaces*, Math. Ann. **295** (3), (1993), pp. 513-525;

6) J. P. Perez & A. Ros, *The space of properly embedded minimal surfaces with finite total curvature*, Indiana Univ. Math. J. **45** 1, (1996), pp.177-204.

c) Minimal surfaces of finite topology properly embedded in E (Euclidean 3-space).(6 lectures in English)

Prof. Harold ROSENBERG (Univ. Paris VII, Paris, France)

Abstract:

We will prove that a properly embedded minimal surface in E of finite topology and at least two ends has finite total curvature. To establish this we first prove that each annular end of such a surface M can be made transverse to the horizontal planes

(after a possible rotation in space), [Meeks-Rosenberg]. Then we will prove that such an end has finite total curvature [Pascal Collin]. We next study properly embedded minimal surfaces in E with finite topology and one end. The basic unsolved problem is to determine if such a surface is a plane or helicoid when simply connected. We will describe partial results. We will prove that a properly immersed minimal surface of finite topology that meets some plane in a finite number of connected components, with at most a finite number of singularities, is of finite conformal type. If in addition the curvature is bounded, then the surface is of finite type. This means M can be parametrized by meromorphic data on a compact Riemann surface. In particular, under the above hypothesis, M is a plane or helicoid when M is also simply connected and embedded. This is work of Rodriquez- Rosenberg, and Xavier. If time permits we will discuss the geometry and topology of constant mean curvature surfaces properly embedded in E.

References

1) H. Rosenberg, *Some recent developments in the theory of properly embedded minimal surfaces in E*, Asterique **206**, (1992), pp. 463-535;

2) W.Meeks & H. Rosenberg, *The geometry and conformal type of properly embedded minimal surfaces in E*, Invent. **114**, (1993), pp.625-639;

3) P. Collin, *Topologie et courbure des surfaces minimales proprement plongées dans E*, Annals of Math. **145**, (1997), pp. 1-31

4) H. Rosenberg, *Minimal surfaces of finite type*, Bull. Soc. Math. France **123**, (1995), pp. 351-359;

5) Rodriquez & H. Rosenberg, *Minimal surfaces in E with one end and bounded curvature*, Manusc. Math. **96**, (1998), pp. 3-9.

Applications

Those who want to attend the Session should fill in an application to the C.I.M.E Foundation at the address below, **not later than** May 15, 1999.

An important consideration in the acceptance of applications is the scientific relevance of the Session to the field of interest of the applicant.

Applicants are requested, therefore, to submit, along with their application, a scientific curriculum and a letter of recommendation.

Participation will only be allowed to persons who have applied in due time and have had their application accepted.

CIME will be able to partially support some of the youngest participants. Those who plan to apply for support have to mention it explicitely in the application form

Attendance

No registration fee is requested. Lectures will be held at Martina Franca on July 7, 8, 9, 10, 11, 12, 13, 14, 15. Participants are requested to register on July 6, 1999.

Site and lodging

Martina Franca is a delightful baroque town of white houses of Apulian spontaneous architecture. Martina Franca is the major and most aristocratic centre of the Murgia dei Trulli standing on an hill which dominates the well known Itria valley spotted with Trulli conical dry stone houses which go back to the 15th century. A masterpiece of baroque architecture is the Ducal palace where the workshop will be hosted. Martina Franca is part of the province of Taranto, one of the major centres of Magna Grecia, particularly devoted to mathematics. Taranto houses an outstanding museum of Magna Grecia with fabulous collections of gold manufactures.

Lecture Notes

Lecture notes will be published as soon as possible after the Session.

Arrigo CELLINA
CIME Director

Vincenzo VESPRI
CIME Secretary

Fondazione C.I.M.E. c/o Dipartimento di Matematica U. Dini Viale Morgagni, 67/A - 50134 FIRENZE (ITALY) Tel. +39-55-434975 / +39-55-4237123 FAX +39-55-434975 / +39-55-4222695 E-mail CIME@UDINI.MATH.UNIFI.IT

Information on CIME can be obtained on the system World-Wide-Web on the file HTTP: //WWW.MATH.UNIFI.IT/CIME/WELCOME.TO.CIME.

FONDAZIONE C.I.M.E.
CENTRO INTERNAZIONALE MATEMATICO ESTIVO
INTERNATIONAL MATHEMATICAL SUMMER CENTER

"Direct and Inverse Methods in Solving Nonlinear Evolution Equations"

is the subject of the fifth 1999 C.I.M.E. Session.

The session, sponsored by the Consiglio Nazionale delle Ricerche (C.N.R.), the Ministero dell'Università e della Ricerca Scientifica e Tecnologica (M.U.R.S.T.) and the European Community, will take place, under the scientific direction of Professor Antonio M. Greco (Università di Palermo), at Grand Hotel San Michele, Cetraro (Cosenza), from September 8 to September 15, 1999.

a) Exact solutions of nonlinear PDEs by singularity analysis (6 lectures in English)

Prof. Robert CONTE (Service de physique de l'état condensé, CEA Saclay, Gif-sur-Yvette Cedex, France)

Abstract

1) Criteria of integrability : Lax pair, Darboux and Bäcklund transformations. Partial integrability, examples. Importance of involutions.

2) The Painlevé test for PDEs in its invariant version.

3) The "truncation method" as a Darboux transformation, ODE and PDE situations.

4) The one-family truncation method (WTC), integrable (Korteweg-de Vries, Boussinesq, Hirota-Satsuma, Sawada-Kotera) and partially integrable (Kuramoto-Sivashinsky) cases.

5) The two-family truncation method, integrable (sine-Gordon, mKdV, Broer-Kaup) and partially integrable (complex Ginzburg-Landau and degeneracies) cases.

6) The one-family truncation method based on the scattering problems of Gambier: BT of Kaup-Kupershmidt and Tzitzéica equations.

References

References are divided into three subsets: prerequisite (assumed known by the attendant to the school), general (not assumed known, pedagogical texts which would greatly benefit the attendant if they were read before the school), research (research papers whose content will be exposed from a synthetic point of view during the course).

Prerequisite bibliography.

The following subjects will be assumed to be known : the Painlevé property for nonlinear ordinary differential equations, and the associated Painlevé test.

Prerequisite recommended texts treating these subjects are

[P.1] E. Hille, *Ordinary differential equations in the complex domain* (J. Wiley and sons, New York, 1976).

[P.2] R. Conte, *The Painlevé approach to nonlinear ordinary differential equations, The Painlevé property, one century later*, 112 pages, ed. R. Conte, CRM series in mathematical physics (Springer, Berlin, 1999). Solv-int/9710020.

The interested reader can find many applications in the following review, which should not be read before [P.2] :

[P.3] A. Ramani, B. Grammaticos, and T. Bountis, *The Painlevé property and singularity analysis of integrable and nonintegrable systems*, Physics Reports 180 (1989) 159–245.

A text to be avoided by the beginner is Ince's book, the ideas are much clearer in Hille's book.

There exist very few pedagogical texts on the subject of this school.

A general reference, covering all the above program, is the course delivered at a Cargèse school in 1996 :

[G.1] M. Musette, *Painlevé analysis for nonlinear partial differential equations, The Painlevé property, one century later*, 65 pages, ed. R. Conte, CRM series in mathematical physics (Springer, Berlin, 1999). Solv-int/9804003.

A short subset of [G.1], with emphasis on the ideas, is the conference report

[G.2] R. Conte, *Various truncations in Painlevé analysis of partial differential equations*, 16 pages, Nonlinear dynamics : integrability and chaos, ed. M. Daniel, to appear (Springer? World Scientific?). Solv-int/9812008. Preprint S98/047.

Research papers.

[R.2] J. Weiss, M. Tabor and G. Carnevale, *The Painlevé property for partial differential equations*, J. Math. Phys. 24 (1983) 522–526.

[R.3] Numerous articles of Weiss, from 1983 to 1989, all in J. Math. Phys. [singular manifold method].

[R.4] M. Musette and R. Conte, *Algorithmic method for deriving Lax pairs from the invariant Painlevé analysis of nonlinear partial differential equations*, J. Math. Phys. 32 (1991) 1450–1457 [invariant singular manifold method].

[R.5] R. Conte and M. Musette, *Linearity inside nonlinearity: exact solutions to the complex Ginz-burg-Landau equation*, Physica D 69 (1993) 1–17 [Ginzburg-Landau].

[R.6] M. Musette and R. Conte, *The two-singular manifold method, I. Modified KdV and sine-Gordon equations*, J. Phys. A 27 (1994) 3895–3913 [Two-singular manifold method].

[R.7] R. Conte, M. Musette and A. Pickering, *The two-singular manifold method, II.* Classical Boussinesq system, J. Phys. A 28 (1995) 179–185 [Two-singular manifold method].

[R.8] A. Pickering, *The singular manifold method revisited*, J. Math. Phys. 37 (1996) 1894–1927 [Two-singular manifold method].

[R.9] M. Musette and R. Conte, *Bäcklund transformation of partial differential equations from the Painlevé-Gambier classification, I. Kaup-Kupershmidt equation*, J. Math. Phys. 39 (1998) 5617–5630. [Lecture 6].

[R.10] R. Conte, M. Musette and A. M. Grundland, *Bäcklund transformation of partial differential equations from the Painlevé-Gambier classification, II. Tzitzéica equation*, J. Math. Phys. 40 (1999) to appear. [Lecture 6].

b) Integrable Systems and Bi-Hamiltonian Manifolds (6 lectures in English)

Prof. Franco MAGRI (Università di Milano, Milano, Italy)

Abstract

1) Integrable systems and bi-hamiltonian manifolds according to Gelfand and Zakharevich.

2) Examples: KdV, KP and Sato's equations.

3) The rational solutions of KP equation.

4) Bi-hamiltonian reductions and completely algebraically integrable systems.

5) Connections with the separabilty theory.

6) The τ function and the Hirota's identities from a bi-hamiltonian point of view.

References

1) R. Abraham, J.E. Marsden, *Foundations of Mechanics*,Benjamin/Cummings, 1978

2) P. Libermann, C. M. Marle, *Symplectic Geometry and Analytical Mechanics*, Reidel Dordrecht, 1987

3) L. A. Dickey, *Soliton Equations and Hamiltonian Systems*, World Scientific, Singapore, 1991, Adv. Series in Math. Phys Vol. 12

4) I. Vaisman, *Lectures on the Geometry of Poisson Manifolds*, Progress in Math., Birkhäuser, 1994

5) P. Casati, G. Falqui, F. Magri, M. Pedroni (1996), *The KP theory revisited. I,II,III,IV.* Technical Reports, SISSA/2,3,4,5/96/FM, SISSA/ISAS, Trieste, 1995

c) Hirota Methods for non Linear Differential and Difference Equations (6 lectures in English)

Prof. Junkichi SATSUMA (University of Tokyo, Tokyo, Japan)

Abstract

1) Introduction;

2) Nonlinear differential systems;

3) Nonlinear differential-difference systems;

4) Nonlinear difference systems;

5) Sato theory;

6) Ultra-discrete systems.

References.

1) M.J.Ablowitz and H.Segur, *Solitons and the Inverse Scattering Transform*, (SIAM, Philadelphia, 1981).

2) Y.Ohta, J.Satsuma, D.Takahashi and T.Tokihiro, " Prog. Theor. Phys. Suppl. No.94, p.210-241 (1988)

3) J.Satsuma, *Bilinear Formalism in Soliton Theory*, Lecture Notes in Physics No.495, Integrability of Nonlinear Systems, ed. by Y.Kosmann-Schwarzbach, B.Grammaticos and K.M.Tamizhmani p.297-313 (Springer, Berlin, 1997).

d) Lie Groups and Exact Solutions of non Linear Differential and Difference Equations (6 lectures in English)

Prof. Pavel WINTERNITZ (Universitè de Montreal, Montreal, Canada) 3J7

Abstract

1) Algorithms for calculating the symmetry group of a system of ordinary or partial differential equations. Examples of equations with finite and infinite Lie point symmetry groups;

2) Applications of symmetries. The method of symmetry reduction for partial differential equations. Group classification of differential equations;

3) Classification and identification of Lie algebras given by their structure constants. Classification of subalgebras of Lie algebras. Examples and applications;

4) Solutions of ordinary differential equations. Lowering the order of the equation. First integrals. Painlevè analysis and the singularity structure of solutions;

5) Conditional symmetries. Partially invariant solutions.

6) Lie symmetries of difference equations.

References.

1) P. J. Olver, *Applications of Lie Groups to Differential Equations*, Springer,1993,

2) P. Winternitz, *Group Theory and Exact Solutions of Partially Integrable Differential Systems*, in Partially Integrable Evolution Equations in Physics, Kluwer, Dordrecht, 1990, (Editors R.Conte and N.Boccara).

3) P. Winternitz, in "*Integrable Systems, Quantum Groups and Quantum Field Theories*", Kluwer, 1993 (Editors L .A. Ibort and M. A. Rodriguez).

Applications

Those who want to attend the Session should fill in an application to the C.I.M.E Foundation at the address below, **not later than May 30**, 1999.

An important consideration in the acceptance of applications is the scientific relevance of the Session to the field of interest of the applicant.

Applicants are requested, therefore, to submit, along with their application, a scientific curriculum and a letter of recommendation.

Participation will only be allowed to persons who have applied in due time and have had their application accepted.

CIME will be able to partially support some of the youngest participants. Those who plan to apply for support have to mention it explicitely in the application form.

Attendance

No registration fee is requested. Lectures will be held at Cetraro on September 8, 9, 10, 11, 12, 13, 14, 15. Participants are requested to register on September 7, 1999.

Site and lodging

The session will be held at Grand Hotel S. Michele at Cetraro (Cosenza), Italy. Prices for full board (bed and meals) are roughly 150.000 italian liras p.p. day in a single room, 130.000 italian liras in a double room. Cheaper arrangements for multiple lodging in a residence are avalaible. More detailed informations may be obtained from the Direction of the hotel (tel. +39-098291012, Fax +39-098291430, email: sanmichele@antares.it.

Further information on the hotel at the web page www.sanmichele.it

Lecture Notes

Lecture notes will be published as soon as possible after the Session.

Arrigo CELLINA
CIME Director

Vincenzo VESPRI
CIME Secretary

Fondazione C.I.M.E. c/o Dipartimento di Matematica U. Dini Viale Morgagni, 67/A - 50134 FIRENZE (ITALY) Tel. +39-55-434975 / +39-55-4237123 FAX +39-55-434975 / +39-55-4222695 E-mail CIME@UDINI.MATH.UNIFI.IT

Information on CIME can be obtained on the system World-Wide-Web on the file HTTP: //WWW.MATH.UNIFI.IT/CIME/WELCOME.TO.CIME.

Lecture Notes in Mathematics

For information about Vols. 1–1525
please contact your bookseller or Springer-Verlag

Vol. 1526: J. Azéma, P. A. Meyer, M. Yor (Eds.), Séminaire de Probabilités XXVI. X, 633 pages. 1992.

Vol. 1527: M. I. Freidlin, J.-F. Le Gall, Ecole d'Eté de Probabilités de Saint-Flour XX – 1990. Editor: P. L. Hennequin. VIII, 244 pages. 1992.

Vol. 1528: G. Isac, Complementarity Problems. VI, 297 pages. 1992.

Vol. 1529: J. van Neerven, The Adjoint of a Semigroup of Linear Operators. X, 195 pages. 1992.

Vol. 1530: J. G. Heywood, K. Masuda, R. Rautmann, S. A. Solonnikov (Eds.), The Navier-Stokes Equations II – Theory and Numerical Methods. IX, 322 pages. 1992.

Vol. 1531: M. Stoer, Design of Survivable Networks. IV, 206 pages. 1992.

Vol. 1532: J. F. Colombeau, Multiplication of Distributions. X, 184 pages. 1992.

Vol. 1533: P. Jipsen, H. Rose, Varieties of Lattices. X, 162 pages. 1992.

Vol. 1534: C. Greither, Cyclic Galois Extensions of Commutative Rings. X, 145 pages. 1992.

Vol. 1535: A. B. Evans, Orthomorphism Graphs of Groups. VIII, 114 pages. 1992.

Vol. 1536: M. K. Kwong, A. Zettl, Norm Inequalities for Derivatives and Differences. VII, 150 pages. 1992.

Vol. 1537: P. Fitzpatrick, M. Martelli, J. Mawhin, R. Nussbaum, Topological Methods for Ordinary Differential Equations. Montecatini Terme, 1991. Editors: M. Furi, P. Zecca. VII, 218 pages. 1993.

Vol. 1538: P.-A. Meyer, Quantum Probability for Probabilists. X, 287 pages. 1993.

Vol. 1539: M. Coornaert, A. Papadopoulos, Symbolic Dynamics and Hyperbolic Groups. VIII, 138 pages. 1993.

Vol. 1540: H. Komatsu (Ed.), Functional Analysis and Related Topics, 1991. Proceedings. XXI, 413 pages. 1993.

Vol. 1541: D. A. Dawson, B. Maisonneuve, J. Spencer, Ecole d' Eté de Probabilités de Saint-Flour XXI - 1991. Editor: P. L. Hennequin. VIII, 356 pages. 1993.

Vol. 1542: J.Fröhlich, Th.Kerler, Quantum Groups, Quantum Categories and Quantum Field Theory. VII, 431 pages. 1993.

Vol. 1543: A. L. Dontchev, T. Zolezzi, Well-Posed Optimization Problems. XII, 421 pages. 1993.

Vol. 1544: M.Schürmann, White Noise on Bialgebras. VII, 146 pages. 1993.

Vol. 1545: J. Morgan, K. O'Grady, Differential Topology of Complex Surfaces. VIII, 224 pages. 1993.

Vol. 1546: V. V. Kalashnikov, V. M. Zolotarev (Eds.), Stability Problems for Stochastic Models. Proceedings, 1991. VIII, 229 pages. 1993.

Vol. 1547: P. Harmand, D. Werner, W. Werner, M-ideals in Banach Spaces and Banach Algebras. VIII, 387 pages. 1993.

Vol. 1548: T. Urabe, Dynkin Graphs and Quadrilateral Singularities. VI, 233 pages. 1993.

Vol. 1549: G. Vainikko, Multidimensional Weakly Singular Integral Equations. XI, 159 pages. 1993.

Vol. 1550: A. A. Gonchar, E. B. Saff (Eds.), Methods of Approximation Theory in Complex Analysis and Mathematical Physics IV, 222 pages, 1993.

Vol. 1551: L. Arkeryd, P. L. Lions, P.A. Markowich, S.R. S. Varadhan. Nonequilibrium Problems in Many-Particle Systems. Montecatini, 1992. Editors: C. Cercignani, M. Pulvirenti. VII, 158 pages 1993.

Vol. 1552: J. Hilgert, K.-H. Neeb, Lie Semigroups and their Applications. XII, 315 pages. 1993.

Vol. 1553: J.-L- Colliot-Thélène, J. Kato, P. Vojta. Arithmetic Algebraic Geometry. Trento, 1991. Editor: E. Ballico. VII, 223 pages. 1993.

Vol. 1554: A. K. Lenstra, H. W. Lenstra, Jr. (Eds.), The Development of the Number Field Sieve. VIII, 131 pages. 1993.

Vol. 1555: O. Liess, Conical Refraction and Higher Microlocalization. X, 389 pages. 1993.

Vol. 1556: S. B. Kuksin, Nearly Integrable Infinite-Dimensional Hamiltonian Systems. XXVII, 101 pages. 1993.

Vol. 1557: J. Azéma, P. A. Meyer, M. Yor (Eds.), Séminaire de Probabilités XXVII. VI, 327 pages. 1993.

Vol. 1558: T. J. Bridges, J. E. Furter, Singularity Theory and Equivariant Symplectic Maps. VI, 226 pages. 1993.

Vol. 1559: V. G. Sprindžuk, Classical Diophantine Equations. XII, 228 pages. 1993.

Vol. 1560: T. Bartsch, Topological Methods for Variational Problems with Symmetries. X, 152 pages. 1993.

Vol. 1561: I. S. Molchanov, Limit Theorems for Unions of Random Closed Sets. X, 157 pages. 1993.

Vol. 1562: G. Harder, Eisensteinkohomologie und die Konstruktion gemischter Motive. XX, 184 pages. 1993.

Vol. 1563: E. Fabes, M. Fukushima, L. Gross, C. Kenig, M. Röckner, D. W. Stroock, Dirichlet Forms. Varenna, 1992. Editors: G. Dell'Antonio, U. Mosco. VII, 245 pages. 1993.

Vol. 1564: J. Jorgenson, S. Lang, Basic Analysis of Regularized Series and Products. IX, 122 pages. 1993.

Vol. 1565: L. Boutet de Monvel, C. De Concini, C. Procesi, P. Schapira, M. Vergne. D-modules, Representation Theory, and Quantum Groups. Venezia, 1992. Editors: G. Zampieri, A. D'Agnolo. VII, 217 pages. 1993.

Vol. 1566: B. Edixhoven, J.-H. Evertse (Eds.), Diophantine Approximation and Abelian Varieties. XIII, 127 pages. 1993.

Vol. 1567: R. L. Dobrushin, S. Kusuoka, Statistical Mechanics and Fractals. VII, 98 pages. 1993.

Vol. 1568: F. Weisz, Martingale Hardy Spaces and their Application in Fourier Analysis. VIII, 217 pages. 1994.

Vol. 1569: V. Totik, Weighted Approximation with Varying Weight. VI, 117 pages. 1994.

Vol. 1570: R. deLaubenfels, Existence Families, Functional Calculi and Evolution Equations. XV, 234 pages. 1994.

Vol. 1571: S. Yu. Pilyugin, The Space of Dynamical Systems with the C^0-Topology. X, 188 pages. 1994.

Vol. 1572: L. Göttsche, Hilbert Schemes of Zero-Dimensional Subschemes of Smooth Varieties. IX, 196 pages. 1994.

Vol. 1573: V. P. Havin, N. K. Nikolski (Eds.), Linear and Complex Analysis – Problem Book 3 – Part I. XXII, 489 pages. 1994.

Vol. 1574: V. P. Havin, N. K. Nikolski (Eds.), Linear and Complex Analysis – Problem Book 3 – Part II. XXII, 507 pages. 1994.

Vol. 1575: M. Mitrea, Clifford Wavelets, Singular Integrals, and Hardy Spaces. XI, 116 pages. 1994.

Vol. 1576: K. Kitahara, Spaces of Approximating Functions with Haar-Like Conditions. X, 110 pages. 1994.

Vol. 1577: N. Obata, White Noise Calculus and Fock Space. X, 183 pages. 1994.

Vol. 1578: J. Bernstein, V. Lunts, Equivariant Sheaves and Functors. V, 139 pages. 1994.

Vol. 1579: N. Kazamaki, Continuous Exponential Martingales and *BMO*. VII, 91 pages. 1994.

Vol. 1580: M. Milman, Extrapolation and Optimal Decompositions with Applications to Analysis. XI, 161 pages. 1994.

Vol. 1581: D. Bakry, R. D. Gill, S. A. Molchanov, Lectures on Probability Theory. Editor: P. Bernard. VIII, 420 pages. 1994.

Vol. 1582: W. Balser, From Divergent Power Series to Analytic Functions. X, 108 pages. 1994.

Vol. 1583: J. Azéma, P. A. Meyer, M. Yor (Eds.), Séminaire de Probabilités XXVIII. VI, 334 pages. 1994.

Vol. 1584: M. Brokate, N. Kenmochi, I. Müller, J. F. Rodriguez, C. Verdi, Phase Transitions and Hysteresis. Montecatini Terme, 1993. Editor: A. Visintin. VII. 291 pages. 1994.

Vol. 1585: G. Frey (Ed.), On Artin's Conjecture for Odd 2-dimensional Representations. VIII, 148 pages. 1994.

Vol. 1586: R. Nillsen, Difference Spaces and Invariant Linear Forms. XII, 186 pages. 1994.

Vol. 1587: N. Xi, Representations of Affine Hecke Algebras. VIII, 137 pages. 1994.

Vol. 1588: C. Scheiderer, Real and Étale Cohomology. XXIV, 273 pages. 1994.

Vol. 1589: J. Bellissard, M. Degli Esposti, G. Forni, S. Graffi, S. Isola, J. N. Mather, Transition to Chaos in Classical and Quantum Mechanics. Montecatini Terme, 1991. Editor: 2S. Graffi. VII, 192 pages. 1994.

Vol. 1590: P. M. Soardi, Potential Theory on Infinite Networks. VIII, 187 pages. 1994.

Vol. 1591: M. Abate, G. Patrizio, Finsler Metrics – A Global Approach. IX, 180 pages. 1994.

Vol. 1592: K. W. Breitung, Asymptotic Approximations for Probability Integrals. IX, 146 pages. 1994.

Vol. 1593: J. Jorgenson & S. Lang, D. Goldfeld, Explicit Formulas for Regularized Products and Series. VIII, 154 pages. 1994.

Vol. 1594: M. Green, J. Murre, C. Voisin, Algebraic Cycles and Hodge Theory. Torino, 1993. Editors: A. Albano, F. Bardelli. VII, 275 pages. 1994.

Vol. 1595: R.D.M. Accola, Topics in the Theory of Riemann Surfaces. IX, 105 pages. 1994.

Vol. 1596: L. Heindorf, L. B. Shapiro, Nearly Projective Boolean Algebras. X, 202 pages. 1994.

Vol. 1597: B. Herzog, Kodaira-Spencer Maps in Local Algebra. XVII, 176 pages. 1994.

Vol. 1598: J. Berndt, F. Tricerri, L. Vanhecke, Generalized Heisenberg Groups and Damek-Ricci Harmonic Spaces. VIII, 125 pages. 1995.

Vol. 1599: K. Johannson, Topology and Combinatorics of 3-Manifolds. XVIII, 446 pages. 1995.

Vol. 1600: W. Narkiewicz, Polynomial Mappings. VII, 130 pages. 1995.

Vol. 1601: A. Pott, Finite Geometry and Character Theory. VII, 181 pages. 1995.

Vol. 1602: J. Winkelmann, The Classification of Three-dimensional Homogeneous Complex Manifolds. XI, 230 pages. 1995.

Vol. 1603: V. Ene, Real Functions – Current Topics. XIII, 310 pages. 1995.

Vol. 1604: A. Huber, Mixed Motives and their Realization in Derived Categories. XV, 207 pages. 1995.

Vol. 1605: L. B. Wahlbin, Superconvergence in Galerkin Finite Element Methods. XI, 166 pages. 1995.

Vol. 1606: P.-D. Liu, M. Qian, Smooth Ergodic Theory of Random Dynamical Systems. XI, 221 pages. 1995.

Vol. 1607: G. Schwarz, Hodge Decomposition – A Method for Solving Boundary Value Problems. VII, 155 pages. 1995.

Vol. 1608: P. Biane, R. Durrett, Lectures on Probability Theory. Editor: P. Bernard. VII, 210 pages. 1995.

Vol. 1609: L. Arnold, C. Jones, K. Mischaikow, G. Raugel, Dynamical Systems. Montecatini Terme, 1994. Editor: R. Johnson. VIII, 329 pages. 1995.

Vol. 1610: A. S. Üstünel, An Introduction to Analysis on Wiener Space. X, 95 pages. 1995.

Vol. 1611: N. Knarr, Translation Planes. VI, 112 pages. 1995.

Vol. 1612: W. Kühnel, Tight Polyhedral Submanifolds and Tight Triangulations. VII, 122 pages. 1995.

Vol. 1613: J. Azéma, M. Emery, P. A. Meyer, M. Yor (Eds.), Séminaire de Probabilités XXIX. VI, 326 pages. 1995.

Vol. 1614: A. Koshelev, Regularity Problem for Quasilinear Elliptic and Parabolic Systems. XXI, 255 pages. 1995.

Vol. 1615: D. B. Massey, Le Cycles and Hypersurface Singularities. XI, 131 pages. 1995.

Vol. 1616: I. Moerdijk, Classifying Spaces and Classifying Topoi. VII, 94 pages. 1995.

Vol. 1617: V. Yurinsky, Sums and Gaussian Vectors. XI, 305 pages. 1995.

Vol. 1618: G. Pisier, Similarity Problems and Completely Bounded Maps. VII, 156 pages. 1996.

Vol. 1619: E. Landvogt, A Compactification of the Bruhat-Tits Building. VII, 152 pages. 1996.

Vol. 1620: R. Donagi, B. Dubrovin, E. Frenkel, E. Previato, Integrable Systems and Quantum Groups. Montecatini Terme, 1993. Editors:M. Francaviglia, S. Greco. VIII, 488 pages. 1996.

Vol. 1621: H. Bass, M. V. Otero-Espinar, D. N. Rockmore, C. P. L. Tresser, Cyclic Renormalization and Auto-morphism Groups of Rooted Trees. XXI, 136 pages. 1996.

Vol. 1622: E. D. Farjoun, Cellular Spaces, Null Spaces and Homotopy Localization. XIV, 199 pages. 1996.

Vol. 1623: H.P. Yap, Total Colourings of Graphs. VIII, 131 pages. 1996.

Vol. 1624: V. Brınzanescu, Holomorphic Vector Bundles over Compact Complex Surfaces. X, 170 pages. 1996.

Vol.1625: S. Lang, Topics in Cohomology of Groups. VII, 226 pages. 1996.

Vol. 1626: J. Azéma, M. Emery, M. Yor (Eds.), Séminaire de Probabilités XXX. VIII, 382 pages. 1996.

Vol. 1627: C. Graham, Th. G. Kurtz, S. Méléard, Ph. E. Protter, M. Pulvirenti, D. Talay, Probabilistic Models for Nonlinear Partial Differential Equations. Montecatini Terme, 1995. Editors: D. Talay, L. Tubaro. X, 301 pages. 1996.

Vol. 1628: P.-H. Zieschang, An Algebraic Approach to Association Schemes. XII, 189 pages. 1996.

Vol. 1629: J. D. Moore, Lectures on Seiberg-Witten Invariants. VII, 105 pages. 1996.

Vol. 1630: D. Neuenschwander, Probabilities on the Heisenberg Group: Limit Theorems and Brownian Motion. VIII, 139 pages. 1996.

Vol. 1631: K. Nishioka, Mahler Functions and Transcendence.VIII, 185 pages.1996.

Vol. 1632: A. Kushkuley, Z. Balanov, Geometric Methods in Degree Theory for Equivariant Maps. VII, 136 pages. 1996.

Vol.1633: H. Aikawa, M. Essén, Potential Theory – Selected Topics. IX, 200 pages.1996.

Vol. 1634: J. Xu, Flat Covers of Modules. IX, 161 pages. 1996.

Vol. 1635: E. Hebey, Sobolev Spaces on Riemannian Manifolds. X, 116 pages. 1996.

Vol. 1636: M. A. Marshall, Spaces of Orderings and Abstract Real Spectra. VI, 190 pages. 1996.

Vol. 1637: B. Hunt, The Geometry of some special Arithmetic Quotients. XIII, 332 pages. 1996.

Vol. 1638: P. Vanhaecke, Integrable Systems in the realm of Algebraic Geometry. VIII, 218 pages. 1996.

Vol. 1639: K. Dekimpe, Almost-Bieberbach Groups: Affine and Polynomial Structures. X, 259 pages. 1996.

Vol. 1640: G. Boillat, C. M. Dafermos, P. D. Lax, T. P. Liu, Recent Mathematical Methods in Nonlinear Wave Propagation. Montecatini Terme, 1994. Editor: T. Ruggeri. VII, 142 pages. 1996.

Vol. 1641: P. Abramenko, Twin Buildings and Applications to S-Arithmetic Groups. IX, 123 pages. 1996.

Vol. 1642: M. Puschnigg, Asymptotic Cyclic Cohomology. XXII, 138 pages. 1996.

Vol. 1643: J. Richter-Gebert, Realization Spaces of Polytopes. XI, 187 pages. 1996.

Vol. 1644: A. Adler, S. Ramanan, Moduli of Abelian Varieties. VI, 196 pages. 1996.

Vol. 1645: H. W. Broer, G. B. Huitema, M. B. Sevryuk, Quasi-Periodic Motions in Families of Dynamical Systems. XI, 195 pages. 1996.

Vol. 1646: J.-P. Demailly, T. Peternell, G. Tian, A. N. Tyurin, Transcendental Methods in Algebraic Geometry. Cetraro, 1994. Editors: F. Catanese, C. Ciliberto. VII, 257 pages. 1996.

Vol. 1647: D. Dias, P. Le Barz, Configuration Spaces over Hilbert Schemes and Applications. VII. 143 pages. 1996.

Vol. 1648: R. Dobrushin, P. Groeneboom, M. Ledoux, Lectures on Probability Theory and Statistics. Editor: P. Bernard. VIII, 300 pages. 1996.

Vol. 1649: S. Kumar, G. Laumon, U. Stuhler, Vector Bundles on Curves – New Directions. Cetraro, 1995. Editor: M. S. Narasimhan. VII, 193 pages. 1997.

Vol. 1650: J. Wildeshaus, Realizations of Polylogarithms. XI, 343 pages. 1997.

Vol. 1651: M. Drmota, R. F. Tichy, Sequences, Discrepancies and Applications. XIII, 503 pages. 1997.

Vol. 1652: S. Todorcevic, Topics in Topology. VIII, 153 pages. 1997.

Vol. 1653: R. Benedetti, C. Petronio, Branched Standard Spines of 3-manifolds. VIII, 132 pages. 1997.

Vol. 1654: R. W. Ghrist, P. J. Holmes, M. C. Sullivan, Knots and Links in Three-Dimensional Flows. X, 208 pages. 1997.

Vol. 1655: J. Azéma, M. Emery, M. Yor (Eds.), Séminaire de Probabilités XXXI. VIII, 329 pages. 1997.

Vol. 1656: B. Biais, T. Björk, J. Cvitanic, N. El Karoui, E. Jouini, J. C. Rochet, Financial Mathematics. Bressanone, 1996. Editor: W. J. Runggaldier. VII, 316 pages. 1997.

Vol. 1657: H. Reimann, The semi-simple zeta function of quaternionic Shimura varieties. IX, 143 pages. 1997.

Vol. 1658: A. Pumarino, J. A. Rodrıguez, Coexistence and Persistence of Strange Attractors. VIII, 195 pages. 1997.

Vol. 1659: V, Kozlov, V. Maz'ya, Theory of a Higher-Order Sturm-Liouville Equation. XI, 140 pages. 1997.

Vol. 1660: M. Bardi, M. G. Crandall, L. C. Evans, H. M. Soner, P. E. Souganidis, Viscosity Solutions and Applications. Montecatini Terme, 1995. Editors: I. Capuzzo Dolcetta, P. L. Lions. IX, 259 pages. 1997.

Vol. 1661: A. Tralle, J. Oprea, Symplectic Manifolds with no Kähler Structure. VIII, 207 pages. 1997.

Vol. 1662: J. W. Rutter, Spaces of Homotopy Self-Equivalences – A Survey. IX, 170 pages. 1997.

Vol. 1663: Y. E. Karpeshina; Perturbation Theory for the Schrödinger Operator with a Periodic Potential. VII, 352 pages. 1997.

Vol. 1664: M. Väth, Ideal Spaces. V, 146 pages. 1997.

Vol. 1665: E. Giné, G. R. Grimmett, L. Saloff-Coste, Lectures on Probability Theory and Statistics 1996. Editor: P. Bernard. X, 424 pages, 1997.

Vol. 1666: M. van der Put, M. F. Singer, Galois Theory of Difference Equations. VII, 179 pages. 1997.

Vol. 1667: J. M. F. Castillo, M. González, Three-space Problems in Banach Space Theory. XII, 267 pages. 1997.

Vol. 1668: D. B. Dix, Large-Time Behavior of Solutions of Linear Dispersive Equations. XIV, 203 pages. 1997.

Vol. 1669: U. Kaiser, Link Theory in Manifolds. XIV, 167 pages. 1997.

Vol. 1670: J. W. Neuberger, Sobolev Gradients and Differential Equations. VIII, 150 pages. 1997.

Vol. 1671: S. Bouc, Green Functors and G-sets. VII, 342 pages. 1997.

Vol. 1672: S. Mandal, Projective Modules and Complete Intersections. VIII, 114 pages. 1997.

Vol. 1673: F. D. Grosshans, Algebraic Homogeneous Spaces and Invariant Theory. VI, 148 pages. 1997.

Vol. 1674: G. Klaas, C. R. Leedham-Green, W. Plesken, Linear Pro-p-Groups of Finite Width. VIII, 115 pages. 1997.

Vol. 1675: J. E. Yukich, Probability Theory of Classical Euclidean Optimization Problems. X, 152 pages. 1998.

Vol. 1676: P. Cembranos, J. Mendoza, Banach Spaces of Vector-Valued Functions. VIII, 118 pages. 1997.

Vol. 1677: N. Proskurin, Cubic Metaplectic Forms and Theta Functions. VIII, 196 pages. 1998.

Vol. 1678: O. Krupková, The Geometry of Ordinary Variational Equations. X, 251 pages. 1997.

Vol. 1679: K.-G. Grosse-Erdmann, The Blocking Technique. Weighted Mean Operators and Hardy's Inequality. IX, 114 pages. 1998.

Vol. 1680: K.-Z. Li, F. Oort, Moduli of Supersingular Abelian Varieties. V, 116 pages. 1998.

Vol. 1681: G. J. Wirsching, The Dynamical System Generated by the 3n+1 Function. VII, 158 pages. 1998.

Vol. 1682: H.-D. Alber, Materials with Memory. X, 166 pages. 1998.

Vol. 1683: A. Pomp, The Boundary-Domain Integral Method for Elliptic Systems. XVI, 163 pages. 1998.

Vol. 1684: C. A. Berenstein, P. F. Ebenfelt, S. G. Gindikin, S. Helgason, A. E. Tumanov, Integral Geometry, Radon Transforms and Complex Analysis. Firenze, 1996. Editors: E. Casadio Tarabusi, M. A. Picardello, G. Zampieri. VII, 160 pages. 1998.

Vol. 1685: S. König, A. Zimmermann, Derived Equivalences for Group Rings. X, 146 pages. 1998.

Vol. 1686: J. Azéma, M. Émery, M. Ledoux, M. Yor (Eds.), Séminaire de Probabilités XXXII. VI, 440 pages. 1998.

Vol. 1687: F. Bornemann, Homogenization in Time of Singularly Perturbed Mechanical Systems. XII, 156 pages. 1998.

Vol. 1688: S. Assing, W. Schmidt, Continuous Strong Markov Processes in Dimension One. XII, 137 page. 1998.

Vol. 1689: W. Fulton, P. Pragacz, Schubert Varieties and Degeneracy Loci. XI, 148 pages. 1998.

Vol. 1690: M. T. Barlow, D. Nualart, Lectures on Probability Theory and Statistics. Editor: P. Bernard. VIII, 237 pages. 1998.

Vol. 1691: R. Bezrukavnikov, M. Finkelberg, V. Schechtman, Factorizable Sheaves and Quantum Groups. X, 282 pages. 1998.

Vol. 1692: T. M. W. Eyre, Quantum Stochastic Calculus and Representations of Lie Superalgebras. IX, 138 pages. 1998.

Vol. 1694: A. Braides, Approximation of Free-Discontinuity Problems. XI, 149 pages. 1998.

Vol. 1695: D. J. Hartfiel, Markov Set-Chains. VIII, 131 pages. 1998.

Vol. 1696: E. Bouscaren (Ed.): Model Theory and Algebraic Geometry. XV, 211 pages. 1998.

Vol. 1697: B. Cockburn, C. Johnson, C.-W. Shu, E. Tadmor, Advanced Numerical Approximation of Nonlinear Hyperbolic Equations. Cetraro, Italy, 1997. Editor: A. Quarteroni. VII, 390 pages. 1998.

Vol. 1698: M. Bhattacharjee, D. Macpherson, R. G. Möller, P. Neumann, Notes on Infinite Permutation Groups. XI, 202 pages. 1998.

Vol. 1699: A. Inoue, Tomita-Takesaki Theory in Algebras of Unbounded Operators. VIII, 241 pages. 1998.

Vol. 1700: W. A. Woyczyński, Burgers-KPZ Turbulence, XI, 318 pages. 1998.

Vol. 1701: Ti-Jun Xiao, J. Liang, The Cauchy Problem of Higher Order Abstract Differential Equations, XII, 302 pages. 1998.

Vol. 1702: J. Ma, J. Yong, Forward-Backward Stochastic Differential Equations and Their Applications. XIII, 270 pages. 1999.

Vol. 1703: R. M. Dudley, R. Norvaiša, Differentiability of Six Operators on Nonsmooth Functions and p-Variation. VIII, 272 pages. 1999.

Vol. 1704: H. Tamanoi, Elliptic Genera and Vertex Operator Super-Algebras. VI, 390 pages. 1999.

Vol. 1705: I. Nikolaev, E. Zhuzhoma, Flows in 2-dimensional Manifolds. XIX, 294 pages. 1999.

Vol. 1706: S. Yu. Pilyugin, Shadowing in Dynamical Systems. XVII, 271 pages. 1999.

Vol. 1707: R. Pytlak, Numerical Methods for Optical Control Problems with State Constraints. XV, 215 pages. 1999.

Vol. 1708: K. Zuo, Representations of Fundamental Groups of Algebraic Varieties. VII, 139 pages. 1999.

Vol. 1709: J. Azéma, M. Émery, M. Ledoux, M. Yor (Eds), Séminaire de Probabilités XXXIII. VIII, 418 pages. 1999.

Vol. 1710: M. Koecher, The Minnesota Notes on Jordan Algebras and Their Applications. IX, 173 pages. 1999.

Vol. 1711: W. Ricker, Operator Algebras Generated by Commuting Projections: A Vector Measure Approach. XVII, 159 pages. 1999.

Vol. 1712: N. Schwartz, J. J. Madden, Semi-algebraic Function Rings and Reflectors of Partially Ordered Rings. XI, 279 pages. 1999.

Vol. 1713: F. Bethuel, G. Huisken, S. Müller, K. Steffen, Calculus of Variations and Geometric Evolution Problems. Cetraro, 1996. Editors: S. Hildebrandt, M. Struwe. VII, 293 pages. 1999.

Vol. 1714: O. Diekmann, R. Durrett, K. P. Hadeler, P. Maini, H. L. Smith, Mathematics Inspired by Biology. Martina Franca, 1997. Editors: V. Capasso, O. Diekmann. VII, 268 pages. 1999.

Vol. 1715: N. V. Krylov, M. Röckner, J. Zabczyk, Stochastic PDE's and Kolmogorov Equations in Infinite Dimensions. Cetraro, 1998. Editor: G. Da Prato. VIII, 239 pages. 1999.

Vol. 1716: J. Coates, R. Greenberg, K. A. Ribet, K. Rubin, Arithmetic Theory of Elliptic Curves. Cetraro, 1997. Editor: C. Viola. VIII, 260 pages. 1999.